INTEGRATING SCALE IN REMOTE SENSING AND GIS

Taylor & Francis Series in
Remote Sensing Applications

Series Editor
Qihao Weng

Indiana State University
Terre Haute, Indiana, U.S.A.

Integrating Scale in Remote Sensing and GIS, *Dale A. Quattrochi, Elizabeth A. Wentz, Nina Siu-Ngan Lam, and Charles W. Emerson*

Remote Sensing Applications for the Urban Environment, *George Z. Xian*

Remote Sensing of Impervious Surfaces in Tropical and Subtropical Areas, *Hui Lin, Yuanzhi Zhang, and Qihao Weng*

Global Urban Monitoring and Assessment through Earth Observation, *edited by Qihao Weng*

Remote Sensing of Natural Resources, *edited by Guangxing Wang and Qihao Weng*

Remote Sensing of Land Use and Land Cover: Principles and Applications, *Chandra P. Giri*

Remote Sensing of Protected Lands, *edited by Yeqiao Wang*

Advances in Environmental Remote Sensing: Sensors, Algorithms, and Applications, *edited by Qihao Weng*

Remote Sensing of Coastal Environments, *edited by Yeqiao Wang*

Remote Sensing of Global Croplands for Food Security, *edited by Prasad S. Thenkabail, John G. Lyon, Hugh Turral, and Chandashekhar M. Biradar*

Global Mapping of Human Settlement: Experiences, Data Sets, and Prospects, *edited by Paolo Gamba and Martin Herold*

Hyperspectral Remote Sensing: Principles and Applications, *Marcus Borengasser, William S. Hungate, and Russell Watkins*

Remote Sensing of Impervious Surfaces, *edited by Qihao Weng*

Multispectral Image Analysis Using the Object-Oriented Paradigm, *Kumar Navulur*

Urban Remote Sensing, *edited by Qihao Weng and Dale A. Quattrochi*

INTEGRATING SCALE IN REMOTE SENSING AND GIS

Dale A. Quattrochi

Elizabeth A. Wentz

Nina Siu-Ngan Lam

Charles W. Emerson

CRC Press
Taylor & Francis Group
Boca Raton London New York

CRC Press is an imprint of the
Taylor & Francis Group, an **informa** business

CRC Press
Taylor & Francis Group
6000 Broken Sound Parkway NW, Suite 300
Boca Raton, FL 33487-2742

First issued in paperback 2019

ISBN-13: 978-1-4822-1826-8 (hbk)
ISBN-13: 978-0-367-86899-4 (pbk)

British Library Cataloguing-in-Publication Data
A catalogue record for this book is available from the British Library

Library of Congress Cataloging-in-Publication Data
Names: Quattrochi, Dale A., editor. Title: Integrating scale in remote sensing and GIS / edited by Dale A. Quattrochi, Elizabeth A. Wentz, Nina Siu-Ngan Lam, and Charles W. Emerson. Description: London : Routledge, 2017. | Includes bibliographical references and index. Identifiers: LCCN 2016028202| ISBN 9781482218268 (hardback : alk. paper) | ISBN 9781482218275 (ebook) Subjects: LCSH: Geographic information systems. | Remote sensing. Classification: LCC G70.212 .I565 2017 | DDC 910.285--dc23 LC record available at https://lccn.loc.gov/2016028202

Visit the Taylor & Francis Web site at
http://www.taylorandfrancis.com

and the CRC Press Web site at
http://www.crcpress.com

Contents

List of Figures

Series Foreword

Scale is an essential, key issue in remote sensing, geographic information systems (GIS), and geographical and landscape analyses. In 1997, Drs. Dale Quattrochi and Michael Goodchild published a seminal work entitled "Scale in Remote Sensing and GIS," in which methods for analysis of characteristics, elements, and errors associated with scaling and multiscaling were laid out. The past two decades witnessed new developments in Earth Observation Technology, such as high spatial resolution, hyperspectral sensors and data, lidar sensing, and increasing availability of freely accessible image data (Weng 2012). In alignment with these developments in data collection and access is the invention of new remote sensing imaging processing techniques, such as object-oriented image analysis, data fusion, data mining, artificial neural network, and methods for better use of temporal dimension of image data. On the other front, GIS has re-invented itself with increasingly powerful computer, network, and telecommunication technologies (Weng 2009). GIS has altered the way we view, use, and disseminate geospatial data not only in data handling, but also in data collection (e.g., volunteered geographic information and social media), storage, and processing (e.g., cloud computing). These developments in remote sensing and GIS data and technology prompt us to rethink how scale, the ubiquitous issue, impacts our understanding of landscape properties, patterns, and processes, and geographical analyses. In this context, I am pleased to learn that Dr. Dale Quattrochi and his colleagues had the vision and energy to edit a book on this important topic. Together with recent books on the same topic (Weng 2014; Zhang et al. 2014), readers, hopefully, would gain more insights into why the subject of scale in remote sensing and GIScience is still pertinent today.

This book intends to demonstrate the evolution in the concept of scale within the purview of multispatial, multitemporal, and multispectral data. In the selected 13 chapters, four themes have been examined, including scale and multiscaling issues, physical scale, human scale, and health/social scale. I agree with the editors that this book will better elucidate the underpinnings of scale as a foundation for geographic analysis and will illustrate how scale has vast implications on remote sensing and GIS integration.

This book is the 15th volume in the Taylor & Francis Series in Remote Sensing Applications, and 2017 marks the 10th anniversary of the series. As envisioned in 2007, the books in the series can contribute to advancements in theories, methods, techniques, and applications of remote sensing in varying fields. In fact, as seen today, many of the books in the series have served well as references for professionals, researchers, and scientists, and as textbooks for teachers and students throughout the world. I hope that the publication of this book will promote a wider and deeper appreciation of scale issues in analyzing geospatial data and will facilitate modeling of space, time, and space–time in landscape and geographic complexity. Finally, my hearty congratulations to Dr. Quattrochi and his colleagues at this important "moment" in the history of remote sensing and GIScience.

Qihao Weng
Hawthorn Woods, Indiana

References

Quattrochi, D.A. and M.F. Goodchild. 1997. *Scale in Remote Sensing and GIS.* Boca Raton, FL: CRC Press, p. 432.

Weng, Q. 2009. *Remote Sensing and GIS Integration: Theories, Methods, and Applications.* New York: McGraw-Hill Professional, p. 397. (Chinese translation has been published by Science Press, Beijing, in September 2011.)

Weng, Q. 2012. *An Introduction to Contemporary Remote Sensing.* New York: McGraw-Hill Professional, p. 320.

Weng, Q. 2014. *Scale Issues in Remote Sensing.* Hoboken, NJ: Wiley, p. 352.

Zhang, J., Atkinson, P. and M.F. Goodchild. 2014. *Scale in Spatial Information and Analysis.* Boca Raton, FL: CRC Press, p. 367.

Preface

Scale as a fundamental concept in remote sensing and geographic information science (GIScience) has been much described, much discussed—and perhaps even much maligned—but it still matters as a pervasive and overarching driver for integrating multiscaled and multisource data for spatial analysis. The predecessor to this book, entitled *Scale in Remote Sensing and GIS*, edited by Dale A. Quattrochi and Michael Goodchild and published by CRC Press in 1997, laid a framework for developing a better understanding of the characteristics, elements, and errors associated with bringing together multiscaled remote sensing data sets into a geographic information systems (GIS) construct. In the almost 10 years since the book was published, great leaps have occurred in spatial, spectral, and temporal resolutions in the sensors available for Earth observation. Concomitant with the improvements in remote sensing characteristics, GIS technology and models have progressed into methods that can be employed to simulate the complex nature of space, time, and space–time interactions. These attributes are a fulcrum for the multidisciplinary field of GIScience, which embraces technological advances in geospatial data capture, analysis, modeling, and output. The pervasive question in both GIScience and remote sensing is, how can we successfully, robustly, and accurately integrate and apply data sets collected at varying spatial, temporal, and spectral scales? It is the intent of this book to not only reiterate the importance of scale as it relates to remote sensing and GIS but also elucidate the key role scale plays in the applications of remote sensing data and GIScience.

This book examines the attributes and characteristic of multiscaled data from an applications perspective wherein we explore such challenges as the following: What are the implications of scale invariance and transformations of scale in aggregating or disaggregating data? What are the ways to best measure the impact of scale or how is the observation of processes affected by changes in scale? How can we measure the degree to which processes are manifested at different scales? What are the implications of scale in process models and how are models affected by using data at multiple scales? What tools can be used to support multiscale databases and associated modeling and analysis? This book explores these issues within an applications purview for four thematic areas: scale issues as related to multiple time and space scales; scale attributes and natural resources; scale and urban analysis; and scale as it impacts human health and social attributes. This book's purpose, therefore, is to extend and update the tenets established in the 1997 volume and to reemphasize the importance of scale issues in combining remote sensing data obtained at different spatial, temporal, and spectral scales as a key aspect of GIScience. Even though scale has been much discussed throughout the spatial sciences literature, scale still matters, and it is the aim of this book to illustrate why this is so, particularly for the applications of multiscaled remote sensing data as an integral component of GIScience.

Dale A. Quattrochi
Elizabeth A. Wentz
Nina Siu-Ngan Lam
Charles W. Emerson

Editors

Dale A. Quattrochi is a geographer and senior research scientist with the NASA George C. Marshall Space Flight Center, Earth Science Office in Huntsville, Alabama. His research has focused on the analysis of multiscaled remote sensing data for GIS integration, the use of NASA satellite and airborne remote sensing data for analysis of land cover/land use changes, particularly as related to the urban environment, thermal remote sensing of the urban heat island effect, and on the applications of NASA data and models to public health issues. He is the coeditor of three books published by CRC Press: *Scale in Remote Sensing and GIS* (1997), *Thermal Remote Sensing in Land Surface Processes* (2004), and *Urban Remote Sensing* (2007). Dr. Quattrochi is the recipient of numerous awards, including the American Association of Geographers Remote Sensing Specialty Group Outstanding Achievement Award (1999), the NASA Medal for Exceptional Scientific Achievement (2001), the Ohio University College of Arts and Sciences Distinguished Alumni Award (2002), and the American Meteorological Society Helmut E. Landsberg Award (2015). He received his BS from Ohio University, his MS from the University of Tennessee, and his PhD from the University of Utah, all in geography.

Elizabeth A. Wentz is dean of social science in the College of Liberal Arts and Sciences, associate director of the Institute for Social Science Research, and professor in the School of Geographical Sciences and Urban Planning at Arizona State University, Tempe, Arizona. Her research focuses on the development and implementation of geographic technologies designed to establish better understanding of the urban environment. In particular, she has been involved in geographic tool development, urban remote sensing, and urban environmental analysis. Her research record includes more than 35 peer-reviewed publications in high-caliber journals and has primarily been funded through (single PI and collaborative projects) from NIH, USDA, NASA, and the NSF. Her research is highly collaborative with researchers across a broad range of social, physical, and computational disciplines through collaborative research both and ASU and other academic institutions including the University of Rhode Island, Yale University, The Polytechnic University of Lausanne in Switzerland, and NASA. In 2015–2016, she served as president of the University Consortium for Geographic Information Science. She earned her PhD in geography from the Pennsylvania State University, her MA in geography from The Ohio State University, and her BS in mathematics from The Ohio State University.

Nina Siu-Ngan Lam received her BSSc in geography from the Chinese University of Hong Kong in 1975 and her MS and PhD in geography from the University of Western Ontario in 1976 and 1980. Dr. Lam is currently a professor and an E.L. Abraham Distinguished Professor in the Department of Environmental Sciences at Louisiana State University. She was chair of the department (2007–2010), director of the National Science Foundation's Geography and Spatial Sciences Program (1999–2001), and president of the University Consortium for Geographic Information Science (UCGIS, 2005). Dr. Lam has authored or coauthored over 90 referred journal articles and book chapters, including a book titled *Fractals in Geography*. Other topics on which she is published include spatial interpolation, scale and uncertainties, cancer mortality, the spread of AIDS, environmental justice issues, disaster resilience, and coupled natural–human system modeling. Dr. Lam has been principal investigator or co-principal investigator on over 40 externally funded research projects. She teaches courses in GIS, remote sensing, and spatial modeling and has served as major advisor of 2 post-doctoral associates, 17 PhD students, and 29 master's students.

Charles W. Emerson received his BS degree in geography from the University of Georgia and MA and PhD degrees in geography from the University of Iowa. He has been a faculty member at Southwest Missouri State University for 3 years and has been at Western Michigan University since 1999. His research focuses on quantitative analysis of remotely sensed imagery using geostatistical techniques and fractals, integration of biophysical measurements with socioeconomic data, and using remotely sensed imagery from satellites and unmanned aerial vehicles to assist paleontological surveys.

Contributors

Sean C. Ahearn
Department of Geography
Center for the Advanced Research
  of Spatial Information
Hunter College of the City
  University of New York
New York, New York

Michael Alonzo
Department of Environmental Science
American University
University of California
Washington, DC

Michael P. Bishop
Department of Geography
Center for Geospatial Science,
  Applications and Technology
Texas A&M University
College Station, Texas

Pietro Ceccato
International Research Institute for
  Climate and Society
The Earth Institute
Columbia University
New York, New York

Guillaume Chabot-Couture
Institute for Disease Modeling
Intellectual Ventures
Bellevue, Washington

Minjuan Cheng
Department of Earth and
  Environmental Systems
Center for Urban and
  Environmental Change
Indiana State University
Terre Haute, Indiana

Edwin Chow
Department of Geography
Texas State University
San Marcos, Texas

Keith C. Clarke
Department of Geography
University of California Santa
  Barbara
Santa Barbara, California

Stephen Connor
Department of Geography
School of Environmental Sciences
University of Liverpool
Liverpool, UK

Phillip E. Dennison
Department of Geography
University of Utah
Salt Lake City, Utah

Tufa Dinku
International Research Institute for
  Climate and Society
The Earth Institute
Columbia University
New York, New York

Iliyana D. Dobreva
Department of Geography
Center for Geospatial Science,
  Applications and Technology
Texas A&M University
College Station, Texas

Kenneth L. Dudley
Biodiversity and Biocomplexity
    Unit
Okinawa Institute of Science and
    Technology
Onna, Okinawa, Japan

Dolores Jane Forbes
Geographic Research & Innovation
    Staff
Geography Division
U.S. Census Bureau
Boca Raton, Florida

Gordon M. Green
Department of Geography
Center for the Advanced Research
    of Spatial Information
Hunter College of the City
    University of New York
New York, New York

Themistocles Herekakis
Institute for Astronomy, Astrophysics,
    Space Applications & Remote
    Sensing
National Observatory of Athens
Penteli, Greece

Emmanuela Ieronymidi
Institute for Astronomy, Astrophysics,
    Space Applications & Remote
    Sensing
National Observatory of Athens
Penteli, Greece

Ian J. Irmischer
Geography and Environmental
    Engineering
United States Military Academy
West Point, New York

Bandana Kar
Department of Geography and
    Geology
University of Southern Mississippi
Hattiesburg, Mississippi

Iphigenia Keramitsoglou
Institute for Astronomy, Astrophysics,
    Space Applications & Remote
    Sensing
National Observatory of Athens
Penteli, Greece

Charalampos Kontoes
Institute for Astronomy, Astrophysics,
    Space Applications & Remote
    Sensing
National Observatory of Athens
Penteli, Greece

Andrew Kruczkiewicz
International Research Institute for
    Climate and Society
The Earth Institute
Columbia University
New York, New York

Jerrod Lessel
International Research Institute for
    Climate and Society
The Earth Institute
Columbia University
New York, New York

Shaun Lovejoy
Department of Physics
McGill University
Montréal, Quebec, Canada

Wenge Ni-Meister
Geography Department
Hunter College of The City
    University of New York
New York, New York

Ioannis Papoutsis
Institute for Astronomy,
    Astrophysics, Space Applications
    & Remote Sensing
National Observatory of Athens
Penteli, Greece

Dar Roberts
Department of Geography
University of California Santa Barbara
Santa Barbara, California

Alex de Sherbinin
Center for International Earth
    Science Information Network
    (CIESIN)
The Earth Institute
Columbia University
New York, New York

Alexandra Sweeney
International Research Institute for
    Climate and Society
The Earth Institute
Columbia University
New York, New York

Madeleine C. Thomson
International Research Institute for
    Climate and Society
The Earth Institute
Columbia University
New York, New York

Qihao Weng
Department of Earth and
    Environmental Systems
Center for Urban and
    Environmental Change
Indiana State University
Terre Haute, Indiana

Erin B. Wetherley
Department of Geography
University of California Santa
    Barbara
Santa Barbara, California

Introduction

In 1997, the book entitled *Scale in Remote Sensing and GIS* was published by CRC Press with Dale Quattrochi and Michael Goodchild as editors. Much has happened in the fields of remote sensing and geographic information systems (GIS) technology since the publication of that book, but the concept of scale as an integrating factor between remote sensing data and GIS still exists as the foundation upon which both of these technologies are grounded. The amount and types of remote sensing data that are now currently available over what was extant in 1997 has, for most intents and purposes, exponentially increased due to the launching of a number of National Aeronautics and Space Administration (NASA) Earth observing satellite systems as well as satellites launched by the European Space Agency, other international entities, and commercial remote sensing firms. Remote sensing satellite data now exist in an array of different spatial, spectral, and temporal resolutions that were not available at the writing of the 1997 book. Earth observing missions launched since then, such as the NASA Terra and Aqua missions launched in 1999 and 2002, respectively, with the Moderate-Resolution Imaging Spectroradiometer, the Advanced Spaceborne Thermal Emission and Reflection Radiometer, and the Visible Infrared Imaging Radiometer Suite instrument onboard the National Oceanic and Atmospheric Administration National Polar-Orbiting Operational Environmental Satellite System (now called the *Suomi National Polar-Orbiting Partnership* or *NPP space platform*) launched in 2011 provide multiscaled data that have been used extensively in multidisciplinary areas of environmental research. These satellites have produced a literal explosion in the amount of remote sensing data that are available for integration into GIS for analysis, modeling, and mapping.

Advances in geographic information science (GIScience) have progressed at a pace at least as quickly as that for remote sensing. Technologies such as data mining and machine learning procedures have greatly enhanced both the speed and power at which data are processed and analyzed *via* GIS. Enhancements in Internet and wireless and satellite communications, which were nascent in 1997, now allow for real-time robust analysis of GIS and remote sensing–based data. Web-based tools, such as Google Earth, whose use by academics and the general public has become ubiquitous, have provided ready access to geospatial data for "common" consumption. Together, remote sensing and GIS have come together to produce data combinations that are used to approach and solve increasingly complex problems. In concert with this, the interoperability of GIS has resulted in standards that address incompatibilities in data formats, software products, spatial conceptions, quality standards, modeling efficiencies, and associated issues that relate to the robustness of remote sensing data and GIS integration.

Although tremendous insight has been gained by research on the science of scale *via* the plethora of information since the 1997 book was published, the applications of space and time scale theory, attributes, and variables in GIScience and remote sensing still need further exploration. It is here where the elucidation of the original points of focus in the 1997 book within the purview of applied perspective can be addressed as organic real-world problems:

- *Invariants of scale*: What measures of scale are invariant of the geographic detail in data that survive manipulation, such as conversion from analog to digital form or transformation of coordinates? What properties of physical and human systems are invariant with respect to scale?

- *The ability to change scale*: What kinds of transformations of scale are available to aggregate or disaggregate data in ways that are logical, rigorous, and well grounded in theory? Can we develop a generic set of methods for disaggregating coarse data and aggregating fine data, in ways that are compatible with our understanding of Earth system processes and climate change?

- *Measures of the impact of scale*: Is it possible to implement methods that assess the impact of scale change, through measures of information loss or gain? How is the observation of processes affected by changes of scale, and how can we measure the degree to which processes are manifested at different scales?

- *Scale as a parameter in process models*: How is scale represented in the parameterization of process models, and how are models affected by the use of data at inappropriate scales?

- *Implementation of multiscale approaches*: What is the potential for integrated tools to support multiscale databases and associated modeling and analysis? Can tools provide a compatible framework for multiscale data? What are the problems that must be overcome in integrating data from different scales?

Thus, a great overall challenge in both GIScience and remote sensing is, how do we successfully and robustly integrate and apply multiple kinds of remote sensing data collected at varying spatial, temporal, and radiometric scales? It is the intent of this book to explore this challenge by delving into these facets of scale and illustrate why, despite the growing repository of literature on scale, the issue still matters, particularly in the applications of multiscaled remote sensing data in GIScience.

In GIScience and remote sensing, scale has four common usages in either the spatial or temporal domain. From smallest to largest, the *measurement scale* is the resolution used to define an object or image such as the pixel resolution in remote sensing image. The *operational scale* is the extent at which a process operates in the environment, and it usually involves a number of pixels to identify the process. The *observational* (or geographic) *scale* refers to the spatial extent of the study and would most likely include a larger area covering a number of processes. In addition, the fourth usage, the *cartographic* or *map scale*, refers to the ratio between ground measurements and representation on a map. Another form of geographic scale is a hierarchical or informal structure of human-derived constructs such as nations or neighborhoods. Similarly, this has the inference of extent, perspective, and detail, which in part are scale, but this social construction leads to potential biases.

The further challenge is that few studies address the implications of map scale or social construction of scale on domain outcomes. Few researchers investigate simultaneously whether the drivers and outcomes are replicated through different scale lenses. These implications range from accuracy assessment, modeling results, social theory implications, and solutions to scale problems. Researchers have attempted to determine the "best" or most optimal spatial scale (measurement) for a specific application, often concluding that larger (geographical) scales and higher resolution scale perform more favorably (Iverson et al. 1997). For a modeling application involving soil moisture correlated with slope and aspect, for example, Iverson et al. (1997) concluded that a digital elevation model derived from a 1:24,000 scale map performed well but that smaller scales such as 1:250,000 were ineffective at providing the slope and aspect information needed. Studies have shown, however, that higher levels of detail sometimes lead to diminishing returns. Building on the framework outlined by Ruddell and Wentz (2009), Table 0.1 identifies the many of concepts of scale examined in this book.

This book addresses the importance of scale from the various frames in Table 0.1. The chapters ultimately show how multiple scale issues combined with the application of geospatial technologies in natural resources, urbanization, and human health provide better insight into solving some of the most complex problems facing society today.

The Book's Contents

This book contains 13 chapters, grouped into four thematic groups: (1) scale issues and multiple scaling; (2) physical scale as applied to natural resources; (3) urban scale; and (4) human health/social scale. The overall integration of these chapters into the book offers a good perspective on the importance of scale in remote sensing and elucidates the progress that has been made since publication of the 1997 book in the application of scale within the broader discipline of GIScience.

The first chapter in the book, by Keith Clarke, provides a review and speculation on the past, present, and future of research within the disciplines of

TABLE 0.1

Scale Concepts

Concept	Definition	Implications
Precision and accuracy	The number of significant digits in observational data	There remains a level of uncertainty with quantifying observations
Minimum mapping unit and support	The smallest area detectable in a vector map and the smallest pixel with detectable information in a raster format	No other smaller area can be quantified or analyzed with any level of reliability
Social construction of scale	Scale is a reflection of social behavior rather than a strict hierarchical organization	Represents an organizing principle that involves the role of capitalism and is malleable
The modifiable area unit problem	The observation that choices in zonal data impact results	Statistical and visual biases occur when zone boundaries change
Ecological fallacy	The logical error of assigning a value or judgment to an individual based on observations of the collective	Policies or practices targeted at individual units inappropriately
Fractals	The observation that complex patterns repeat at multiple scales of analysis	Measuring the level of repeatability (fractal dimension) is considered a solution to scale problems
Economies of scale	Advantages that occur when size or scale of operation are consolidated	Highlighting where aggregating leads to more optimal use of resources

geography and GIScience on the nature of space, time, and space–time, or what might be called *spatiotemporal geography*. The review is not intended to be comprehensive; indeed, there are several excellent extant review papers to which the reader is referred. Instead, the principal computational, geographic, and philosophical issues and challenges are the focus. Central to these is the geographic (and cartographic) concept of scale, which has profound implications for research in space–time, as in geography generally. The essay covers first space and then time, and then examines their integration before concluding with some observations about contemporary research.

Dolores Forbes then presents intriguing concepts on the relationships of complexity and geographic scale. The aim of the chapter is to encourage the analysis of geographic scale for examining complex systems. Forbes first describes the nature of complex systems and how determining the causal factors of a complex system is compounded when the system of interest is also directly connected to distinctly different complex systems. Forbes discusses the relationship of geographic scale to complex systems and how the dynamic attributes of complex systems change through time. She then describes how in complex systems, the dynamic nature of equilibrium points or the thresholds that indicate a shift between states or scale domains within the system are indicative of the ability of a system to adapt, which in turn leads to response. Here, self-organization is crucial to achieving equilibrium, which is the ability of the system to configure itself through feedback loops and which makes complex systems unpredictable or leads to sudden changes. Scale is key for understanding how complex systems reorganize. This applies to geography, which has scale as its foundation, and complexity theory provides a theoretical and practical background for examination of complex systems that aligns well with geographic practice. Complexity theory, therefore, focuses in part on the limitations to understanding complex systems and these limitations are well known in geographic analysis. Hence, there is a clear need for modeling and mapping tools that address complexity through multiscale analysis in space and time to capture the scale dependencies and scale domains of natural and social processes.

Shaun Lovejoy in Chapter 3 expounds upon complexity by examining issues related to scaling geocomplexity in remote sensing data. As the resolution of remotely sensed data improves, we increasingly zoom into complex radiance fields that reveal hierarchies of structures within structures. Traditional approaches effectively isolate narrow ranges of scales and seek deterministic, mechanistic models and explanations. However, most of the variabilities (e.g., variance) are found to lie in the scaling "background" processes, and these are conveniently modeled with fractal sets and multifractal fields. In this chapter, Lovejoy focuses on recent developments in scaling processes and analyses that are of particular significance to remote sensing. On the empirical side, he notes the increasing availability of global scale data sets including their evolutions in time that require space–time formalisms and analysis techniques. Specific examples discussed here include global scale visible, infrared, and microwave (passive and active) atmospheric fields, ocean color, topography, soil moisture, and vegetation indices. New theoretical developments include advances in the theory of radiative transfer in multifractal media, improved (real space) fluctuation analyses (especially Haar fluctuations), as well as improvements in the older multifractal trace moment technique that now allows for a direct determination of the outer scale and direct confirmation of the multiplicative (cascading) nature of the intermittency.

Chapter 4 by Bandana Kar and Edwin Chow describes how the revolutionary advances that have occurred in geospatial technologies have enabled the collection and generation of large amounts of geospatial data at varying space and time resolutions, as well as analysis of multisourced data at varying scales. The availability of such an overwhelming volume of data enables near-real-time analysis but requires fusion of spatial and nonspatial data available at multiple spatial and temporal scales. Likewise, multiscaled data need to be resampled to a specific scale for ease of analysis. The variance associated with each scale of analysis needs to be modeled to ensure appropriate interpretation of accuracy by upscaling or downscaling. This chapter explores these issues and focuses on the following three objectives: (1) to discuss fusion techniques used for multisourced data and for integration of spatial and temporal data;

(2) to discuss the analytical issues associated with data fusion and multiscaling analysis; and (3) to discuss the geovisualization techniques and existing issues with visual representation of data fusion and multiscaling outcomes.

On a related theme in Chapter 5, Edwin Chow and Bandana Kar discuss how geospatial data derived from many sources at varying spatial and temporal resolutions are stored in different models. During the process of data integration, varying amounts of errors are introduced through many pathways that affect the quality of the final outcomes. Therefore, it is important to identify the sources of uncertainty and their corresponding error in order to quantify the confidence level of modeled products and improve accuracy. As scale affects accuracy assessment, it is important to have reliable estimates of the accuracy of individual data sources. In the case of multisource data fusion, the estimated accuracy for each input data will impact the estimated accuracy of the final product due to error propagation. The accuracy of an individual data source or product can be determined by the following: (1) empirical validation against reference data (i.e., ground truth) and (2) near-real-time error propagation based on *a priori* covariance estimation. If the first method is used, the reference data should have a higher level of accuracy in order to determine the quality of the fused product. In the second method, an understanding of the error sources and their dissemination will help in quantifying the confidence level associated with the accuracy of the fused product. Depending on the usage, it is essential to determine the best techniques to represent error, including full error covariance, summary statistics such as horizontal circular error, vertical linear error, error ellipsoids, and absolute and relative metrics. The purpose of this chapter is to review the current understanding of the sources, causes, and representation of error when integrating remote sensing and GIS data sets.

In the second section of the book, which focuses on physical scale as applied to natural resources, Charalampos Kontoes and coauthors (Chapter 6) discuss remote sensing techniques for wildland fire management. Wildfires play a key role in structuring vegetation and landscapes worldwide and represent a significant threat to humans in fire-prone lands. In the Mediterranean region, where wildfires impact large areas and cause significant damage, climate change scenarios indicate an increase in fire risk, leading to high fire frequency and to extended fire seasons. Earth observation techniques and processing chains for the early detection and systematic monitoring and management of wildfires on an operational basis are of paramount importance for the effective deployment of fire suppression resources, the protection of human lives and infrastructural assets, and the promotion of evidence-based decision-making. Remote sensing and GIS provide powerful tools for acquisition, storage, retrieval, manipulation, and visualization of geospatial data; however, there are inherent trade-offs with respect to the spatial resolution and the revisit time of satellite acquisitions. This calls for the development of complex near-real-time data analysis at fine resolution scales, with powerful downscaling methods *via* the integration of multisource spatial data, including satellite imagery, meteorological data, and geospatial information.

Kontoes et al. describe the design, architecture, implementation, and evaluation of a real-time, automatic, robust, Internet-based, and fully operational fire detection, fire monitoring, burnt area mapping, and damage assessment system that serves the entire Greek territory. The system utilizes multiscale and multitemporal satellite data from geostationary coarse resolution to high-resolution polar-orbiting satellites and employs novel image-processing algorithms. The various modules have been wrapped into a single platform, named *FireHub*. The performance of FireHub was evaluated against real wildfire data using the historical data of the Hellenic Fire Brigade. The described solution contributes substantially to judicious wildland fire management and is fully exploited operationally by local fire management authorities without further processing.

In Chapter 7, Michael Bishop and Iliyana Dobreva discuss issues of scale and complexity in geomorphometry and mountain geodynamics. Mountain topography is the result of highly scale-dependent interactions among climatic, surface, and tectonic processes. Moreover, the complexity of mountain geodynamics is governed by process–form relationships, polygenetic evolution, and the operational scale dependencies of feedback mechanisms and systems coupling. Geomorphometry can be used to study the complexities of mountain geodynamics, as digital elevation models can be used to quantitatively characterize topographic parameters and assess process–form relationships. GIS-based spatial analysis, however, does not effectively address numerous issues of scale that permit us to go beyond empirical analysis. Subjective interpretation of scale dependencies and empirical thresholds and indices dominate our analysis of topography. Consequently, the objective of this chapter is to identify, characterize, and discuss various concepts of scale in order to address notoriously difficult issues in understanding surface processes, landforms, and mountain geodynamics. Specifically, they examine scale and complexity from numerous perspectives including space–time representation, spatial extent and parameter magnitude, anisotropy, structure and hierarchical organization, semantic modeling, indeterminate boundaries, and spatiotemporal dynamics. Their conceptual and quantitative treatments of these concepts of scale demonstrate the importance of morphometric information, its linkage to process and polygenetic topographic evolution, and the significance of a systems perspective that highlights the role of topography in characterizing the spatiotemporal scale dependencies of various aspects of mountain geodynamics.

In the last chapter in this section (Chapter 8), Gordon Green and colleagues explore some of the potentials of on-demand processing using downscaling of forest canopy structure as an example. Distributed processes now dominate the computational landscape, from cloud computing to mobile apps. On-demand access to spatial data is now routine. Web mapping services allow users to efficiently view large data sets by accessing small subsets on demand, and dynamic services are now a part of familiar applications like route navigation, weather forecasting, and traffic monitoring. Most products based on remotely sensed data are published in a static form, but on-demand processing offers some promising new possibilities, particularly related to problems of scale.

The third section of the book focuses on issues of multiscale analysis of urban areas and extraction of urban roads from combined data sets of high-resolution satellite imagery and lidar data. Dar Roberts and his coauthors (Chapter 9) examine how spectral mixing models can be used for multiscale analysis of urban areas. The clear relationship between urban surface cover and environmental response has prompted considerable research on the use of remote sensing to map vegetated and impervious surface fractions. Urban environments are challenging in that they are composed of extremely diverse surface types localized into objects that are smaller than the spatial resolution of most space-borne, and some airborne, systems. Airborne systems can provide data at scales from several centimeters to a few meters and thus can resolve fine-scale objects. High spatial resolution imagery's spectral range and resolution is frequently insufficient to discriminate urban materials such as soils from senesced grass or tile roofs from other roof types. Spectral variability significantly impacts the ability to discriminate materials in the urban environment. The combination of high spectral diversity and coarse spatial resolution relative to object size has led to a suite of specialized analysis approaches designed to address limitations of remotely sensed imagery and provide the necessary information on urban form and composition, in particular green cover and impervious surface fractions. A common approach is spectral mixture analysis (SMA), in which a mixed pixel is decomposed into fractional estimates of pure, subpixel components, often called *endmembers*. SMA is highly appropriate as a tool for estimating vegetated and impervious surface fractions in the urban environment because it is designed to retrieve subpixel fractions from images where the spatial resolution is coarser than objects in the environment. The authors use a modification of SMA called *multiple endmember spectral mixture analysis* (MESMA), which accounts for endmember variability by allowing the number and types of endmembers to vary on a per-pixel basis, to better identify urban land covers at multiple spatial scales. Here, the authors highlight the potential value of MESMA for mapping an urban environment by analyzing Advanced Visible and Infrared Spectrometer (AVIRIS) data acquired at 7.5-m spatial resolution. They produced a classification based on estimates of six urban material classes: nonphotosynthetic vegetation (NPV), green vegetation (GV), rocks, soils and sand, impervious ground, and roofs. Fractional cover of GV, NPV, paved, rock, roof, and soil was based on two-, three-, and four-endmember models. This study illustrates some of the potentials of MESMA, in the application of this technique to high spatial resolution AVIRIS data acquired over the cities of Santa Barbara and Goleta, California. They used MESMA both as a classifier and a tool for mapping fractional cover. MESMA proved to be extremely versatile in mapping urban land cover. Use of multiple endmembers and multiple levels of complexity provides a high level of detail, enabling mapping of a variety of impervious surface types.

The second chapter in this section (Chapter 10), by Minjuan Cheng and Qihao Weng, focuses on extracting roads from remote sensing data sources, which is an important task in urban remote sensing. Automatic feature extraction from remote sensing imagery reduces the need to perform time-consuming field surveys and is also easy to update in a timely manner. In recent years, various sources of remote sensing data have been widely used in feature extraction. This chapter explores the use of IKONOS data in conjunction with lidar data for urban road extraction in Indianapolis, Indiana, with a geographic object–based image analysis method. The data source, extraction process, and effectiveness were examined and compared by using an IKONOS image alone and by the combined use of IKONOS and lidar data. Results indicate that the extraction with the combined data set obtained higher accuracy through an analysis of four accuracy measurements. The chapter further explains why the combined data set can greatly improve the extraction quality and what sources of errors that were encountered.

The last section of this book (Chapter 11) looks at some very interesting applications of scaling remote sensing data for human health and social attributes. The first chapter in this section by Pietro Ceccato and colleagues provides observations on a number of the major human infectious diseases that plague the developing world that are sensitive to interseasonal and interdecadal changes in environment and climate. Monitoring variations in environmental conditions such as rainfall, temperature, water bodies, and vegetation helps decision makers at public health agencies to assess the risk levels of vector-borne disease epidemics and evaluate the impact of control measures. The International Research Institute for Climate and Society at Columbia University has developed products based on remotely sensed data to monitor those changes and provide the information directly to the decision makers. This chapter presents recent developments that use remote sensing and available ground-based observations to monitor climate variability and environmental conditions for applications in public health, specifically vector-borne diseases.

In the next chapter (Chapter 12), Guillaume Chabot-Couture provides an intriguing examination of scale in disease transmission, surveillance, and modeling. Certain infectious diseases spread over continents while others can be found only in remote focus areas. For example, the influenza virus can infect people anywhere in the world; vector-borne diseases like malaria, dengue, and yellow fever can be carried from the tropics to temperate zones by infected travelers who act as reservoirs. In contrast with these widespread diseases, poliomyelitis and Guinea worm are now found only in a few areas of Africa and Asia because of long-running global campaigns to eradicate them. As a starting point, he defines the scale of an infectious disease as the geographic distance and the time period over which significant changes in prevalence and/or incidence take place. Although infectious diseases can vary over widely different spans of time and space, scale can be discussed in terms of three main factors: the human mixing scale (a function of proximity and mobility), the scale of environmental determinants of transmission (e.g., airborne, waterborne, or vector-borne transmission), and the scale of disease control efforts (e.g., vaccination, treatment distribution, or disease prevention awareness). Today, efforts to map diseases of global health importance are expanding. The tools of digital epidemiology, specifically, remote sensing, GIS, low-cost handheld GPS receivers, and computer disease models, are increasingly enabling data collection on an unprecedented scale. If these efforts can be used to both increase the quality and the quantity of data available to fight diseases, we may be able to radically reduce the morbidity and the mortality that communicable diseases create around the world. And if we are persistent, we may even permanently wipe out some of these diseases from the planet.

Lastly, the chapter by Alex de Sherbinin (Chapter 13) describes lessons learned in assessing remote sensing and socioeconomic data integration in studies performed by the NASA Socioeconomic Data and Applications Center (SEDAC) located at Columbia University. Many of the core research questions of the Anthropocene are spatial in nature and require spatial data integration to provide answers: (1) Where are the populations most vulnerable to environmental changes located? (2) How do global environmental changes affect people, ecosystems, or production systems in a given location? and (3) What are the impacts of human activities in the coastal zone, mountainous areas, or drylands? This chapter provides examples of the integration of remotely sensed biophysical data with population and other socioeconomic data that illustrate the benefits of spatial data integration. The examples are organized into sections based on the ways in which population grids are used in research, followed by some examples of integration using other types of socioeconomic data. The chapter also addresses challenges in integrating data developed at different scales and for different purposes through sharing lessons learned from 20 years of operating SEDAC.

Insights from the Book

Scale can be a problematic concept that requires matching the remote sensing data used with the purpose of a particular study for analysis within GIScience applications. In this book, we try to provide supporting information on why the attributes and challenges of scale, although they have been examined in numerous journal articles and books, still matter with respect to advancing the integration of remote sensing data in GIScience. By presenting the 13 chapters here focusing on the four themes (scale issues, physical scale, human scale, and health/social scale), we hope that we have better elucidated the underpinnings of scale as a foundation for geographic analysis. We see through the chapters in this book how scale has vast implications on remote sensing data and GIScience integration by examining the following topics: (1) the past, present, and future of research within the disciplines of geography and geoscience within space, time, and space–time; (2) what the challenges and attributes are of geographic scale in examining complex systems and scaling geocomplexity in remote sensing data; (3) the revolutionary advances in geospatial technologies that have enabled the collection and generation of large amounts of geospatial data at varying space and time resolutions, and how geospatial data derived from many different sources at varying spatial and temporal resolutions can be stored in different models; (4) examples of how multiscaled remote sensing data can be used in applications related to the physical environment and geomorphometry and on-demand processing of forest canopy and structure; (5) how multiscaling as an integral factor in the classification of urban surface types is enhanced through Multispectral Endmember Spectral Mixture Analysis (MESMA), and how extraction of urban road networks using multiscaled remote sensing data is improved through Geographic Object Based Image Analysis (GEOBIA); and (6) how scaling of remote sensing data can be used for applications to human health and analysis of social attributes and vulnerabilities. Thus, it is our hope that this book will provide more insight into why the subject of scaling in remote sensing and GIScience is still an important topic for research, particularly in the applications of multiscaled data. Scale and its associated complexities are a fulcrum for GIScience. Research into the relationships of space, time, and spectral scales will become even more important as new remote sensing systems are brought on line, as will finding out how data from these sensors can best be integrated within GIScience from an application perspective.

References

Louis R. Iverson, Martin E. Dale, Charles T. Scott and Anantha Prasad, 1997. A GIS-derived integrated moisture index to predict forest composition and productivity of Ohio forests, USA. *Landscape Ecology*, 12:331–348.

Darren Ruddell and Elizabeth A. Wentz, 2009. Multi-tasking: Scale in geography. *Geography Compass*, 3:681–697.

1

On Scale in Space, Time, and Space–Time

Keith C. Clarke and Ian J. Irmischer

CONTENTS

Introduction

This chapter is a review and speculation on the past, present, and future of research within the disciplines of geography and geographic information science on the nature of space, time, and space–time, or what might be called *spatiotemporal geography*. The review is not intended to be comprehensive; indeed, there are several excellent extant review papers to which the reader is referred. Instead, the principal computational, geographic, and philosophical issues and challenges are the focus. Central to these is the geographic (and cartographic) concept of scale, which has profound implications for research in space–time, as in geography generally. The chapter covers first space, then time, and then examines their integration before concluding with some observations about contemporary research.

A Geography of Space

A fundamental primitive of geography from a descriptive perspective is a single measurement of position (Goodchild 1999; Goodchild et al. 2007). For much of the history of the discipline, space was sufficiently described by two independent values, a longitude and a latitude or x and y. Understanding of Earth's three-dimensional form and geodesy added a third value, elevation or height, requiring a geodetic model or ellipsoid in addition to the reference prime meridian, equator, and poles. Using three values allows any position on, below, or above the Earth to be measured and described fully. Adding an attribute allows geography to also state what is where, and attributes can be identifiers, names, classes, or measurements themselves. Such tuples (x, y, z, A) can represent point locations, centroids of areas, or sequences and surface models, lines, polygons, or even fields and volumes (Clarke 2015). Yet geographical description remains incomplete without a time stamp (x, y, z, A, t). Maps require dates, features require time fixes and durations, and no two measurements can be truly synchronous. A time associated with a space "locates" the place on a timeline, gives it a fix in both linear and circular time, and functions as a sort of attribute. Time is not, however, simply another spatial dimension, as is often supposed in four (or more)-dimensional geographic information systems (GIS; Raper 2000). This difference is because of time's unique nature, its variation in only one true direction. While we can regard time as a simple relative measure and measure it with extraordinary precision, time's "arrow" moves only forward. Descriptive geography becomes increasingly vague as one moves backward in time, but it becomes highly uncertain as we move forward (Goldstein et al. 2003). Indeed the best model of the future is the present, for two reasons: it is easier to ignore change than to predict or forecast it, and the change signal may not be larger than the expected error in the current and projected mapped distributions (Pontius and Lippitt 2006). Only since the advent of the global positioning system (GPS) has the systematic collection of lifelines, individual people's movements captured as traces of positions and time, been effective. Such data are now routinely used for tracking, behavioral analysis, crowd study, animal movements, and in managing fleets of vehicles. Increasingly, the source of such data is cellular telephones equipped with GPS, and the data are gathered from social media websites or customer databases.

Geography was characterized for much of its history as largely ignoring time (Kellerman 1989). Maps reflected the reality of the time at which they were made or showed thematic data at a snapshot in time or over a duration. Early mapping methods included flow maps and small multiples (Tufte 1983). Derivatives and rates can be displayed using static methods such as choropleth maps, but it was not until animation (Acevedo and Masuoka 1997) and the Internet (Peterson 1995) that dynamics found their way into the suite of cartographic methods (Kossoulakou and Kraak 1992; Mitasova et al. 1995). A notable exception is the work of Tobler (1970).

Meanwhile, geography was slowly assimilating the conceptual frameworks necessary for adding the time dimension. While Sauer (1941) had stressed the need for a historical approach to geography, the contributions of Hagerstrand cannot be understated, with the original contribution coming from the study of migration using spatial potentials, waves of innovation, and a focus on individual movements centered around space–time (Hägerstraand 1970). Among his key ideas was the time–space prism, a means of representing the set of possible movement paths for an individual given constraints on their movement, in which two-dimensional space was shown as a perspective map and time formed a vertical axis. Hagerstrand's concepts were quickly refined, expanded, and disseminated by Pred (1977) and Thrift (1977). Miller (1991) produced a formal computational geometric implementation of Hagerstrand's time prism, for use in GIS and in analysis. This implementation was later used by Kwan (1998) to examine the constraints on human movement present in tracking data. Later, major contributions by Langran (1992) and Peuquet and Duan (1995) led to an event-based temporal model that was consistent with first-generation GIS. In this model, the simplistic frame-based or snapshot model as used in raster GIS, animation, and cartography was replaced by the event-based spatiotemporal data model. Here, time essentially became an attribute of map features and so could be used to sort and analyze them or to construct a map geometry at any specific time. Various methods emerged in GIS for handling spatiotemporal data and even processes enumerated by Goodchild (2013), who laid out seven means by which GIS have dealt with dynamics to date. In complexity theory, time is an inherent part of the concerts of emergence and phase transitions (Holland 1998) and key to the temporal behavior of cellular automata (Goodchild 2013).

Space–time geography remains an active field of study. Various research initiatives have taken place, such as that of the National Center for Geographic Information and Analysis (Egenhofer and Golledge 1998) and that reported by Yuan and Hornsby (2007). Key recent review papers include those authored by Andrienko et al. (2003), which comprehensively explored the visual analytics of space–time geography; Long and Nelson (2013), which focused on space–time analytical methods; Goodchild (2013) in the context of GIS; and a revisit of the time prism with a new focus on decision-making (Forer et al. 2007). Recent trends are the overlap of space–time geography with research in computer graphics and machine vision in the context of modeling multiple moving objects (Hornsby and Cole 2007); links to the study of history (Goodchild 2008); and the continued search for new conceptual models (Yi et al. 2014). The topic has received considerable attention at recent meetings of the Association of American Geographers, with multiple paper tracks and sessions now an annual event.

A Geography of Time

Why should the concept of time be different from a geographical perspective? Philosophically, there are two different approaches to the conceptualization of time. Newton's realist view of time holds that time is part of the underlying structure of the whole physical universe—a dimension that is independent of events and in which specific events occur in an order or sequence, as along a timeline. An alternative approach is that the concept of time does not refer to any kind of finite bucket chain that events and objects pass along as in a timeline but instead is part of a conceptual structure with which humans comprehend, sequence, organize, and compare events. This approach, called the *Leibnizian view*, assumes that time is irrelevant outside of the events and things that constitute the world and our experience and consequently that it is not measurable. Newton's view assumes that space and time exist independent of their content, whereas Leibniz defined *space* and *time* by the objects and events they contain. Einstein's theories of relativity led to the important consideration that time depends on other natural forces and noted the importance of the speed of light as a universal constant. Time is also one of seven fundamental physical quantities in the International System of Units, in which it is used to define other quantities, for example velocity, and so forms a basis for measurement.

A critical property of time is the human ability to divide it into even and uneven periods or eras, each with a duration. This ability is inherent in human language and semantics, with the use of common language terms such as *before*, *during*, and *after*. Breaks in the time periods are events, which can be instantaneous, transitional, or cyclical. Examples are the history of wars, geological eras, and the cycle of US elections. Cyclical time implies a granularity or resolution. Time can be instantaneous (strontium clocks are reported to be accurate to within 10^{-16} seconds), clock-related, daily, weekly, monthly, seasonal, annual, decadal, or any metacombination up to the geological era or universal history. Periodic time has its own set of analytical methods, time series analysis based on harmonic, Fourier, and wavelet analysis. These methods allow any time-variant quantity to reveal whatever cycles are internally present and other structures.

Time measures can be precise or approximate. A time stamp from GPS records instantaneous time, but similarly time can be measured by approximation or generalization. Archaeology often measures time using proxies: styles of building, pottery, or type of stone tools. Geological eras also have fluid boundaries and depend on the presence of minerals and fossils in rocks. This includes approximate time, which allows for overlap and slow change. Lastly, the sampling theorem that applies in space (Tobler 2000) also applies in time. Sampling at any regular time interval *t* means that events with shorter durations will be invisible. This is particularly true of decennial censuses, webcams, and weather records.

The implication of these similarities is that space and time are similar from a scaling perspective. Just as polygons can be merged into super-classes, durations can be merged into longer periods, for example geological eras, reigns of monarchs (e.g., Egyptian history), and periods of civilization (e.g., the Middle Ages). The same issues of boundaries, overlap, uncertainty, and approximation apply. Nevertheless, this is only true in retrospective, that is, with time running backwards. No such scaling applies to the future, in spite of the fact that cyclical time allows the finite buckets to be easily created (e.g., next week, next year, the twenty-second century). Division also applies, for example, a major event (the assassination of Archduke Ferdinand in Sarajevo on June 28, 1914) can be divided into almost unlimited subevents with their own sequence, durations, and sub-durations. The conclusion is that time and space do indeed behave as quantifiable dimensions, three for space and one for time, but only theoretically for the future. Therefore, all we know about cartographic generalization—including selection, simplification, combination, and displacement—also applies to time, at least in the Newtonian model. These apply to measurement dependencies, such as interpolation, extrapolation, and inference. However, whereas geographic space has the independent variables x and y and the constrained variable z (due to the Earth's surface), t is not independent of location. Its constraints (Forer et al. 2007) are similar to those regarding spatial similarity as stated by Tobler's first law (Sui 2004). The temporal equivalent might be that history is bunk, but recent events are more related than past ones. A spatiotemporal equivalent would hold that events that occur nearby and at about the same time are more related than others. This has obvious implications in geography and GIScience for crime analysis, landscape hazards, human geography, and tracking applications.

A Geography of Space and Time

Relatively few studies within geography take an integrated view of space and time, what might be called a *spatiotemporal* or *space–time* view. The geographical circumstances under which such systems are desirable were enumerated by Worboys (1994). Although the ease of time measurement is now ubiquitous, geography and GIScience are still struggling with conceptual models of time that can be implemented as computational systems. Even a precise ontology is lacking, although there are many useful contributions (Galton 2001; Tilly 1995). Once such an ontology exists, geography will be able to move beyond description and mapping (though these will remain important) to analysis, forecasting, and prediction with an authority lacking today in what might be called *space–time intelligent systems* (Jacquez et al. 2005). In fact, the promise of large numbers of new systems, products, devices, and applications once such an understanding is in place is large. Intelligent scheduler apps that handle appointments and travel, delivery systems that truly support just-in-time production and planning, and data mining applications that identify and act on particular circumstances will become simple commodities.

Long and Nelson (2013), in their treatment of spatiotemporal analysis, identified seven methods for dealing with such data. These methods are time geography, path description, similarity indices, pattern and cluster analysis, comparison of individual and group dynamics, and analysis of spatial fields and of spatial range. Long and Nelson considered issues of scaling, noting the levels of instantaneous, interval, episodic, and global time; the problems of sampling to capture events; and the fact that analytic methods depend on granularity and data volume, just as in spatial analysis. Goodchild (2013) posed three questions worthy of geographic space–time research: the visualization of space–time data, the measurement and portrayal of uncertainty in such data, and the meaning of scale in time, space, and space–time.

The critical nature of spatiotemporal scaling is a function of the defining measures. This discussion began with the spatial and temporal primitive, but space–time primitives that unite these dimensions are velocity, extent, flux, granularity, form, and process. It is necessary to invoke a process, given the Leibnitz time model. The process can take multiple forms, but it is essentially a reaction describing the movement of objects or changes in static values at a place. For simplicity, these objects will be termed "*d*-objects"—that is, they exist in a delta or derivative space. The simplest is a single point moving in space (trajectory). More complex processes are multiple moving points, changing fields, changing features, or evolving patterns. Among each of these, velocity is a vector, with magnitude and direction, but also a speed or rate in which time is part of the denominator. The extent is the spatial footprint of the process. Key to the process is how long the *d*-object takes to cover the extent. A 1-km^2 zone, for example, has a diagonal of 1.4142 km—a car moving at 60 km/h (1 km/min) could cross this space in 1 minute 25 seconds. Any spatiotemporal measure with a temporal sampling frame of less than this would miss the car entirely and any less than 42.5 seconds would not permit the calculation of a velocity at all. Flux corresponds to the quantity associated with a vector; it is both the rate of flow of matter or energy associated with the *d*-object and a quantity expressing the strength of a field of force in a given area. This latter quantity is what earlier spatial analytic research called a *potential* or *propensity for spatial interaction*. It could be the mass of the car, the energy it expends in crossing the square kilometer, or the forces of gravity, air resistance, and friction it must overcome to move (a reaction, but equal to the forward force). As measurements, these values are subject to the same scaling phenomena as geographic data: the sampling theorem's limitations, the hierarchical nature of the measurement systems, uncertainty, and the ratio of granularity to extent.

Form and process are different entirely. While they are to a great extent determined by the type and number of *d*-objects, it is possible to recognize processes that repeat often in geography. Yi et al. (2014) used the aggregate descriptors of core points, footprint borders, and composite borders as spatial *d*-objects. They noted that change can take place to objects based on their identity. Existential states are existing, nonexisting with history, non-existing without history, and transitional, that is, changes in identity. Spatial objects can then join or combine, distribute, be destroyed, reappear and disappear, and be born. A full set of relations was proposed for group moving objects as follows: concurrence (objects moving at the same velocity and in the same direction), trend-setting (anticipating the motion of other objects), flocking (a cluster moving as concurrent objects), leadership/following (when the motion of one *d*-object is assumed by others), convergence (defined as objects passing through the same region), encounter (when objects meet), and convoy (when objects move together in sequence). Although these behaviors were created to deal with flow patterns in the South China Sea from a large empirical data set, nevertheless there are many geographic phenomena that share this behavior, including humans and vehicles.

Some time–space behaviors are well known, such as the pattern of events and outcomes that are associated with tornados, which reoccur with subtle differences in fluxes, velocities, and locations (McIntosh and Yuan 2005). These are the processes that produce spatial distributions over time: dispersal, diffusion, agglomeration, clustering, navigation, random movement, bifurcation, and attraction. Other processes characterize ecological and biological phenomena: home range following, competition, invasion, succession, evolution, and retreat. Others are derived from physics: dissolution, precipitation, agglomeration, reaction–diffusion, coalescence, collision, avoidance, and repulsion. The processes creating these patterns are only visible over time sequences, usually when holding space constant. Just as interesting are fixed process sequences varying in space (e.g., hurricanes) but differing in some way. In many cases, these sequences can be described by language, narratives, and semantics (e.g., collective political events in historical England as "action phases"; Tilly 1995). In all cases, these processes leave a spatial pattern that varies in space–time, a sort of spatiotemporal signature. Such signatures are available for quantification and description, and the linking of their forms over time can reveal information about sequence that could lead to more general theory. Of interest is whether these processes follow geographical rules and laws and whether they are predictable.

A final observation on the scaling of space–time relates to the time-prism model and decision-making. Forer et al. (2007) in their work on human movement identified the needs associated with time–space trajectories. They saw an appointment in time (and space) as a contract—the voluntary acceptance of a constraint on one's position at a specific time (e.g., to board an airplane)—that is characterized by opportunity, purpose, the location of an individual, mode of travel, coupling windows, and risk avoidance. Naturally, these are scale-dependent but reflect the asymmetry of time. For example, it is acceptable to arrive at a train platform any amount of time before the departure of a train, but not after. Forer et al. hinted that the form of the contract is information. Such information is increasingly abstract and virtual, for example, purchasing an electronic ticket for a plane flight. The importance of such a contract is focal on a specific time (becoming more critical as the time approaches) and space (more spatially constrained closer to the departure time). Such a contract implies a spatiotemporal link between the future and the present that takes the form of information. One can imagine a vast array of time–space prisms associated with each decision and contract that forms part of our daily activities, a suite of time–space contracts arriving as information flows from the future. Some of these are routine and fixed for long periods (workplace, home); others are more temporary (travel, recreation) and still others more random. In other words, they are scaled. While software tools such as Oculus' GeoTime, USA, are already available for visualizing the dynamics of space–time prisms and multiple moving objects at a specific scale, including all or even some of these contracts seems overwhelming, especially given the increasing availability of tracking data from social media. No software or method yet exists that can integrate process information across spatial and temporal scales. Data mining methods will need to encapsulate knowledge about contracts, process, fluxes, and predictable spatial forms if they are to be effective in the future.

Conclusion

In this chapter, the literature and approaches within Geography and GIScience to spatial, temporal, and spatiotemporal data have been surveyed. The increasing availability of time-stamped data (from remote sensing with multitemporal and repeat images, from GPS, and from sensors and mobile devices) means that at present technology is far ahead of theory in dealing with space–time. In spite of some impressive developments in geography dating from the classic work of Hagerstrand and his successors, it seems that the discipline will be swimming in data with little new in the way of theory for some time to come. Space–time methods have the potential not just to increase explanation, understanding, and prediction but are likely to guide technology towards new markets and beneficial applications to humankind. In spite of several recent concerted efforts to garner the research, it may take a coordinated, centrally led, transdisciplinary effort to make further progress. Contributions will come from philosophy, mathematics, physics, semantics, and computer science in addition to geography and GIScience. When such theory emerges, the now-vague analogies between dimensions may be clarified and the future may become more lucid.

References

Acevedo, W. and Masuoka, P. (1997). Time-series animation techniques for visualizing urban growth. *Computers & Geosciences* 23(4): 423–436.

Andrienko, N., Andrienko, G. and Gatalsky, P. (2003). Exploratory spatio-temporal visualization: an analytical review. *Journal of Visual Languages and Computing* 14(6): 503–541.

Clarke, K. C. (2015). *Maps and Web Mapping*. Upper Saddle River, NJ: Pearson eBook.

Egenhofer, M. J. and Golledge, R. G. (1998). *Spatial and Temporal Reasoning in Geographic Information Systems*. Oxford: Oxford University Press.

Forer, P., Huisman, O. and McDowall, C. (2007). Dynamic prisms and 'instant access': linking opportunities in space to decision making in time. In *Societies and cities in the age of instant access*, Miller H. J. (ed.), Berlin, Germany: Springer.

Galton, A. P. (2001). Space, time, and the representation of geographical reality. *Topoi* 20(2): 173–187. doi:10.1023/A:1017913008827.

Goldstein, N. C., Candau, J. T. and Clarke, K. C. (2003). Approaches to simulating the 'March of ... and Mortar'. *Computers, Environment and Urban Systems* 28(1–2): 125–147.

Goodchild, M. F. (1999). Measurement-based GIS. In: *International Symposium on Spatial Data Quality*, Department of Land Surveying and Geo-Informatics, Hong Kong Polytechnic University, Hong Kong.

Goodchild, M. F. (2008). Combining space and time: New potential for temporal GIS. In: *Placing History: How Maps, Spatial Data, and GIS Are Changing Historical Scholarship*, A. K. Knowles and A. Hillier (eds.). Redlands, CA: ESRI Press, pp. 179–198.

Goodchild, M. F., Yuan, M. and Cova, T. J. (2007). Towards a general theory of geographic representation in GIS. *International Journal of Geographical Information Science* 21(3): 239–260.

Goodchild, M. F. (2013). Prospects for a space–time GIS. *Annals of the Association of American Geographers* 103(5): 1072–1077.

Hägerstraand, T. (1970). What about people in regional science? *Papers in Regional Science* 24(1): 7–24. doi:10.1111/J.1435-5597.1970.TB01464.X.

Holland, J. H. (1998). *Emergence: From Chaos to Order*. Oxford: Oxford University Press.

Hornsby, K. S. and Cole, S. (2007). Modeling Moving Geospatial Objects from an Event-based Perspective. *Transactions in GIS* 11: 555–573.

Jacquez, G. M., Goovaerts, P. and Rogerson, P. (2005). Space-time intelligence systems: Technology, applications and methods. *Journal of Geographical Systems* 7(1): 1–5. doi:10.1007/s10109-005-0146-7.

Kellerman, A. (1989). *Time, Space and Society: Geographical-Societal Perspectives*. Boston, MA: Kluwer Academic Publishers.

Kossoulakou, A. and Kraak, M.-J. (1992). Spatio-temporal maps and cartographic communication. *Cartographic Journal* 29(2): 101–108.

Kwan, M.-P. (1998). Space-time and integral measures of individual accessibility: A comparative analysis using a point-based network. *Geographical Analysis* 30(3): 191–216.

Langran, G. (1992). *Time in Geographic Information Systems*. Boca Raton, FL: CRC Press.

Long, J. A. and Nelson, T. A. (2013), Measuring Dynamic Interaction in Movement Data. *Transactions in GIS* 17: 62–77.

McIntosh, J. and Yuan, M. (2005). Assessing similarity of geographic processes and events. *Transactions in GIS* 9(2): 223–245.

Miller, H. J. (1991). Modelling accessibility using space-time prism concepts within geographical information systems. *International Journal of Geographical Information Systems* 5(3): 287–302.

Mitasova, H., Mitas, L., Brown, W. M., et al. (1995). Modelling spatially and temporally distributed phenomena: New methods and tools for GRASS GIS. *International Journal of Geographical Information Systems* 9(4): 433–446.

Peterson, M. P. (1995). *Interactive and Animated Cartography*. Englewood Cliffs, NJ: Prentice Hall.

Peuquet, D. J. and Duan, N. (1995). An event-based spatiotemporal data model (ESTDM) for temporal analysis of geographical data. *International Journal of Geographical Information Systems* 9(1): 7–24. doi:10.1080/02693799508902022.

Pontius, R. G., Jr. and Lippitt, C. D. (2006). Can error explain map differences over time? *Cartography and Geographic Information Science* 33(2): 159–171.

Pred, A. (1977). The choreography of existence: Comments on Hägerstrand's time-geography and its usefulness. *Economic Geography* 53(2): 207–221. doi:10.2307/142726.

Raper, J. R. (2000). *Multidimensional Geographic Information Systems*. London: Taylor and Francis.

Sauer, C. O. (1941). Foreword to historical geography. *Annals of the Association of American Geographers* 31(1): 1–24. doi:10.2307/2560961.

Sui, D. Z. (2004). Tobler's first law of geography: A big idea for a small world. *Annals of the Association of American Geographers* 94: 269–277.

Thrift, N. (1977). *An Introduction to Time-Geography*. Norwich, UK: Concepts and Techniques in Modern Geography. Geo Abstracts, University of East Anglia.

Tilly, T. (1995). *Popular Contention in Great Britain, 1758–1834*. Cambridge, MA: Harvard University Press.

Tobler, W. R. (1970). A computer movie simulating urban growth in the Detroit region. *Economic Geography* 46: 234–240.

Tobler, W. R. (2000). "The development of analytical cartography: A personal note," special content issue on the nature of analytical cartography. *Cartography and Geographic Information Science* 27(3): 189–194.

Tufte, E. R. (1983). *The Visual Display of Quantitative Information*. Cheshire, CT: Graphics Press.

Yi, J., Du, Y., Liang, F., Zhou, C., Wu, D. and Mo, Y. (2014). A representation framework for studying spatiotemporal changes and interactions of dynamic geographic phenomena. *International Journal of Geographical Information Science* 28 (5): 1010–1027.

Yuan, M. and Hornsby, K. S. (2007). *Computation and Visualization for Understanding Dynamics in Geographic Domains: A Research Agenda*. Boca Raton, FL: CRC Press.

Worboys, M. F. (1994). A unified model for spatial and temporal information. *The Computer Journal* 37(1): 26–34. doi:10.1093/comjnl/37.1.26.

2

Complexity and Geographic Scale

Dolores Jane Forbes

CONTENTS

Introduction

This chapter came about from a struggle to engage with the complex problem of emergent super algal blooms in an estuary, the northern Indian River Lagoon (IRL) (Figure 2.1), on the east central coast of Florida. Climate change was initially deemed responsible for the emergent condition, but estuaries are complex "systems of systems," with perhaps unknown elements that interact in as yet unknown ways. The list of possible causes is lengthy and crosses disciplinary boundaries, including nutrient overloading from land use change, increasing human population and economic growth, overfishing, extreme weather events, and phytoplankton biology and metabolism.

By definition, an *estuary* is a body of water where freshwater from rivers and streams mixes with saltwater from the sea: "the link between land and sea" (Van Arman 1987, 1). The problem of determining causal factors in a complex system is compounded when the system of interest is also directly connected to distinctly different complex systems. Questions about approaches

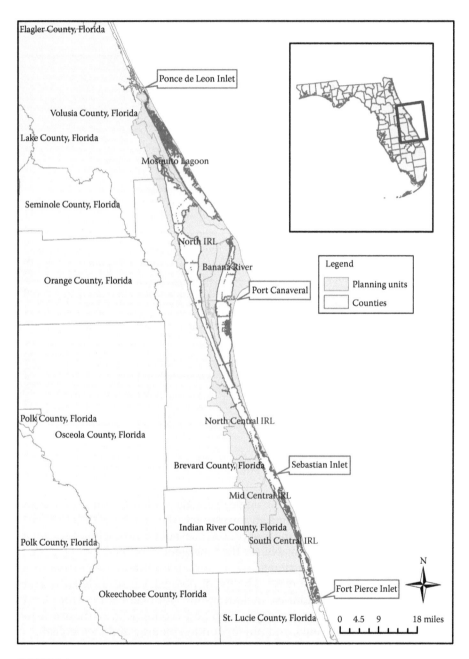

FIGURE 2.1
Northern Indian River Lagoon Estuary, Cape Canaveral Basin, Florida.

for separating cause and effect in such interconnected complex systems led to an exploration of the complexity approach, an approach increasingly being embraced by a wide range of academics and professionals for dealing with pressing social–ecological challenges (Rogers et al. 2013).

Learning about the complexity approach involved readings across disciplines including complexity, resilience, sustainability, ecology, and geography as well as physics and biology. Geographic subdisciplines were also useful, including land systems science, geomorphology, and landscape ecology, among many others. The literature on scale is also interdisciplinary. The definition of complexity, like the definition of scale, is also contested. The results of these explorations of complexity and scale are presented here, where an argument is constructed to emphasize that scale is critical to engaging with and understanding complex geographic systems.

Geography is the study of the world in which we live, composed of both living and nonliving elements and entities. Geographers "synthesize the understanding of physical and human systems derived from the pure and human sciences and analyze and describe the interaction and dynamics of these systems" (Richards 2002, 99). It is possible to conceptualize the geographic domain, the Earth and its atmosphere, as a complex system of systems. Doing so focuses on the interrelationships and the interconnectedness between systems and between internal system elements and external elements. This conceptualization aligns with geographic practice.

Geography also offers a long history of research on scale, as well as a deep understanding of and expertise in systems that span spatial and temporal scales (Manson 2009, 77). Scale, and more specifically geographic scale, is the foundational reference for disciplines that study the Earth, its processes, and inhabitants, and is important to understanding complex systems (Manson 2008). Scale identification techniques are employed for analyses of Earth systems and processes, including observing and measuring natural and disturbed systems, designing and analyzing experiments, and setting management objectives and policies (Gardner 1998). Scale-identification techniques allow for the inclusion of the human with the physical (Sayre 2009), separation of impacting factors (Wilbanks 2002), and better understanding of system dynamics by locating relationships, change, thresholds, and dissimilarities in space and time.

Every discipline defines its own terms, which is an aspect of science that can frustrate evaluations of transdisciplinary approaches to problems. The definition of *scale* used here is a basic one from ecology: *scale* is the physical dimension of a thing (O'Neill and King 1998); this is *scale* as a descriptor of space. Defining *scale* as the measure of something extends from space to include time, so a *timescale* is defined as an interval, a measure of time, in the way that spatial scale is an interval or a real measure of the Earth. *Geographic scale* thus defined anchors processes and observations to place and time.

Complexity science as the study of complex systems is, like geography, a broad and multidisciplinary topic. Complexity is not confined to any one discipline, nor is it entirely or solely about algorithms, chaos theory, and solving systems of nonlinear equations. The aim of this paper is to encourage geographers to engage with complexity science by illustrating the importance of geographic scale for examining complex systems. Rather than attempting to measure the value of complexity in terms of geography, perhaps the better question is what value geographic practice has for the study of complex systems. Examining complexity from the perspective of geographic scale yields insight into geography's ability to engage with and advance understanding of Earth's complex systems.

This chapter describes complex topics from a pragmatic approach in what is hoped is an accessible style. This makes the topic suitable to geographers of varying levels of expertise. Terms are explicitly defined in everyday language. It is hoped that readers can connect to the material with or without specialized disciplinary knowledge and that they might draw on their own observations of the world to connect to and support (or critique) the ideas discussed. This chapter is also intended to be of interest to both human and physical geographers.

Systems

Conceptualizing the Earth as a complex system of systems is an attempt to conceptually model a complex reality. Doing so focuses attention on the interrelationships and the interconnectedness between systems and other systems, between system components or elements and other components. The Earth as a complex system of systems (Cilliers 1998) requires defining what is meant by the term *system*. We often talk about natural systems, ecosystems, political systems, and environmental systems as types of systems defined by specific processes performed by entities. As a broad and basic definition, systems are composed of elements (or entities) that interact and are interrelated. General classifications for systems are *simple, complicated*, and *complex*. Academic and practical disciplines define the distinctions between these classes differently.

Disciplinary classification of systems differentiates between simple, complicated, and complex based on the number of parts in the system, the amount of information needed to describe the system, or how accurately the system can be predicted. Simple systems have few parts, require little information to describe them, and are fully predictable and "easily" modeled or solved using mathematical or statistical principles (Figure 2.2). Complicated systems have more parts than simple systems, require more information to describe them, are fully predictable, and therefore can be fully modeled, although perhaps not as easily or simply as a simple system. Complex systems also have many parts, require vast amounts of information to describe them,

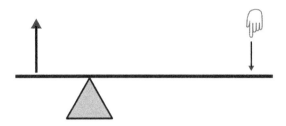

FIGURE 2.2
A lever as an example of a simple system.

are not predictable or not fully predictable, and are therefore difficult to model accurately or cannot be fully modeled.

These definitions highlight a specific problem in classifying systems using imprecise terms such as a threshold for the number of parts, the exact definition of *easily understood*, or the "least" amount of information needed to describe the system. This leads to questions as to the exact number of elements that differentiate a complicated system with fewer parts from a complex system with many parts.

The scale at which a system is viewed determines the number of parts perceived. A tide pool observed from a distance appears to be a simple system: a puddle of water left behind by the tide that can be reasonably measured and described. As the distance between the observer and the tide pool diminishes, additional features and elements might be observed in the tide pool. Using a microscope to enhance human visual acuity leads to the realization that what appears to be just seawater contains microscopic life; there are many more parts of the system than were seen at a distance.

Classifying a system as complex as opposed to merely complicated based on the number of elements is therefore problematic given that the number of parts observed is scale dependent. Holling (1994, 598) bases the classification of a system as complex entirely on perspective: "The complexity of a system is in the eye of the beholder." For example, the aerial photo of the Pelican Island National Wildlife Refuge in Figure 2.3b shows a portion of the IRL estuary from an oblique angle above the surface of the water. Decreasing the distance or changing the angle between the observer and the estuary will change the number of elements perceived. In the image, individual birds are white specks on the image and are difficult to see. Images taken closer to the island might reveal additional birds or additional species. From an underwater perspective, the number of elements in the system again changes. The image itself sets a boundary as to what is included and seen. Anything to the left or right, bottom or top of the image frame is excluded.

Determining the number of parts in a system is therefore dependent on the scale of observation, the distance from which the system is observed, and the perspective chosen as the frame of reference (Holling 1994). The image of Pelican Island in Figure 2.3 does not reveal all species or all elements within

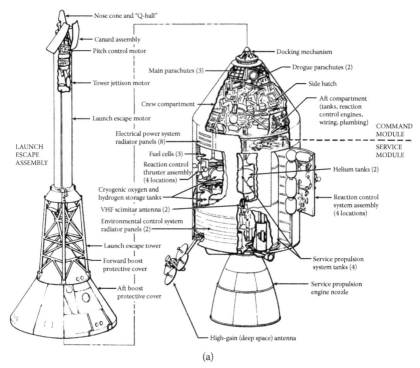

(a)

(b)

FIGURE 2.3
A comparison between a complicated system and a complex system. (a) The Apollo modules
and (b) An aerial photo of the Pelican Island National Wildlife Refuge.

the estuarine basin (Figure 2.1). Furthermore, the image depicts only a very small portion of the northern IRL basin shown in Figure 2.1. Pelican Island is located just south of the Sebastian Inlet (Figure 2.1). Pelican Island therefore does not describe the entire northern IRL basin. The greater northern IRL (Figure 2.1) varies widely as to plant, land, and marine species; water depth; economic activity and human population; agriculture; tidal variation; and salinity, as just a few examples.

Weinberg (1975), a computer scientist with an interest in systems theory, developed a systems classification of small-number, middle-number, and large-number systems. Small-number systems can be completely described by a system of 2^n equations, where n is the number of system parts or elements. Small-number approaches include classical Newtonian dynamics or differential equations for predator–prey models (King 2012, 187). The mathematical expression devised by Weinberg (1975) defines a specific threshold between simple and complicated/complex for system classification. Yet determining n (the number of parts) is dependent on both scale and perspective, as in the example tide pool or the examples in Figure 2.3. Something else is needed to differentiate complicated systems from complex systems.

Large-number systems in the Weinberg classification have a large number of parts and can be described by the statistical properties of the collective set of elements, making these systems predictable. Large-number systems have a large number of elements that are "alike" such that they can be summed or averaged without loss of accuracy; they are predictable and can be modeled with reasonable accuracy. Middle-number systems are systems that have too many parts to be described using small-number approaches, and too few parts to be accurately described by statistics. An estuary is considered to be a middle-number system using this method of classification.

The distinction made by Weinberg between large-number and middle-number systems then has to do with the heterogeneity of the parts, not the number of parts. For example, a volume of air is a large-number system comprised of millions of like molecules that, without loss of accuracy, can be assumed to be the same and therefore summed or averaged (O'Neill and King 1998). The volume of air is predictable in the sense that different volumes of air can be calculated with reasonable accuracy in relation to volumetric measures. This calculation is not possible for "middle-number" systems, which are comprised of a large number of heterogeneous parts. Assumptions of heterogeneity, averages, or sums cannot necessarily be imposed on middle-number systems without loss of accuracy.

It is important to note that heterogeneity of the parts in a system is also dependent on scale. The example of the tide pool suggests this. In Figure 2.3a showing the Apollo modules, it is understood that only the major systems are shown in the drawing. Individual subsystems, each of which are designed to perform a different, specific task, are much more complicated than as depicted in the drawing. The heterogeneity of the parts of the complicated system is also scale dependent. To render all parts in these complicated systems would require very detailed drawings. Intuitively, we understand the heterogeneity of the parts in the estuary shown in Figure 2.3b.

Distinguishing a complicated system from a complex system therefore cannot be done using either the number of parts or the heterogeneity of the parts. Both of these system or element attributes are scale dependent. What then distinguishes the Apollo module in Figure 2.3 on the left from the complex estuary on the right is the predictability of the outcomes of the system (Meadows 2008). We know from historical analysis that the Apollo program was successful in meeting its goals, including scientific exploration and, more importantly, bringing the astronauts safely back to the Earth's surface. The Apollo systems behaved as designed (Apollo 13 notwithstanding) and this behavior was planned for and predicted. The portion of the estuary shown Figure 2.3b may behave in an unpredictable manner. The emergence of super algal blooms in an estuary where they have not been known to occur is only one example of this unpredictability. It is this unpredictable quality that differentiates complex from complicated systems.

Homogeneity of system parts, where all the parts are similar and where this similarity is scale and perspective invariant (e.g., a large-number system), lends some measure of predictability to the system. It is this predictability that differentiates large-number systems from middle-number systems and complicated systems from complex systems. The elements of a complex system are sufficiently heterogeneous that the ability to accurately predict the behavior of a complex system may be impaired. Because the number of parts is also scale dependent, the distinction between *complicated* and *complex* ultimately resides in the predictability of system behavior (Meadows 2008).

The Weinberg system classification was originally embraced to support hierarchy theory, the somewhat arbitrary imposition of hierarchical "levels," for examination of complex ecological systems. However, hierarchy theory has been shown to be problematic as a general approach to complex systems analysis (O'Neill and King 1998). The problem is that when the resolution (the sampling intervals) or the extent of the observation set is changed (in space or time), the data essentially change (O'Neill and King 1998). O'Neill and King (1998) argued that hierarchical levels should be extracted from data, not imposed. Any change to the data set, or a change in the spatiotemporal extent, or a change in the sampling intervals, may result in the need for new levels, a different hierarchy.

Complex Systems

Geoscientists assign Earth systems to both the complicated and complex classes (Kastens et al. 2009). Earth systems are complex in that they exhibit nonlinear interactions, multiple stable states, fractal and chaotic behavior, self-organized criticality, and non-Gaussian distributions of outputs (Kastens et al. 2009). Earth systems are also *complicated* in the ordinary sense of the word where multiple mechanical, chemical, biological, and anthropogenic processes may be active and interacting at the same time and place (Kastens et al. 2009). It is possible to connect the ordinary sense of *complicated* to this discussion by saying that some portion of Earth system processes may be predictable.

In Earth systems, different processes may be occurring in the same place at the same time, or they may also occur at different spatial and temporal scales, based on the entity performing the interaction that defines the process. For example, the scale of a fish's activities may differ from the scale of a dolphin's activities, which differs from the scale of a human's activities, all of which can occur simultaneously in an estuary. The fish, dolphin, and human may also be interacting and their activities may be overlapping in space and time.

Complex systems exhibit these multiple processes operating at numerous spatial scales, inclusive of both social and natural activities. To Cilliers (1956–2011), an electrical engineer with an interest in systems who became a Professor in Complexity and Philosophy (https://www.linkedin.com/in/paul-cilliers-78899331) at University of Stellenbosch in South Africa, complex systems are usually associated with living things such as a bacterium, the brain, social systems, or natural language (Cilliers 1998). Cilliers (1998) asserted that another characteristic of complex systems that distinguishes them from complicated systems is that not only do the system elements interact, but this interaction is dynamic; interactions between elements in complex systems change over time. The Apollo modules as a complicated system were engineered specifically such that interactions between elements did not change, ensuring that the mission met its goals.

Dynamism

Complex systems characterized as having changing interactions over time aligns with consideration of Earth systems as dynamic and heterogeneous to space as well as time. Complex systems have a history; not only do they evolve through time, but the past is co-responsible for their present behavior, so time must be considered (Carpenter et al. 2009; Cilliers 1998). Change occurs in space over time within the geographic domain, sometimes "quickly," sometimes "slowly." Humans regularly observe changes in space over time as cities expand or decay, political power shifts from one interest group to another, or super algal blooms appear in an estuary where they have not previously been known to occur.

The dynamical nature of complex systems is also scale dependent with respect to time, where "the same ecological dynamics may be considered transient or in steady state, depending on the scale of observation" (Wu and Loucks 1995, 446). Observations made over short timescales may suggest that little has changed. Observations made over geologic timescales show that much may have changed. The perception of change or lack of change is only apparent relative to the spatial and temporal scales at which the system is observed.

The dynamic nature of complex systems also means that complex systems operate under conditions far from equilibrium with respect to time (Cilliers 1998). Equilibrium can be thought of as a lack of change where there is no interaction, no motion, no development, no flow, or no inputs to the system. Equilibrium is equivalent to "death" (Cilliers 1998) in the same way that humans die without inputs of air, water, and food. There has to be a constant flow of energy inputs to the system (in some form) to maintain the organization of the system, to ensure its survival, and to fight entropy (Cilliers 1998).

Holling et al. (1995) characterized the dynamics of complex systems as shifts or jumps between states. The definition of a *state* for a complex system used here is a scale that exhibits stable structure, interrelatedness, and interconnectedness. Another way of considering a *stable* state is to consider a stable state as a specific location at a specific time in which process–pattern relationships are consistent (Meentemeyer 1989; Stallins 2006; Wiens 1989). Examples of systemic states for the Earth at geologic timescales can be an Earth without life, an Earth without humans, a hothouse Earth, or a snowball Earth (Kastens et al. 2009). Wiens (1989) termed these scale regions where process–patterns are stable *scale domains*. The term used here refers to scales where process–patterns, interrelatedness, and interconnectedness (and therefore structure) are stable.

Shifts or jumps between states may occur quickly or over long periods of time due to the dynamic nature of complex systems. One example of "slow" shifts between states (relative to human lifespan) is the Florida Everglades in South Florida, where in the early twentieth century artificial control and diversion of surface waters was imposed to prevent loss of human life from devastating floods. Doing so saved unknown human lives but over time led to significant changes in the Everglades as well as in Florida Bay, the estuary between the Everglades and the sea. It took many decades for the initial changes to be realized as adverse conditions in both Florida Bay and the Everglades. Sudden shifts in state can occur as well; one tragic example is the mudslide in Oso, Washington, in 2014.

Shifts and jumps between states that can occur at different time intervals, combined with the inherent heterogeneity of complex systems, mean that complex systems exist in multiple "stable" states that vary across space and time (May 1977). These multiple stable states may exist at multiple scales. For example, a lake may be stable as to water volume at the spatial scale of the entire lake on an annual basis. At a finer timescale, the lake volume is greater after precipitation but over time is balanced out by evaporation, leading to a predictable volume of water when sampled on an annual basis. Scale domains define these spatiotemporal extents within a complex system, where process–patterns and relationships are predictable. The scale domain for the lake's water volume is the entire lake over a year's time.

A scale domain thus describes a stable state for a specific area over a specific time where process–patterns are consistent, homogeneous, or persistent over the timescale of interest (Meentemeyer 1989; Stallins 2006; Wiens 1989).

Adjacent scale domains (in space and/or time) are separated by transitions, or thresholds (Wiens 1989). These thresholds define the end or beginning of a scale domain and mark the transition from one stable state with respect to a different stable state.

Thresholds or scale breaks between domains indicate zones of variability where there is a change in the dominant processes or the state of the system. These thresholds or breaks may reflect "fundamental shifts in underlying processes or controlling factors" (Wu and Li 2006, 12), so identification of scale domains can assist in determining possible factors leading to changes in complex systems (Figure 2.4).

The growing acceptance that complex systems consist of multiple stable states leads to complex systems characterized as nonequilibrial when considered in their entirety. As Figure 2.4 illustrates, for some time periods, there are multiple scale domains. Figure 2.4 also illustrates multiple thresholds

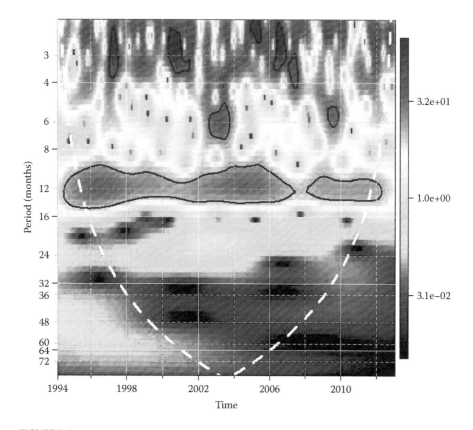

FIGURE 2.4
Plot of wavelet transform coefficients for dissolved oxygen observations (mg/L) in the Mosquito Lagoon, northern Indian River Lagoon, Cape Canaveral Basin, Florida.

between scale domains. It is important to note that Figure 2.4 illustrates only a single "element" (dissolved oxygen) of the complex estuary.

The recognition of complex systems as nonequilibrial systems (Figure 2.4) led to the discovery that descriptions of such systems can be made using nonlinear equations. The link between *far from equilibrium* and *nonlinearity* was made by the physicist Prigogine; see Prigogine and Stengers (1984) for a nontechnical review (Capra and Luisi 2014). Prigogine realized that open systems maintain themselves in varying states of balance but far from equilibrium.

Self-Organization

In complex systems, the dynamic nature of equilibrial points (the thresholds that indicate a shift between states or between scale domains) is indicative of the ability of the system to adapt, which in turn leads to response (Garmestani et al. 2009) in an endless cycle of adaptation–response through time (Levin 1998). The systems renew themselves, produce new structures, break down, and rebuild (Capra and Luisi 2014). This property of self-organization, the ability of the system to configure itself (Garmestani et al. 2009) through feedback loops, further distinguishes complex systems from complicated systems. It is this self-organizing ability of complex systems that can make them unpredictable or lead to sudden "surprises." Here again, scale is key for understanding how complex systems reorganize, where "even small disturbances such as fires have the capacity to reorder entities and relationships throughout an ecosystem, causing it to move through cycles of destruction and rejuvenation" (Manson 2009, 71), a type of cascading effect of change occurring in a small area leading to changes over a larger area.

Self-organization is not limited to biological or ecological systems. Social systems also self-organize, leading to changes in structure and changes in relationships (Levin 1998). There is growing acknowledgment and acceptance that scales are produced (Marston 2000; Sayre 2009; Swyngedouw 1997) over time. Marston (2000, 220) argued this point in saying that "scale is not necessarily a preordained hierarchical framework for ordering the world—local, regional, national, and global. It is instead a contingent outcome of the tensions that exist between structural forces and the practices of human agents." Marston also argued against arbitrary imposition of hierarchical levels in alignment with O'Neill and King (1998) as discussed previously, with a separate but supporting argument based on scales being produced through human practice, process, and activity. Examples of self-organizing social systems abound, including the reorganization of Europe after the Second World War, the recent devastation and reorganization in the Middle East from war and in Syria specifically due to drought, the 2004 Indian Ocean earthquake and tsunami and its effects on Southeast Asia, and the effects of Hurricane Katrina on both New Orleans and Houston.

Self-Organized Criticality

Self-organized criticality is a theory introduced by Bak and Chen (1991) that states that complex systems may display unexpected, often unpredictable behavior in the context of the self-organizing behavior of complex systems (Cilliers 1998), just as super algal blooms appeared unexpectedly in the northern IRL. Cilliers (1998) extended this theory to conclude that it is not the properties of the individual elements of a complex system that describe the system, but their relationships that lead to this self-organizing behavior. Cilliers (1998) specifically used the term *interaction*. The properties or characteristics of the complex system emerge as a result of interactions between related system components or entities (Cilliers 1998).

Interaction means "to be or become involved with, to interact with, or to communicate with someone or something." Entities or components of a complex system are therefore related if they interact. In other words, interactions occurring between entities indicate their relatedness and their connectedness. Interactions occurring over time can be termed *processes*. These processes can take the form of activities that transfer "information" (communication), or materials, or energy between entities. Plants interacting with the sun do so through the process of photosynthesis, where the plant converts solar energy to a form of energy the plant can utilize to bloom, grow, or reproduce.

Complex systems are dynamic as to interaction according to Cilliers (1998), as stated previously. Relationships between entities within complex systems change over time (Levin 1998). As the relationships change, processes may also change. For example, complex systems like an estuary cycle matter and use energy. Energy inputs drive the flow of matter within and between elements and their environment. These energy inputs are also dynamic. Solar irradiation from the sun changes over daily and seasonal (annual) cycles, so processes in estuaries, for example, are constantly changing in response to changes in these energy inputs over time. As the inputs change, the processes change, the relationships change, and the scales at which processes occur also change. Complex systems examined in terms of the relatedness of the elements are to conceptualize these systems as a network, a conceptual model based on the relatedness of the elements. The relatedness of network elements is also scale dependent (Leitner 2004). Changing the scale of observation of the network also alters the ability to perceive relatedness or unrelatedness because relationships between systems entities occur at specific scales and not at other scales. The dynamic nature of relationships means they are also timescale dependent.

Geography has historically been the study of the distribution, in space, of the physical features of the Earth and its atmosphere (physical geography) and of human activity in the distribution of populations and resources, land use, and industries (human geography). Addressing the dynamic nature of these distributions (patterns) in complex systems requires addressing the

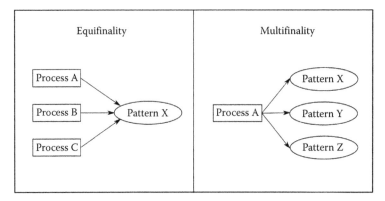

FIGURE 2.5
The equifinality and multifinality problem.

dynamic nature of the processes that lead to the resulting patterns in space. Focusing on pattern alone may be a problem due to the *equifinality problem* (von Bertalanffy 1969), which states that many processes can give rise to the same (or similar) pattern, so that the relationship between process and pattern is many to one (Manson and O'Sullivan 2006). In other words, more than one process may produce the same end pattern (Cliff and Ord 1975). The problem is also that many different patterns can arise from a single process; there is no unique relationship in general between patterns and process (Manson and O'Sullivan 2006) (Figure 2.5).

The equifinality problem (Figure 2.5) says that a pattern observed in the present may have resulted from several possible past processes; the multi-finality problem says that a process observed in the present may result in several possible patterns in the future.

Self-Actualization

The self-actualization property of complex systems results from this ability of a complex system to self-organize. Self-actualization means that in the system there are more possibilities than can be actualized (Cilliers 1998). This can be considered in human terms where it is possible to change careers, body shape, and other attributes by changing the processes that lead to these different outcomes. When a given process may result in many different patterns (Figure 2.5), in the absence of probabilistic certainty for a given process, patterns alone may not distinguish the process that preceded it. For the example estuary, it is the "information," energy or material exchanged by related entities through processes, that is of critical importance for understanding the system and the emergence of super algal blooms.

When patterns observed in a static satellite image or a map of a landscape can represent (but not necessarily describe) fixed and dynamic processes due to the nonequilibrial nature of complex systems, when the observed patterns of a complex system can be static, fixed, dynamic, flexible, and responsive at the same time (Golley 2000), it becomes clear that pattern alone cannot assist in understanding the complexity of geographic systems (Manson and O'Sullivan 2006). Manson and O'Sullivan (2006) emphasized that understanding patterns requires matching the spatial and temporal scales of the processes responsible for those patterns (Swetnam et al. 1999).

The importance of process scales over patterns alone is still not yet reflected in the major tool of analysis used in geography and many other disciplines: the geographic information systems (GIS) (Gunderson et al. 2007). In GIS, the primary tool of analysis is the static map because of the emphasis that GIS places on the spatial-only representation of the world (Yuan 2001). This leads to analyses that default to analyses of spatial patterns (Gunderson et al. 2007), ignoring the temporal dimension and the dynamic nature of land-scapes. Partially, this is due to the relative lack of long-term data sets, but it is also driven by GIS technologies that make spatial pattern analyses simple (Gunderson et al. 2007) and temporal or space–time analyses difficult.

In emphasizing the importance of process scales for analysis of complex systems, it is also important to note that it is not necessarily true that "process scale, observation scale, and modeling (working) scale require different definitions" (Blöschl and Sivapalan 1995, 251); these are merely different scales resulting or generated from different processes. Operational scales are scales of process enacted by an entity (or related group of entities) through its activities, such as the commute to work by a human or the migratory domain of birds. The act of observing is also a process, one that establishes a relationship between the observer and what is observed. Operational and observational scales are both scales of process, just different processes; both are related to perspective. Operational scales are relative to the perspective of the entity performing the activity. Observational scales are relative to the perspective of the observer.

For example, examining the work commute of a single employee is choosing to examine the commute process. This is different from examining the work commute for all employees of a given company, which is different from examining the work commute for all workers within a city or a region. These choices indicate not only different observational scales chosen for analysis but also different operational scales conducted by the individual or group commuting to work.

Perspective can also be the choice of scales for observation, such as the boundary of an estuary or the resolution of a satellite image; choosing to analyze a landscape using an aerial photo taken at an oblique angle instead of overhead; setting the distance between sampling areas along a transect; selecting a portion of a landscape to examine; or the choice of analyzing a single social group relative to "everything else." These are all examples of choices made for scales of observation, and the perspective is reflected in the choice.

The Nature–Society Divide

If *complex systems* are defined as "systems that self-organize over time, are composed of heterogeneous elements that dynamically interact, and where the relationships between elements are also dynamic," it is important to also note that interactions can and do occur between natural and social elements. The emergent condition in the northern IRL estuary may be due to interactions between natural and social processes. Imposing an artificial boundary between natural and social processes prior to analysis may hinder identification of these causes. For these and other reasons, ecologists have urged greater integration of the social and natural sciences (Sayre 2005), yet academic disciplines, spatial analysis, and quantitative techniques are often structured to the divisions between human and physical geography (Jones and Gomes 2010). Given that it is impossible or nearly so to find a social realm completely devoid of nonhuman life or processes, and vice versa, "the distinction between natural and social sciences is beginning to seem meaningless" (Santos 2007, 15). Acceptance of Earth systems as complex systems composed of interconnected and interacting elements further encourages removal of the division between nature and society.

Dividing the social from the natural allows for the construction of analytical objects that are "natural" separate from "social." Doing so makes research and analyses simplistic and therefore easier. Dividing the social from the natural in some ways converts an interconnected complex system to a merely complicated system, in the way that printed maps as representations of Earth convert complex reality to a bounded set of points, lines, polygons, images, labels, networks, or features.

Despite encouragement to remove the division between the social and the natural, attempts to integrate nature with society have led to a confusing profusion of hyphenated terms (human–environmental interaction, nature–society, social–ecological systems) that imply inclusiveness but nonetheless maintain the distinction, requiring explication and differentiation of meaning to prevent confusion. None of the many approaches and frameworks that have been broached seems to have generalized practical use.

It may be that a general framework will always be elusive. If the only "true" representation of reality is reality itself and complex systems cannot be wholly (or fully) modeled, then a complex reality cannot by any means be adequately described by general theory. Geography is frequently criticized for lacking in general theory yet geographic systems are complex, the geographic research domain is heterogeneous in space and time, and this heterogeneity is scale dependent. General theory only applies when physical laws are present that are "fixed" and immutable (at specific scales, it should be noted); complex geographic systems are the opposite of fixed and immutable.

It is not the aim of this chapter to argue for or against the separation of nature from society but rather to highlight the issues that arise from considering them apart and to suggest a scale solution. Identifying and describing process scales and the related scale domains and scale thresholds generated by natural, physical, or social entities, individually or in groups, assist in integrating nature and society by localizing activities and processes in space–time. If only natural processes are examined and described, the highly localized results ease synthesis with results from other physical or social studies. It is possible to discover and describe scale domains within physical observations (Figure 2.4). The resulting process scales, scale domains, and thresholds, if not immediately explained using common knowledge, suggest connections to other physical or social processes by localizing process scales, scale domains, and thresholds in space and time. This can be accomplished (Figure 2.4) without prior separation or division between nature and society, and it has the added benefit of focusing future research efforts to a very specific time and place (Forbes 2014). The threshold observed in the 12-month process scale in Figure 2.4 was not explained using common knowledge. However, the threshold is highly localized in space (29°0′29.00″N 80°54′34.00″W) as well as time (2007–2008).

Framing Practice

To Cilliers (1998), complex systems such as the brain, living organisms, and social–ecological systems must be studied as intact systems. Simple or merely complicated systems, in contrast, can be taken apart and put together again without losing anything (Cilliers et al. 2013).

One way of subsetting a geographic system for analysis is choosing a boundary as study extent. What is termed *framing* here in one sense equates to this choice of a scale, spatial and temporal, as a study extent or boundary. Boundaries chosen reflect a value-based choice (Heylighen et al. 2007), but this value-based choice may also be an explicit choice based on known scales of a process. For example, the example estuary (Figure 2.1) is bounded by the elevation of the land that surrounds it, a dynamic boundary relative to precipitation, evaporation, and tidal change. The section of the estuary being studied is also contained within three different counties (Figure 2.1), each of which govern and enforce policy related to the estuary. Each of these counties enacts different policies and performs different governance activities, which can be examined in relation to processes occurring within the estuary. Another process scale is the St. Johns River Water Management District, with its scale domain and its own policies, processes, and management activities. The northern IRL is entirely contained within the water management district, but is not the only water body managed by the district. The results within the estuary can be compared to other water bodies within the district's greater area of responsibility.

A second way of framing practice involves selecting certain system variables to examine and excluding other variables from the analysis. The list of chosen variables can be expanded or contracted based on prior knowledge or through analytical techniques to determine statistical basis or explanatory value. A third example is in the selection of stakeholders to interview for determining research needs or objectives. All these choices, from the northern IRL boundary to the selection of stakeholders, are examples of a reductionist approach, which is typical because these choices are tractable and yield results.

The emphasis Cilliers placed on a holistic approach being necessary for analysis of complex systems sometimes leads to the conclusion that he was antireductionist, but it is important to note that Cilliers was focused on determining what makes complex systems difficult to understand, which he termed *critical complexity*. To Cilliers (1998), the problem in the selection of a subset of variables to examine essentially decomposes a complex system and destroys the system properties (Cilliers 1998). What Cilliers was concerned with were the problems that result from this "cutting up" of a system, of which the imposition of a nature–society boundary is just one example. Cilliers asserted that something important is lost in cutting up the system.

The various choices available for bounding a study, as in the previous examples, all refer broadly to framing practices employed for studying geographic systems, cutting up the system through placement of boundaries in space, in time, or within the available data. Cilliers was calling attention to the assumption that this practice is unproblematic (Audouin et al. 2013). Cilliers considered that complex systems are open systems and comprise elements that are dynamically interconnected, so the distinction between the system under study and its environment cannot be predetermined (Audouin et al. 2013).

To understand all interactions, relationships, and processes within a complex system, however, requires understanding an impossibly large number of elements and their interconnections (Audouin et al. 2013; Cilliers 1998). Setting boundaries is therefore necessary; otherwise all research would always have to consider the entire Earth. Setting boundaries is essential as they enable the generation of knowledge (Audouin et al. 2013). This results in a paradox between the need to impose boundaries and the understanding that doing so may pose problems for complex systems research.

This paradox is not limited to the chosen boundary of a study area. "Even if a bounded system could be assumed, another problem would remain: that of the units of observation within the whole" (Sayre 2005, 279). Scale as the size of a thing (Howitt 1998) can also refer to these units of observation, the resolution or grain that determine the precision or detail contained within a representation of the Earth. These units can be related to the gridded resolution of remotely sensed imagery (Lam and Quattrochi 1992) or the chosen spacing of intervals along a transect (Wiens 1989). This resolution is the spacing that defines the threshold between adjacent samples (Wu and Li 2006). Resolution in this sense pertains to both space and time. For time, this is usually referred to as the sampling rate for observations. The dissolved oxygen in Figure 2.4 was sampled at monthly intervals.

These choices for framing, whether as internal framing as resolution or external framing as extent, are choices of perspectives for observation. Cilliers (2005) emphasized the consequences of this cutting up of the system, intentionally or not: all knowledge of the system is valid only in relation to a particular framing practice; if any aspect of this framing practice changes, knowledge of the system changes, in the same way in which different image resolutions of the same remotely sensed area result in differences in what is visible to the human eye. This mirrors what O'Neill and King (1989) stated, which was discussed previously: any change in the data would require a different set of "levels" as a hierarchy; this suggests there is no general case hierarchy and never will be.

The Modifiable Areal Unit Problem

Issues that arise from altering the distance between samples (the resolution) are well known. Arbitrary modification of extents or intervals (resolution) leads to the modifiable areal unit problem (MAUP). While the MAUP is named in terms of space (areal) (Openshaw and Taylor 1979), the problem applies equally to temporal extents (a length of time) and temporal intervals at which samples are collected. The MAUP consists of two closely related aspects: the scale problem (or scale effect) and the zoning problem (or zoning effect) (Wu et al. 2000). The scale problem concerns changes in the extent leading to different results, and the zoning problem results from variations in the results of spatial analysis using different zoning systems (Wu et al. 2000), such as gerrymandering voting districts to influence vote outcomes. The MAUP refers to the likelihood of obtaining different results by varying the internal boundaries (the zoning effect) and the external boundaries (the scale effect) in a study. The MAUP illustrates problems related to different ways of cutting up space (or time).

There are theoretically an infinite number of ways to cut up a spatial extent (Fotheringham and Wong 1991). Different choices for partitioning data into various zones or aggregates affect not only what can be seen or observed but also produce different results in statistical correlation (Openshaw and Taylor 1979). Changing boundaries or resolutions may also result in the observation of different patterns and relationships (Wiens 1989) or the relatedness of elements in a network. Results of aggregate analysis are also strongly dependent on these choices of spatial or temporal scale as extent, partitions, or resolution (Figure 2.6). In Figure 2.6, the same three data points are analyzed

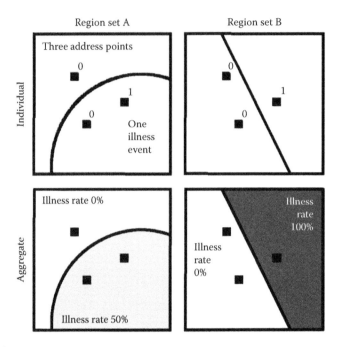

FIGURE 2.6
Illustration of how different data partitions affect illness incidence rates.

individually using two different sub-boundaries (the columns) and then illustrated individually or as aggregate rates (the rows).

The MAUP also makes results sensitive to the resolution at which data are collected (Fotheringham and Wong 1991; Openshaw 1984). Arbitrary scales chosen for collection and analysis tend to reflect hierarchies of spatial scales that are based on human perceptions of what is being analyzed (Wiens 1989). Wiens (1989) described this in terms of ecology: when the focus is on phenomena occurring at a particular scale domain, studies conducted at finer scales fail to include important features of pattern or causal controls of the phenomena; studies at broader scales fail to reveal patterns or relationships because linkages are averaged out or are characteristic only at the phenomena's scale domain, not at the broader scale of observation. A second example is in the wavelet transforms plot in Figure 2.4, where the dominant process scale for dissolved oxygen shown is 12 months. Sampling at yearly intervals instead of monthly intervals would fail to reveal the dominance of this scale or the break in the seasonal cycling that occurs between 2007 and 2008.

Openshaw (1984) stated that defining or creating areal units would be acceptable if they were created using a fixed set of rules with explicit geographically meaningful bases. However, there are no rules for areal aggregation, no standards, and no international conventions to guide the spatial aggregation process (Openshaw 1984). Tobler (1989) noted that Openshaw (1984) cited a paper from 1934 (Gehlke and Biehl 1934); issues related to the MAUP have been known for a long time.

It is now possible to understand the MAUP in terms of the scale-dependent heterogeneity and dynamism of complex systems. The characteristic heterogeneity and dynamism of complex geographic systems suggest there may never be rules, standards, or conventions for spatial (or temporal) aggregation, except through the discovery and description of scale domains. Scale domains by definition are spatial–temporal scales where relationships, patterns, and processes are stable. What both Cilliers and Openshaw were concerned about is that arbitrary choice of partitions of space, time, or data may "leave out" or ignore important relationships. Scale domains describe partitions of space–time in which relationships and processes are stable.

Cao and Lam (1997) provided an excellent and thorough description of the MAUP and concluded that the difficulty in finding a general solution to the MAUP is due to the scale-dependent nature of geographical phenomena. Scale domains that describe spatial–temporal scales in which relationships and processes are stable may represent one possible solution to the issues related to data aggregation or choices of sampling intervals for data collection due to the MAUP. However, scale domains cannot be assumed to describe general laws because they are specific to the location and time, context, perspective, and variables or populations being studied. Figure 2.4 displays data collected at Station IRLV05 in Mosquito Lagoon. Twenty-eight other stations at other locations throughout the northern IRL basin (Figure 2.1) were examined for the same time period (Forbes 2014), and the resulting scale domains and process scales were dissimilar for all 29 stations.

The MAUP and the associated ecological, individualistic, and cross-level fallacies reiterate what Cilliers was saying: knowledge and understanding of the system changes relative to the cutting up of the system. What is seen depends on how it is measured through framing practices that decide the sampling intervals of observation, the resolution of the satellite image, the choice of a spatial or temporal boundary extent, and the choice of stakeholders to interview. Framing practices are choices of what to measure and what not to measure. Framing choices are also choices of perspective; a different perspective may yield different results. Cilliers was emphasizing that as framing practices change, knowledge of the system also changes.

Easterling and Polsky (2004) defined scale as a human construct that locates an observer/modeler relative to a set of objects distributed in space, time, and magnitude. The perspective resulting from the chosen extents, resolutions, or sampling intervals, the framing practice of scientists and analysts, influences the discovery of pattern and process (Easterling and Polsky 2004; Gibson et al. 2000; Wilbanks and Kates 1999). Choices of scale are therefore a choice of perspective, and it can be said that scale as measurement of space and time is one form of perspective. Multiscale analysis is multiperspective analysis in this sense. Choices of perspective can also be nonscalar, as in the example of choosing a subset of stakeholders to interview or variables to collect or examine; both of these are explicit choices of the perspectives considered. Adding or removing a variable or stakeholder or changing the study extent, resolution, or sampling interval alters understanding of the system in the same way that altering the resolution of a satellite image changes what can be seen by the naked eye.

It is understood then that perspective is the key to understanding complex systems (Loehle 2010); scales as measures of space and time are one form of perspective. Cilliers (1998) is not an antireductionist; he emphasized the thoughtful consideration of perspective in framing practices over "strict" reductionism, by emphasizing that knowledge of the system is relative to the perspective chosen. The use of separate terms for "operational" *versus* "observational" scales suggests justification for choosing somewhat arbitrary observation scales. Consideration of critical complexity is to realize that these terms denote different perspectives, and any single perspective of a complex system is limited for all the reasons given. Careful consideration of perspective is needed for complex phenomena.

Consideration of scale as one form of perspective also supports the understanding that "scale is made, and not an ontologically given category" (Marston 2000, 172). Scale becomes a way of framing conceptions of reality (Marston 2000). Given the theoretically infinite possible perspectives for examining a complex system, it is the production of geographical scale, how a scale is produced through process, that is the appropriate research focus (Marston 2000). Indeed, in attempting to unify Prigogine's material approach to complexity (with its focus on the transfer of material, energy, or information between system elements) with the relational approach to complexity (as networks) from Maturana and Varela (1980), Capra and Luisi (2014) concluded that unity is only possible through the consideration of process.

Discussion

A choice of perspective can take many forms, including the decision to examine the perspective of a single entity or group of similar entities relative to "everything else," such as examining dolphin processes and activities within an estuary, selecting sampling intervals for data collection, or examining the social processes performed by a socioeconomic subpopulation. Perspective may also be understood in terms of the process scales generated by the activities of dolphins, social groups, or natural occurrences such as rainfall events. Perspective is chosen through the framing practices of a spatial analyst or the engineered resolution of a satellite image. Knowledge of complex systems entirely depends on the perspective chosen, and scale is one form of perspective. Understanding this leads to solutions and methods for engaging with complex geographic systems. In practice, scale identification techniques can be utilized to determine and describe scales of process and scale domains. The resulting scales as measures of space–time enhance the ability to synthesize multiple perspectives of the same system.

The question put forth here was to examine what value geography may have for complexity research and the answer is hopefully clear. The very foundation of geography is scale, which is appropriate given that scale is a form of perspective and perspective is the foundation for complex systems analysis. Critical complexity theory provides a theoretical and practical background for examination of complex systems that aligns well with geographic practice. Critical complexity focuses in part on the limitations to understanding complex systems, and these limitations are well known to geographers.

Considering critical complexity with scale reveals that geography already engages with complexity with every choice of scale, every map, every image, and every nonscalar choice of perspective. Geographers recognize the characteristic heterogeneity and dynamism of geographic systems. Geographers understand that scales are perceived and valued or created by different actors and that the boundaries chosen, the framing practices employed, the map representations selected reflect different values, for example, different perspectives. The myriad subdisciplines of geography mirror the recognized need for consideration of multiple perspectives for understanding complex geographic phenomena. Geography has already contributed a great deal to complexity science through its existing body of work.

There are benefits to portraying aspects of knowledge as measurements of space and time, and this chapter reiterates and reinforces previous arguments that these portrayals should be based on understanding scales of process, the scales, as measures of space and time, in which activities are carried out. Process scales as spatial and temporal extent anchor process to place and time. Choices of scale for analysis of a process are best served if the chosen scales of observation align with the processes of interest. Scale identification techniques can be employed if process scales are not known.

The complexity of geographic phenomena means that activities can be occurring in the same place at different timescales or occurring simultaneously in time with respect to different places; the dynamics of process scales are important. Multiscale identification techniques are needed to describe the processes occurring in complex, nonequilibrial systems. Wavelet transform coefficients as illustrated in Figure 2.4 are one way to describe and display multiple process scales and scale domains simultaneously. Once identified, the relations between these process scales and scale domains can be investigated in comparison to results from other variables, other perspectives.

The complexity of geographic phenomena may mean that general theory or a general framework may never be possible, but critical complexity suggests approaches through an emphasis on the importance of multiple perspectives for a "holistic" understanding of complex systems. Scales chosen for analysis through framing practices, either as the observational scale or the known scale at which an entity operates, are choices of different perspectives of a complex system. These different perspectives reveal different results, which when combined or synthesized may lead to increased understanding of the whole. If reality is theoretically composed of an infinite number of perspectives or can be decomposed by theoretically infinite ways of cutting up the system, methods are needed to assist in the synthesis of multiple perspectives and the results for multiple process scales. Doing so asymptotically approaches the "reality" of the system.

This leads to the need for tools to visualize scales of process and to visualize multiple perspectives of specific systems in space and time. Figure 2.4 graphically depicts multiple process scales across time but only for a single location in space. Analyzing 29 similar plots in terms of space was challenging. Clustering methods are available to spatially cluster temporal process scales generated from wavelet transform calculations, but graphical illustrations of explicit scale domains in space and time remains a challenge.

These identified needs for analysis, visualization, and synthesis methods for process scales are compounded by the issues inherent in integrating data from multiple sources, such as integrating vector and raster data, integrating spatial and nonspatial data, integrating qualitative and quantitative data, and other data integration issues. Maps unfortunately still look the same as they did in the past despite enormous advances in technology. Perhaps examining the processes that led to how geographic data are currently visualized will assist in devising ways to visualize geographic complexity in ways beyond static maps and the data models or database schemas from the past that are currently still in use.

There is a clear need for modeling and mapping tools that address complexity through multiscale analysis in space and time to capture the scale dependencies and scale domains of processes, networks, and structures (patterns) in space–time. Easterling and Polsky (2004) made a strong case for multiscale modeling; similarly, the seemingly most effective tools to identify process scales employ a "moving window" of varying sizes that scan data at multiple scales (space or time), providing simultaneous results from multiple "perspectives." The wavelet transform computation shown in Figure 2.4 utilizes just such a moving window of varying sizes to describe multiple temporal process scales that can be displayed simultaneously.

There is a further need to question specific space–time dependencies or scale domains as given and presumably transferable from one place to another or one time to another. A "hierarchy" of space–time dependencies or scale domains may not apply everywhere at all times and cannot be assumed in light of the heterogeneity and dynamism of complex systems. Multiscale analysis allows for these dependencies to be explored and affirmed.

Understanding that multiple perspectives are needed for understanding complex systems also leads to recognizing the need to synthesize results across disciplines. Interdisciplinary research integrates the data, tools, perspectives, and theories of two or more disciplines to advance understanding or solve problems. Recognizing the need to incorporate multiple perspectives of problems or systems includes considering perspectives and results from other disciplines. Current geographic tools are recognized and utilized for ease in performing some forms of data integration. Developing additional tools for synthesis of process scales both supports and encourages interdisciplinary or transdisciplinary research.

For the example of the IRL estuary problem from the introduction, the key was to describe process scales and identify scale domains, if they existed. Doing so allowed for connecting physical process scales to unexamined social or ecological processes. This was accomplished through examination of a single perspective as a single physical variable (dissolved oxygen) relative to "everything else." The space–time structure of the observations was described; methods were used to describe temporal process scales (cyclical changes in the observations shown in Figure 2.4) and to identify scale domains (Figure 2.4).

The results were also analyzed relative to multiple spatial scales of management and governance/policy, which revealed specific locations and time periods of interest for further study. Possible causes were identified by connecting scales of processes found with known internal and external processes or phenomena. Not all results were explained using common knowledge, thereby generating hypotheses for future research and informing management as to specific areas of interest at specific time periods. This is especially important given the increasingly limited resources available for research, governance, and management.

The methods used demonstrated a type of reductionism without exclusion or division. Analysis of a single variable situated one variable (one perspective) relative to everything else. Given the complexity of the estuary, additional perspectives are needed. Additional perspectives would provide additional understanding when synthesized with results already completed. The highly localized nature of the resulting scale domains and related thresholds, in both space and time, eases this synthesis.

Conclusions

It is unpredictability that distinguishes complicated systems from complex systems. This unpredictability leads to the "surprises" that emerge from complex systems, such as the rise of fascism in the twentieth century, the Oso, Washington, mudslide, or the Arab Spring. These events are only surprising if the systems from which they arose are considered complicated systems, rather complex.

Treating geographic systems as complicated systems sometimes manifests in statements that current social, political, climatic, or environmental problems are "becoming more complex." Interconnectivity within and between many social and natural systems has certainly increased, with globalization as one example, but defining a system as *more complex* based on the number of connections within the system is as problematic as classifying a system based on the number of parts. Both the number of parts counted or perceived and the number of interconnections between elements in complex systems are dependent on the scale at which the system is examined. A system becoming more complex also suggests that the system was less complex in the past, which perhaps encourages treating a complex reality as a merely complicated system. This also suggests that there is some quantity that distinguishes a system as *more* or *less* complex, and therefore some quantity is needed to define this distinction.

Rather than problems seeming to become more complex, it seems rather that there is a need to understand that the problems themselves are not simple and easily solved. Controlling and diverting surface water in the Everglades in the early twentieth century was really an attempt to treat a complex system as a merely complicated system. Doing so led to unintended consequences. Problems such as flood control or water quality in an estuary have always been complex and systemic. It is the approach to the problems that needs to change, beginning with recognition of the inherent complexity of the systems. The Everglades and Florida Bay as complex systems did self-organize in response to the changes made, but the results were not as expected, were slow to materialize, and were considered less than desirable from many perspectives.

There is therefore a clear need to view the past, present, and future of geographic systems as the dynamic, complex realities that they are. This supports the understanding that a map or a remotely sensed image is one conception, one perspective, of what is essentially a complex reality; maps and images are one representation out of theoretically infinite possible perspectives of the same landscape or object.

Engaging with both critical complexity and geographic scale leads to exploring the ways in which we come to know and derive meaning in the world. The growing understanding of the complexity of the Earth's systems in turn suggests the need for a change in the narratives handed down from the past, such as treating a complex system as a merely complicated system. This suggests changes to practice, but this chapter is not solely an appeal to alter geographic practice. This chapter illustrates how geographers already engage with complex systems and contribute to complexity science. What the discussion in this chapter does suggest is the adoption of a new narrative. This new narrative is based on recognizing the complexity of the systems that support all life on Earth. It is also based on thinking in terms of relatedness between system components for increased understanding of complex systems. This new narrative emphasizes the necessity to consider perspective in research and the importance of process scales over arbitrary choices of scale for analysis.

This new narrative is then a call to see the world not as a simple or complicated machine that can be taken apart and put back together again; something is lost when this is done, as the Everglades and many other areas and events have taught. This new narrative sees the world as a complex system of systems in which processes occur between related elements at multiple scales over time. It emphasizes that engaging with multiple perspectives of these systems leads to understanding and highlights the importance for understanding the dynamics of process scales within and between these systems. It also illustrates and highlights the importance of geographic research and therefore geographic scale for engaging with complex geographic systems.

Acknowledgments

The research on which this chapter was based was supported by the National Aeronautics and Space Administration through the University of Central Florida's NASA Florida Space Grant Consortium. The author thanks the anonymous reviewers who took the time to read and comment on the original draft. This chapter would not have been possible without their encouragement, and their comments contributed significantly to the result. The author is also grateful to Deborah Morcos for her thoughtful and insightful comments on the draft manuscript.

References

Audouin, M., R. Preiser, S. Nienaber, et al. 2013. Exploring the Implications of Critical Complexity for the Study of Social-Ecological Systems. *Ecology and Society* 18 (3): 12.

Bak, P., and K. Chen. 1991. Self-Organized Criticality. *Scientific American (United States)* 264: 1 (January).

Blöschl, G., and M. Sivapalan. 1995. Scale Issues in Hydrological Modelling: A Review. *Hydrological Processes* 9 (3–4): 251–90.

Cao, C., and N. Lam. 1997. Understanding the Scale and Resolution Effects in Remote Sensing and GIS. In *Scale in Remote Sensing and GIS*, edited by D. A. Quattrochi and M. F. Goodchild, 57–72. Boca Raton, FL: CRC Press.

Capra, F., and P. L. Luisi. 2014. *The Systems View of Life: A Unifying Vision*. 1st edition. Cambridge: Cambridge University Press.

Carpenter, S. R., H. A. Mooney, J. Agard, et al. 2009. Science for Managing Ecosystem Services: Beyond the Millennium Ecosystem Assessment. *Proceedings of the National Academy of Sciences of United States of America* 106 (5): 1305–12.

Cilliers, P. 1998. *Complexity and Postmodernism: Understanding Complex Systems*. 1st edition. London: Routledge.

Cilliers, P. 2005. Complexity, Deconstruction and Relativism. *Theory, Culture & Society* 22 (5): 255–67.

Cilliers, P., H. C. Biggs, S. Blignaut, et al. 2013. Complexity, Modeling, and Natural Resource Management. *Ecology and Society* 18 (3): 1.

Cliff, A. D., and J. K. Ord. 1975. Model Building and the Analysis of Spatial Pattern in Human Geography. *Journal of the Royal Statistical Society. Series B (Methodological)* 37 (3): 297–348.

Easterling, W. E., and C. Polsky. 2004. Crossing the Divide: Linking Global and Local Scales in Human-Environment Systems. In *Scale and Geographic Inquiry: Nature, Society, and Method*, edited by E. S. Sheppard and R. B. McMaster, 66–85. Malden, MA: Blackwell.

Forbes, D. J. 2014. Generating Space-Time Hypotheses in Complex Social-Ecological Systems. Ph.D. dissertation, Florida Atlantic University. Ann Arbor: ProQuest/UMI.

Fotheringham, A. S., and D. W. S. Wong. 1991. The Modifiable Areal Unit Problem in Multivariate Statistical Analysis. *Environment and Planning A* 23 (7): 1025–44.

Gardner, R. H. 1998. Pattern, Process, and the Analysis of Spatial Scales. In *Ecological Scale*, edited by D. L. Peterson and V. T. Parker, 17–34. New York: Columbia University Press.

Garmestani, A. S., C. R. Allen, and L. H. Gunderson. 2009. Panarchy: Discontinuities Reveal Similarities in the Dynamic System Structure of Ecological and Social Systems. *Ecology and Society* 14 (1): 15.

Gehlke, C. E., and K. Biehl. 1934. Certain Effects of Grouping upon the Size of the Correlation Coefficient in Census Tract Material. *Journal of the American Statistical Association* 29 (185A): 169–70.

Gibson, C. C., E. Ostrom, and T. Ahn. 2000. The Concept of Scale and the Human Dimensions of Global Change: A Survey. *Ecological Economics* 32 (2): 217–39.

Golley, F. B. 2000. Ecosystem Structure. In *Handbook of Ecosystem Theories and Management*, edited by S. E. Jørgensen and F. Müller. Boca Raton, FL: CRC Press.

Gunderson, L. H., C. R. Allen, and D. Wardwell. 2007. Temporal Scaling in Complex Systems. In *Temporal Dimensions of Landscape Ecology*, edited by J. A. Bissonette and I. Storch, 78–89. New York: Springer.

Heylighen, F., P. Cilliers, and C. Gershenson. 2007. Complexity and Philosophy. In *Complexity, Science, and Society*, edited by J. Bogg and R. Geyer. Oxford: Radcliffe.

Holling, C. S. 1994. Simplifying the Complex: The Paradigms of Ecological Function and Structure. *Futures, Special Issue - Complexity: Fad or Future?* 26 (6): 598–609.

Holling, C. S., D. W. Schindler, B. W. Walker, and J. Roughgarden. 1995. Biodiversity in the Functioning of Ecosystems: An Ecological Synthesis. In *Biodiversity Loss*, edited by C. Perrings, K. Maler, C. Folke, C. S. Holling, and B. Jansson. Cambridge: Cambridge University Press.

Howitt, R. 1998. Scale as Relation: Musical Metaphors of Geographical Scale. *Area* 30 (1): 49–58.

Jones III, J. P., and B. Gomez. 2010. Introduction. In *Research Methods in Geography: A Critical Introduction*, edited by B. Gomez and J. P. Jones III, 1–6. Wiley.

Kastens, K. A., C. A. Manduca, C. Cervato, et al. 2009. How Geoscientists Think and Learn. *Eos, Transactions American Geophysical Union* 90 (31): 265–66.

King, A. W. 2012. Hierarchy Theory: A Guide to System Structure for Wildlife Biologists. In *Wildlife and Landscape Ecology: Effects of Pattern and Scale*, edited by J. A. Bissonette, 185–214. Springer Science & Business Media Springer-Verlag: New York.

Lam, N. S., and D. A. Quattrochi. 1992. On the Issues of Scale, Resolution, and Fractal Analysis in the Mapping Sciences. *The Professional Geographer* 44 (1): 88–98.

Leitner, H. 2004. The Politics of Scale and Networks of Spatial Connectivity: Transnational Interurban Networks and the Rescaling of Political Governance in Europe. In *Scale and Geographic Inquiry: Nature, Society, and Method*, edited by E. S. Sheppard and R. B. McMaster. Malden, MA: Blackwell.

Levin, S. A. 1998. Ecosystems and the Biosphere as Complex Adaptive Systems. *Ecosystems* 1 (5): 431–36.

Loehle, C. 2010. *Becoming a Successful Scientist*. 1st edition. Cambridge: Cambridge University Press.

Manson, S. M. 2008. Does Scale Exist? An Epistemological Scale Continuum for Complex Human–Environment Systems. *Geoforum* 39 (2): 776–88.

Manson, S. M. 2009. Complexity, Chaos and Emergence. In *A Companion to Environmental Geography*, edited by N. Castree, D. Demeritt, D. Liverman, and B. Rhoads, 1st edition. Chichester, UK: Wiley-Blackwell.

Manson, S. M., and D. O'Sullivan. 2006. Complexity Theory in the Study of Space and Place. *Environment and Planning A* 38 (4): 677–92.

Marston, S. A. 2000. The Social Construction of Scale. *Progress in Human Geography* 24 (2): 219–42.

Maturana, H. R., and F. J. Varela. 1980. *Autopoiesis and Cognition: The Realization of the Living*. Springer Science & Business Media, New York.

May, R. M. 1977. Thresholds and Breakpoints in Ecosystems with a Multiplicity of Stable States. *Nature* 269 (5628): 471–77.

Meadows, D. H. 2008. *Thinking in Systems: A Primer*, edited by D. Wright, Sustainability Institute. White River Junction, VT: Chelsea Green Publishing.

Meentemeyer, V. 1989. Geographical Perspectives of Space, Time, and Scale. *Landscape Ecology* 3 (3–4): 163–73.

O'Neill, R. V., and A. W. King. 1998. Homage to St. Michael; Or, Why Are There So Many Books on Scale? In *Ecological Scale*, edited by D. L. Peterson and V. T. Parker, 3–16. New York: Columbia University Press.

Openshaw, S. 1984. *The Modifiable Areal Unit Problem*. Concepts and Techniques in Modern Geography. Norwich, United Kingdom: Geo Books.

Openshaw, S., and P. J. Taylor. 1979. A Million or so Correlation Coefficients: Three Experiments on the Modifiable Areal Unit Problem. In *Statistical Applications in Spatial Sciences*, edited by N. Wrigley, 127–44. London: Pion.

Prigogine, I., and I. Stengers. 1984. *Order out of Chaos: Man's New Dialogue with Nature*. Boulder Colorado: New Science Library.

Richards, A. 2002. Complexity in Physical Geography. *Geography* 87 (2): 99–107.

Rogers, K. H., R. Luton, H. Biggs, et al. 2013. Fostering Complexity Thinking in Action Research for Change in Social-Ecological Systems. *Ecology and Society* 18 (2): 31.

Santos, B. 2007. A Discourse on the Sciences. In *Cognitive Justice in a Global World: Prudent Knowledges for a Decent Life*, edited by B. Santos. Lanham Maryland: Rowman & Littlefield Publishing Group, Inc.

Sayre, N. F. 2005. Ecological and Geographical Scale: Parallels and Potential for Integration. *Progress in Human Geography* 29 (3): 276–90.

Sayre, N. F. 2009. Scale. In *A Companion to Environmental Geography*, edited by N. Castree, D. Demeritt, D. Liverman, and B. Rhoads, 1st edition. Chichester, UK: Wiley-Blackwell.

Stallins, J. A. 2006. Geomorphology and Ecology: Unifying Themes for Complex Systems in Biogeomorphology. *Geomorphology, Linking Geomorphology and Ecology* 77 (3–4): 207–16.

Swetnam, T. W., C. D. Allen, and J. L. Betancourt. 1999. Applied Historical Ecology: Using the Past to Manage for the Future. *Ecological Applications* 9 (4): 1189–1206.

Swyngedouw, E. 1997. Excluding the Other: The Production of Scale and Scaled Politics. In *Geographies of Economies*, edited by Roger Lee and Jane Wills, 167–76. London: Arnold.

Tobler, W. R. 1989. Frame Independent Spatial Analysis. In *The Accuracy of Spatial Databases*, edited by M. F. Goodchild and S. Gopal, 75–79. Boca Raton, FL: CRC Press.

Van Arman, J. A. 1987. Introduction. In *Indian River Lagoon Reconnaissance Report*, edited by J. S. Steward and J. A. Van Arman. West Palm Beach, FL: South Florida Water Management District.

von Bertalanffy, L. 1969. *General System Theory: Foundations, Development, Applications.* Revised edition. New York: George Braziller.

Weinberg, G. M. 1975. *An Introduction to General Systems Thinking.* New York: Wiley.

Wiens, J. A. 1989. Spatial Scaling in Ecology. *Functional Ecology* 3 (4): 385.

Wilbanks, T. J. 2002. Geographic Scaling Issues in Integrated Assessments of Climate Change. *Integrated Assessment* 3 (2): 100–114.

Wilbanks, T. J., and R. W. Kates. 1999. Global Change in Local Places: How Scale Matters. *Climatic Change* 43 (3): 601–28.

Wu, J., D. E. Jelinski, M. Luck, and P. T. Tueller. 2000. Multiscale Analysis of Landscape Heterogeneity: Scale Variance and Pattern Metrics. *Geographic Information Sciences* 6 (1): 6–19.

Wu, J., and H. Li. 2006. Perspectives and Methods of Scaling. In *Scaling and Uncertainty Analysis in Ecology*, edited by J. Wu, K. B. Jones, H. Li, and O. L. Loucks, 17–44. Dordrecht, the Netherlands: Springer.

Wu, J., and O. L. Loucks. 1995. From Balance of Nature to Hierarchical Patch Dynamics: A Paradigm Shift in Ecology. *The Quarterly Review of Biology* 70 (4): 439–66.

Yuan, M. 2001. Representing Complex Geographic Phenomena in GIS. *Cartography and Geographic Information Science* 28 (2): 83–96.

3

Scaling Geocomplexity and Remote Sensing

Shaun Lovejoy

CONTENTS

Introduction

Remotely sensed radiances from planetary surfaces or atmospheres reveal fields of immense complexity with structures typically spanning the range of scales from planetary to submillimetric: 10 or more orders of magnitude. The natural framework for analyzing, modeling, and indeed understanding such hierarchies of structures within structures is scale invariance: fractal sets and multifractal fields. A little over 15 years ago, several colleagues and

I gave a short review of some of the theories relevant to remote sensing, including several examples (Pecknold et al. 1997; see also Lovejoy et al. 2001a, especially the discussion of correlated scaling processes). This chapter is an update discussing some of the advances that have occurred since then.

Resolutions have improved and channels have multiplied. Today, images are often collected at regular intervals so that today remotely sensed data are very much space–time products. As the quantity of remotely sensed data and the range of scales that they span has grown, the data are increasingly of global extent. In some cases such as topography or cloud imagery, composite data analyses spanning up to eight orders of magnitude are possible. In parallel with that, global numerical models may also show wide-range scaling; this includes weather prediction models and reanalysis products. They show that the standard atmospheric state variables are scaling up to planetary scales (typically 5,000 km or larger). This is fortunate because it allows them to be compatible with the scaling of the fields as determined by *in situ* and remotely sensed measurements. Other significant developments include improvements in some of the older data analysis techniques (notably trace moments, in particular their use to explicitly determine the outer scales) but also in the development of new techniques (Haar fluctuations).

However, many phenomena—especially atmospheric—evolve rapidly enough in time that it is important to understand the space–time behavior. With respect to both atmospheric and oceanic phenomena, there has been important progress, notably in clarifying the notion of weather and climate with the realization that there is a third regime (macroweather) that is intermediate between the two (Lovejoy 2013). A consequence is that it turns out that the climate is *not* "what you expect"; it is rather macroweather. Whereas in the weather regime (up to 5–10 days, the lifetime of planetary structures), fluctuations on average increase with scale (the exponent H defined below is >0), in macroweather, on the contrary, they decrease with scale ($H < 0$). In this regime, successive fluctuations tend to cancel each other out: "Macroweather is what you expect." The climate regime starts at much lower frequencies (≈30 years in the industrial epoch) and again, $H > 0$. In the section "Space–Time Scaling: Example of MTSAT," we therefore discuss the space–time development of geostationary thermal IR fields providing a theoretical framework for understanding atmospheric motion vectors (AMVs).

Spatial Scaling Spectra: Some Examples

Spectra

Scale invariance is a symmetry such that when one changes from one scale to another, some aspect is unchanged (conserved). In scale-invariant systems, fluctuations ΔI over distances Δx have power law dependences: $\Delta I(\Delta x) \approx \varphi \, \Delta x^H$, where φ is the driving flux whose statistical mean [φ] does not depend on scale. The exponent H characterizing the fluctuations is scale invariant, whereas the fluctuations themselves are power law functions of scale; they are said to be "scaling." The corresponding (Fourier and power) spectra are of the form $E(k) \approx k^{-\beta}$ where k is a wave number (inverse distance scale) and β is the "spectral exponent."

Spectra have several useful features: they are traditional, familiar, and applicable to any geophysical signal. They are also very sensitive to breaks in scaling and—when these are due to quasiperiodic nonstationarities (such as daily of annual cycles)—of easily separating these from the otherwise scaling "background." The main disadvantages are that (1) their interpretation is not as straightforward as for (real space) fluctuations; (2) they are not well adapted to situations with missing data; and (3) they only characterize the second-order statistical moments so that—unless the signal is quasi-Gaussian—it only gives a very partial characterization of the statistics. In this section, we give a quick tour of some of examples; in the section "Multifractals, Structure Functions," we discuss alternative real space analysis techniques.

Let us recall the basics. We will be dealing mostly with two-dimensional (2D) remotely sensed radiance fields $I(r)$ ($r = (x,y)$ is a position vector), so that we can define the 2D (Fourier) spectrum $P(k)$ by the following:

$$\left\langle \tilde{I}(\underline{k})\tilde{I}(\underline{k}') \right\rangle = \delta(\underline{k}+\underline{k}')P(\underline{k}); \ \tilde{I}(\underline{k}) = \int e^{i\underline{k}\cdot\underline{r}}I(\underline{r})d\underline{r} \tag{3.1}$$

where $\underline{k}$ is the wave vector dual to $\underline{r}$, $\tilde{I}(\underline{k})$ is the Fourier transform of I, and δ is the Dirac delta function. For finite data, the delta function is replaced by a finite difference approximation such that at $\underline{k} = -\underline{k}'$, it is equal to the number of degrees of freedom of the system (the number of pixels) N. For a real signal I, we have $\tilde{I}^{*}(\underline{k}) = \tilde{I}(-\underline{k})$ (the asterisk [*] indicates a complex conjugate) so that:

$$P(\underline{k}) \propto \left\langle \left| \tilde{I}(\underline{k}) \right|^{2} \right\rangle \tag{3.2}$$

where the constant of proportionality is N. If the system is statistically isotropic, then P only depends on the vector norm of k: $P(\underline{k}) = P(k); k = |\underline{k}|$. Now perform an isotropic Fourier space "zoom" $\underline{k} \to \lambda\underline{k}$ (i.e., a standard "blowup") by a factor of $\lambda > 1$ so that in physical space there is an inverse blowup: $\underline{x} \to \lambda^{-1}\underline{x}$. If the system is "self-similar," that is, if it is both isotropic and scaling, then the condition that the smaller scales are related to the larger scales without reference to any characteristic size (i.e., that it is scaling) is that the spectra follow power law relations between large wave numbers $\lambda|\underline{k}|$ and smaller ones $|\underline{k}|$:

$$P(|\lambda\underline{k}|) = P(\lambda|\underline{k}|) = \lambda^{-s}P(|\underline{k}|) \tag{3.3}$$

that is, that the form of P is independent of scale. Equation 3.3 is satisfied by the following scaling law for P:

$$P(\underline{k}) = |\underline{k}|^{-s} \qquad (3.4)$$

We can now obtain the power spectrum $E(k)$ (with $k = |\underline{k}|$) by integrating over all the directions:

$$E(k) = \int_{|\underline{k}'| = k} P(\underline{k}')d\underline{k}' \qquad (3.5)$$

In $d = 2$, the region of integration is the annuli between the radii k and $k + dk$. Therefore, if the process is isotropic in 2D, $E(k) = 2\pi\,kP(k)$. In terms of data analysis, where one has a finite rather than infinite sample size, this angle integration is advantageous because it reduces the noise. In the following examples, we therefore take the power law dependence of the spectrum:

$$E(k) \approx k^{-\beta};\ \beta = s - 1 \qquad (3.6)$$

as evidence for the scaling of the field f, and the exponent β being the "spectral slope." Note that in some areas of geophysics, angle averages ($P(k)$) are used rather than angle integrals ($E(k)$). This has the disadvantage that the resulting spectral exponents will depend on the dimension of space so that, for example, one-dimensional sections will have exponents that differ by 1 from 2D sections. In contrast the angle integrations used here yield the same exponent β in spaces of any dimension (i.e., β is independent of the dimension of space, whereas s is dependent).

When permitted by the data, using the angle integrals over the 2D spectrum $P(k_x, k_y)$ is advantageous because it reduces the spectrum to a function of a single variable (the modulus of the wave vector) while simultaneously improving the statistics. Often the data are easily amenable to integrating over angles—for example geostationary satellites; others such as orbiting satellites commonly have swaths of limited width while being essentially 20,000 km long (i.e., half a circumference—no two points on the Earth can be further apart) so that they are more one dimensional.

Alternatively, it is possible to estimate the 1D spectra $E(k_x)$, $E(k_y)$ by integrating out the conjugate coordinates:

$$E_x(k_x) = \int P(k_x, k_y)dk_y;\ E_y(k_y) = \int P(k_x, k_y)dk_x \qquad (3.7)$$

and for isotropic statistics, $E_x(k) = E_y(k) = E(k)$ (to within constant factors). Of course, if the field is not isotropic, for example if it is scaling but with different exponents in different orthogonal directions, $E_x(k_x) \propto k_x^{-\beta_x}$; $E_y(k_y) \propto k_y^{-\beta_y}$ with $\beta_x \neq \beta_y$, then the isotropic spectrum will in fact display a break (roughly) at the wave number k that satisfies $E_x(k) = E_y(k)$ with one exponent dominating the behavior at the high frequencies and the other at the low frequencies. This is discussed further in the section on space–time scaling in the context of the space–time analysis of satellite data.

The Horizontal

The Atmosphere

Over the last few years, a number of global scale satellite radiances have impressively demonstrated the global scale extent of the scaling. For example, Figure 3.1a and b, taken from over 1,000 orbits of the Tropical Rainfall Measuring Mission (TRMM) satellite, shows that visible, near, and thermal

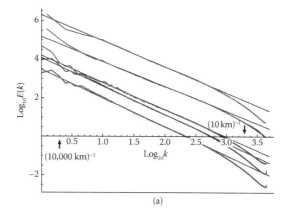

(a)

FIGURE 3.1

(a) Spectra from ≈ 1,000 orbits of the visible and infrared scanner (VIRS) instrument on the Tropical Rainfall Measuring Mission (TRMM) satellite, channels 1–5 (at wavelengths of 0.630, 1.60, 3.75, 10.8, and 12.0 µm, from top to bottom, displaced in the vertical for clarity). The data are for the period January–March 1998 and have nominal resolutions of 2.2 km. The straight regression lines have spectral exponents β = 1.35, 1.29, 1.41, 1.47, and 1.49, respectively, close to the value β = 1.53 corresponding to the spectrum of passive scalars (=5/3 minus intermittency corrections). The units are such that $k = 1$ is the wave number corresponding to the size of the planet $(20,000 \text{ km})^{-1}$. Channels 1 and 2 are reflected solar radiation so that only the 15,600-km sections of orbits with maximum solar radiation were used. The high wave number falloff is due to the finite resolution of the instruments. To understand the figure we note that VIRS Bands 1 and 2 are essentially reflected sunlight (with very little emission and absorption), so that for thin clouds, the signal comes from variations in the surface albedo (influenced by the topography and other factors), whereas for thicker clouds it comes from nearer the cloud top *via* (multiple) geometric and Mie scattering. As the wavelength increases into the thermal IR, the radiances are increasingly due to black body emission and absorption with very little multiple scattering. Whereas at the visible wavelengths we would expect the signal to be influenced by the statistics of cloud liquid water density, for the thermal IR wavelengths it would rather be dominated by the statistics of temperature variations—themselves also close to those of passive scalars (Adapted from Lovejoy, S., et al., *Q. J. Roy. Meteor. Soc.*, 134, 277–300, 2008). *(Continued)*

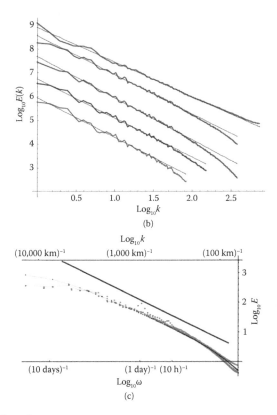

FIGURE 3.1 (Continued)
(b) Spectra of radiances from the Thematic Microwave Imager from the TRMM satellite, ≈1,000 orbits from January through March 1998. From bottom to top, the data are from Channels 1, 3, 5, 6, and 8 (vertical polarizations of 2.8, 1.55, 1.41, 0.81, and 0.351 cm) with spectral exponents β = 1.68, 1.65, 1.75, 1.65, and 1.46, respectively, at resolutions of 117, 65, 26, 26, and 13 km (hence the high wave number cut-offs); each are separated by one order of magnitude for clarity. To understand these thermal microwave results, recall that they have contributions from surface reflectance, water vapor, clouds, and rain. Because the particles are smaller than the wavelengths, this is the Rayleigh scattering regime and as the wavelength increases from 3.5 to 2.8 cm the emissivity/ absorptivity due to cloud and precipitation decreases so that more and more of the signal origi- nates in the lower reaches of clouds and underlying surface. Moreover, the ratio of scattering to absorption increases with increasing wavelength so that at 2.8 cm multiple scattering can be important in raining regions. The overall result is that the horizontal gradients—which influence the spectrum—increasingly reflect large internal liquid water gradients. (c) One-dimensional spectra of Multifunctional Transport Satellite (MTSAT) thermal IR radiances; the Smith prod- uct was developed with similar IR satellite radiance fields. In black: the theoretical spectrum using parameters estimated by regression from the 3-D generalization fo equation 3.7 to include the frequency Greek omega. and taking into account the finite space–time sampling volume. The spectra are $E_x(k_x) \approx k_x^{-\beta x}$, $E_y(k_y) \approx k_y^{-\beta y}$, $E_t(\omega) \approx \omega^{-\beta t}$ with $\beta_x \approx \beta_y \approx \beta_t \approx 1.4\pm0.1$. Other parameters are $L_w \sim$ 20,000 km; $\tau_w \sim$ 20±1 days; $s \approx 3.4\pm0.1$. The straight line is a reference line with slope −1.5 (blue). Pink is the zonal spectrum; orange is the meridional spectrum; blue (with the diurnal spike and harmonic prominent) is the temporal spectrum. See the section titled "Space–Time Scaling: Example of MTSAT" and Figure 3.18 for further analysis and discussion. (Reproduced from Pinel, J., et al., *Atmos. Res.*, 140–141, 95–114, 2014. With permission.)   *(Continued)*

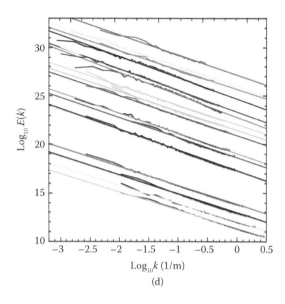

(d)

FIGURE 3.1 (Continued)
(d) The spectra of the 19 (of 38) highest resolution clouds analyzed with a spectral slope $\beta \approx 2$ shown by the reference lines. (Reproduced from Sachs, D., et al., *Fractals*, 10, 253–265, 2002. With Permission.)

infrared as well as passive microwave radiances have nearly perfect scaling up to the largest scales. This is also demonstrated by Multifunctional Transport *Satellite* (MTSAT, also known as "Himawari") geostationary imagery (Figure 3.1c), which shows the results of spectral analysis of a large (roughly $1{,}000^3$ points) data set in (x,y,t) space showing that the (1D) temporal and (horizontal) spatial statistics are nearly identical. This is an isotropic space–time symmetry that we discuss further in "Space–Time Scaling: Example of MTSAT" (we also analyze and display spectra of the same data over various 2D subspaces). Moving to the opposite extreme of small scales, one can evaluate the spectra at small ("cloud") scales, this time looking upwards with a handheld large-format camera (Figure 3.1d), which displays excellent scaling (of the angle integrated) spectrum of individual clouds at visible wavelengths. The theory and numerical modeling of radiative transfer in scaling clouds are reviewed in the section of radiative transfer.

The atmospheric energy budget is essentially determined by the incoming fluxes at visible wavelengths; and the outgoing fluxes at infrared wavelengths, the basic sources and sinks do not introduce characteristic scales. The lower boundary conditions (e.g., the topography and ocean surface) are also scaling (see below), as are the atmospheric dynamical equations. We may therefore expect the state variables (wind, temperature, pressure, humidity) to also be scaling. This is directly demonstrated in Figure 3.2, which shows the spectra of the outputs of atmospheric reanalyses. Reanalyses are data-model hybrid

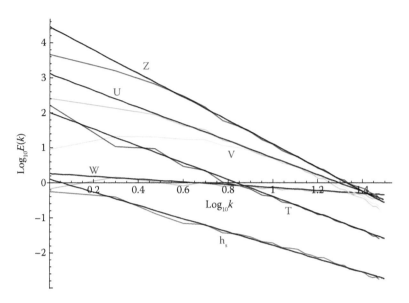

FIGURE 3.2
Comparisons of the spectra of different atmospheric fields from the European Centre for Medium-Range Weather Forecasts (ECMWF) interim reanalysis. Top (red) is the geopotential (β = 3.35); second from the top (green) is the zonal wind (β = 2.40); third from the top (cyan) is the meridional wind (β = 2.40); fourth from the top (blue) is the temperature (β = 2.40); fifth from the top (orange) is the vertical wind (β = 0.4); at the bottom (purple) is the specific humidity (β = 1.6). All are at 700 mb (roughly 3 km altitude) and between ±45° latitude, every day in 2006 at GMT. The scale at the far left corresponds to 20,000 km in the east–west direction, at the far right to 660 km. Note that for these 2D spectra, Gaussian white noise would yield β = –1 (i.e., a positive slope =+1). (Reproduced from Lovejoy, S., and D. Schertzer, *J. Geophys. Res.*, 116, 2011. With permission.)

products in which data of all kinds—increasingly including satellite data—are assimilated into a numerical weather model, which constrains them to satisfy the equations of the atmosphere as embodied in the numerical model.

The Earth's Surface: Topography

The topography is of prime importance for several areas of Earth science, notably as the lower boundary condition for the atmosphere and climate, and also for surface hydrology, oceanography, and hence for hydrosphere–atmosphere interactions. The issue of scaling in topography has an even longer history than it does in atmospheric science, going back over 100 years to when Perrin (1913) eloquently argued that the coast of Brittany was nondifferentiable. Later, Steinhaus (1954) expounded on the nonintegrability of the river Vistula, Richardson (1961) quantified both aspects using scaling exponents, and Mandelbrot (1967) interpreted the exponents in terms of fractal dimensions. Indeed, scaling in the Earth's surface is so prevalent that there are entire scientific specializations such as river hydrology and geomorphology that abound in scaling laws of all types (for a review see Rodriguez-Iturbe and Rinaldo 1997; see Tchiguirinskaia et al. 2000, for a comparison of multifractal and fractal analysis of basins) and that virtually require the topography to be scaling.

Ever since the pioneering power spectrum of Venig-Meinesz (1951) with $\beta \approx 2$, scaling spectra of topography from various regions have routinely been reported with fairly similar exponents (Balmino et al. 1973; Bell 1975 [also with $\beta \approx 2$]; Berkson and Matthews 1983 [$\beta \approx 1.6$–1.8]; Fox and Hayes 1985 [$\beta \approx 2.5$]; Gilbert 1989 [$\beta \approx 2.1$–2.3]; Balmino 1993 [$\beta \approx 2$]; Lavallée et al. 1993; and Gagnon et al. 2006).

Figure 3.3 shows the global scale spectrum of the ETOPO5 data set (a global gridded topography data set including bathymetry at a 5′ arc, i.e., about 10 km), along with those of other higher resolution but regional digital elevation models (DEMs, i.e., gridded topographic maps), land only. These include GTOPO30 (the continental United States at ≈1 km) as well as

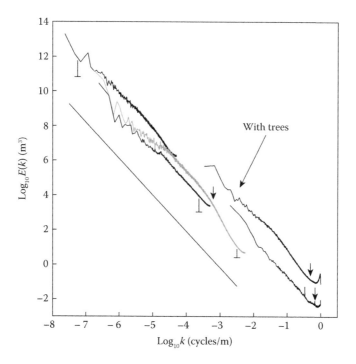

FIGURE 3.3

A log–log plot of the spectral power as a function of wave number for four digital elevation models. From right to left: Lower Saxony, with trees (top), without trees (bottom); United States (in gray), GTOPO30 and ETOPO5. A reference line of slope –2.10 is shown for comparison. The small arrows show the frequency at which the spectra are not well estimated due to their limited dynamical range (for this and scale-dependent corrections; Equation 3.8). (Reproduced from Gagnon, J. S., et al., *Nonlin. Proc. Geophys.*, 13, 541–570, 2006. With permission.)

two other DEMs: the United States at 90-m resolution and part of Saxony in Germany at 50-cm resolution. Overall, the spectrum follows a scaling form with $\beta \approx 2.1$ down to at least ≈ 40 m in scale.

The remarkable thing about the spectra is that the only obvious breaks are near the high wave number (small-scale) end of each data set. In Gagnon et al. (2006), it is theoretically shown that for wave numbers higher than the arrows, the data are corrupted by inadequate dynamical ranges (i.e., the ratio of the largest to the smallest resolvable changes in altitude). The basic idea is straightforward: if the dynamic range is small, then there will be large areas that are nominally at the same altitude, and this leads to spuriously smooth fields. For example, the curve in Figure 3.3 with the largest break in scaling was the DEM at 90-m spatial resolution, which only had a 1-m altitude resolution. This implies that huge swathes of the country had nominally zero gradients and hence overly smooth spectra. To quantify this, denote the minimum and maximum heights by h_{min} and h_{max}. If they are measured in nondimensional digital counts, then for a spectrum $E(k) \approx k^{-\beta}$, we have the following:

$$k_{max}/k_{min} \approx (h_{max} - h_{min})^{2/\beta} \qquad (3.8)$$

The wave numbers at which this formula predicts that the spectra start to be corrupted are shown with arrows in Figure 3.3. This approximate formula works particularly well for the spectrum of the continental United States at 90-m resolution, where it explains the drop in the high frequencies. In fact, the problem of insufficient dynamical range can probably explain many of the scale breaks seen in the literature that are interpreted as characteristic scales of the process. A related cause of spurious high wave number spectral falloffs occurs when data are sampled at a rate higher than the intrinsic sensor resolution (oversampling).

The Ocean Surface: Ocean Color

The ocean surface is particularly important due to its exchanges with the atmosphere. Figure 3.4 shows a particularly striking wide-range scaling result: a swath over 200 km long at 7-m resolution over the St. Lawrence Estuary in eight different narrow visible wavelength channels from the airborne multi-detector electro-optical imaging sensor (MEIS). The use of different channels allows one to determine "ocean color," which itself can be used as a proxy for phytoplankton concentration. For example, the channels fourth and eighth from the top in the figure exhibit nearly perfect scaling over the entire range; these are the channels that are insensitive to the presence of chlorophyll; they give us an indication that over the corresponding range ocean turbulence itself is scaling. In comparison, other channels show a break in the neighborhood of ≈ 200 m in scale; these are sensitive to

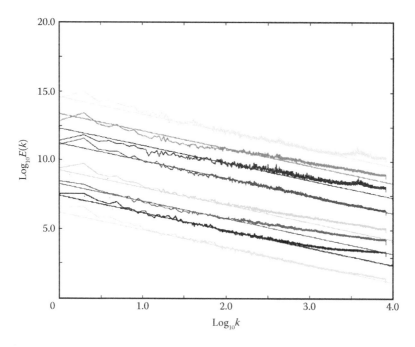

FIGURE 3.4
The ocean, Channels 1–8 offset for clarity, eight visible channels characterizing ocean color, 210-km-long swath, 28,500 × 1,024 pixels, 7-m resolution. The extreme high wave number is (14 m)$^{-1}$. (Adapted from Lovejoy, S., et al., *Inter. J. Remote Sensing*, 22, 1191–1234, 2001a.)

phytoplankton. The latter are "active scalars" undergoing both exponential growth phases ("blooms") as well as being victim to grazing by zooplankton; in Lovejoy et al. (2000), a turbulence theory is developed to explain the break with a zooplankton grazing mechanism.

Another important ocean surface field that has been found to be scaling over various ranges is the sea surface temperature (SST). Relevant studies include *in situ* results such as those of McLeish (1970) (Eulerian, $\beta_T \approx 5/3$), Seuront et al. (1996) (Lagrangian, $\beta_T \approx 2$), and Lovejoy et al. (2000) (towed instruments, $\beta_T \approx 1.63$). Remote sensing using thermal IR images first from aircraft (Saunders 1972) and then from satellites (Deschamps et al. 1981, 1984; Park and Chung 1999) yields, respectively, $\beta_T \approx 5/3$, ≈ 2.2, ≈ 1.9, ≈ 2, $\approx 1.87 \pm 0.25$ out to distances of ≈ 100 km. At larger scales (out to at least ≈ 500 km), Burgert and Hsieh (1989) found $\beta_T \approx 2.1$ from "cloud free" satellite data. The satellite data—even if nominally "cloud free"—are somewhat smoothed by atmospheric effects; hence their β_T values are probably slightly too high. Monthly averaged *in situ* SST data were analyzed by Lovejoy and Schertzer (2013), and the literature was reviewed (see also Table 3.1). They concluded that the scaling with $\beta_T \approx 1.8$ really does continue up to scales of 5,000 km or more, a conclusion that is bolstered by the corresponding spectral and cascade analyses.

TABLE 3.1

A Comparison of Various Horizontal Parameter Estimates: Summarizing the Values of Categories Using Approximate Values

		C_1	α	H	β	L_{eff}	Range of scales[i]
Remote Sensing of the Surface							
Topography[a] (Earth)	Altitude	0.12	1.8	0.7	2.1	20,000	40 m–20,000 km
Topography[b] (Mars)	Altitude	0.10	1.7	0.5	1.8	8,000	10 km–10,000 km
Sea surface temperature[c]	SST	0.12	1.9	0.50	1.8	16,000	500 m–50,000 km
Soil moisture index[d]	MODIS	0.05	2.0	0.14	1.2		512 m–25 km
Vegetation index[d]	MODIS	0.06	2.0	0.16	1.2		
Surface magnetic field[e]	(Aircraft) <10 km	0.14	1.9	0.6	2.0		250 m–20 km
	(Aircraft) >10 km	0.08	2	0.1	1		20 km–1,000 km
Mostly Atmospheric[k]							
State variables[f]	u, v	0.09	1.9	1/3 (0.77)	1.6 (2.4)	(14,000)	280 m–~5,000 km
	w	(0.12)	(1.9)	(−0.14)	(0.4)	(15,000)	
	T	0.11 (0.08)	1.8	0.50 (0.77)	1.9 (2.4)	5,000 (19,000)	
	h	0.09	1.8	0.51	1.9	10,000	
	z	(0.09)	(1.9)	(1.26)	(3.3)	(60,000)	
Precipitation[g]	R	0.4	1.5	0.00	0.2	32,000	4–12,000 km
Passive scalars[h]	Aerosol concentration	0.08	1.8	0.33	1.6	25,000	100 m–125 km

(Continued)

TABLE 3.1 (Continued)

A Comparison of Various Horizontal Parameter Estimates: Summarizing the Values of Categories Using Approximate Values

	C_1	α	H	β	L_{eff}	Range of scales[j]
Radiances[i] Infrared	0.08	1.5	0.3	1.5	15,000	2–15,000 km
Visible	0.08	1.5	0.2	1.5	10,000	2–15,000 km
Passive microwave	0.1–0.26	1.5	0.25–0.5	1.3–1.6	5,000–15,000	20–15,000 km

Notes: When available (and when reliable), the aircraft data were used in precedence over the reanalysis values. In those cases where there was no comparable *in situ* value or when reanalysis was significantly different from the *in situ* value, the latter is given in parentheses. For the estimate of the effective outer scale, L_{eff}, where the anisotropy is significant, the geometric mean of the north–south and east–west estimates is given (the average ratio is 1.6:1 EW/NS, although for the precipitation rate the along-track TRMM estimate was used). Finally, the topography estimate of L_{eff} is based on a single realization (one Earth, one Mars); in both cases, the values pertain to the large-scale scaling regimes: on Mars, >10 km; on Earth, >40 m. See Lovejoy and Schertzer (2013) for an extensive review. Note that the half circumference on Mars (the largest Martian distance) is 10,600 km, compared with 20,000 km on Earth.

[a] From Gagnon et al. (2006); data from GTOPO30 (≈1 km resolution), ETOPO5 (≈10 km resolution), and the continental United States at 90 m and Saxony (Germany) at 50 cm.

[b] From Landais et al. (2015); at scales smaller than 10 km there is also a (nearly) monofractal regime with $H \approx 0.75$.

[c] These values are from both SST data at ≈500 km resolution and also satellite data at ≈500 m resolution; see Table 8.2 of Lovejoy and Schertzer (2013).

[d] From Lovejoy et al. (2007a).

[e] From airborne magnetic field anomaly measurements at 800 m altitude over Canada (two regions). The two different scale regions correspond to a break at the horizontal scale corresponding to the Curie depth (see Lovejoy et al. 2001b; Pecknold et al. 2001). The inner scale (approximately indicated here as 20 km) is thus only a rough average; the Curie depth varies from ≈10 to 50 km.

[f] From instrumented (Gulfstream) aircraft 280 m to 1,100 km (Lovejoy et al. 2010), the values in parentheses are from reanalyses from ≈100 to ≈5,000 km by Lovejoy and Schertzer (2011).

[g] From gauge networks from 200 to 5,000 km, reanalyses (≈100–8,000 km) and from TRMM satellite radar data, 4–12,000 km (Lovejoy et al. 2012).

[h] From airborne lidar backscatter ratios from 100 m to 125 km (horizontal) and 3 m to 5 km (vertical) (Lilley et al. 2004, 2008; Lovejoy et al. 2008).

[i] From >1,000 orbits of TRMM satellite data with visible, near-IR, IR, and passive microwave at varying resolutions (see Figure 3.1a and b). Also from MTSAT (geostationary) satellite data (Figure 3.3c) in the IR, 30–5,000 km, and GOES (Geostationary Operational Environmental Satellite; IR, visible geostationary) satellite data (Lovejoy et al. 1993b; Lovejoy et al. 2001c); there are many other scaling analyses of cloud radiances that used other analysis techniques that give exponents that cannot be used to determine the parameters in this table.

[j] This column gives the range of the scaling observed with the cited references. If the range covered by the cited data covers a well-defined scale break, then the range covered by the data may be larger. Similarly, the scaling range may cover scales beyond those that were empirically studied.

[k] With the exception of the passive scalars from lidar and precipitation (from satellite radar), the radiances are not purely dependent on atmospheric conditions; they depend also on the surface emissivities and temperatures (IR, microwave) and albedos (visible).

Soil Moisture and Vegetation Indices from MODIS

Many surface fields are scaling over wide ranges, particularly as revealed by remote sensing. Figure 3.5a shows six moderate-resolution imaging spectro-radiometer (MODIS) channels at 250-m resolution over Spain (each a 512 × 512 pixel "scene"). The scaling is again excellent except for the single lowest wave number, which is probably an artefact of the contrast enhancement algorithm that was applied to each image before analysis. These channels are used to yield vegetation and surface moisture indices by dividing channel pair differences by their means, so that the scaling is evidence that both vegetation and soil moisture are also scaling.

The vegetation and soil surface moisture indices are standard products described by Rouse et al. (1973) and Lampkin and Yool (2004):

$$\sigma_{\mathrm{VI}} = \frac{I_2 - I_1}{I_2 + I_1}; \sigma_{\mathrm{SM}} = \frac{I_6 - I_7}{I_6 + I_7} \tag{3.9}$$

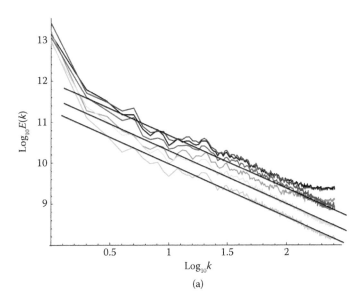

(a)

FIGURE 3.5

(a) Spectra of six bands of MODIS radiances over a 512 × 512 pixel region of Spain (at 250-m resolution; $k = 1$ corresponds to 128 km): $E(k)$ as a function of the modulus of the wave vector. In order from top to bottom at the point $\log_{10}k = 0.7$, the curves are as follows: purple = Band 6, black = Band 1, magenta = Band 7, light green = Band 2, cyan = Band 4, dark green = Band 3. Reference lines have slopes −1.3. The band wavelengths are (in nm): Channel 1: 620–670, Channel 2: 841–876, Channel 3: 459–479, Channel 4: 545–565, Channel 5: 1,230–1,250, Channel 6: 1,628–1,652, Channel 7: 2,105–2,155. These data are used for determining both vegetation and soil moisture indices. Adapted from Figure 3.3a in Lovejoy et al. (2007a).   *(Continued)*

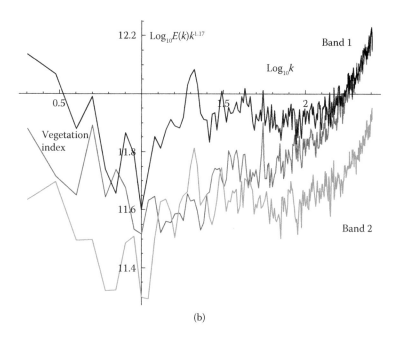

(b)

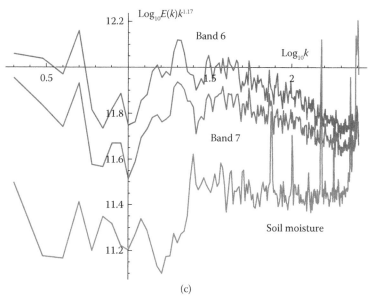

(c)

FIGURE 3.5 (Continued)
(b) The spectra of Bands 1 and 2 and the vegetation index (Equation 3.9) compensated by $k^{1.17}$ so that a spectrum $E(k) \approx k^{-1.17}$ is flat. Note that the variations are less than ± 0.2, whereas the actual spectra vary by almost five orders of magnitude (Figure 3.5a). (c) The spectra of Bands 6 and 7, and the soil moisture index (Equation 3.9) compensated by $k^{1.17}$ so that a spectrum $E(k) \approx k^{-1.17}$ is flat.

where VI and SM are the vegetation and soil surface moisture indices and I is the radiance (the band number is indicated in the subscript); Equation 3.9 is usually stated without reference to any particular scale. We may immediately note that, although the Terra MODIS data have a resolution of 500 m (Bands 1 and 2 were degraded from 250 m) and the variability of the radiances and surface features continues to much smaller scales, the surrogates are defined at a single (subjective) resolution equal to that of the sensor. One of the applications of our analyses was to investigate how the relations between the surrogates and the bands used to define them change with scale: we anticipate that since the scaling properties are different, making the surrogates with data at different resolutions would produce fields with different properties. This is investigated in detail in the section "Soil Moisture and Vegetation Indices."

The Atmosphere, Vertical: Lidar Data

In spite of the fact that gravity acts strongly at all scales, the classical theories of atmospheric turbulence have all been quasi-isotropic in either two or three dimensions. Whereas a few models tentatively predict possible transitions in the horizontal (for example between $k^{-5/3}$ and k^{-3} spectra for a transition from 3D to 2D isotropic turbulence for the wind), in contrast, in the vertical any 2D/3D "dimensional transition" would be even more drastic (Schertzer and Lovejoy 1985a; this is also true for passive scalars such as chaff; see e.g., Lesieur 1987).

The fact that these predicted drastic transitions in horizontal spectra have not been observed was the starting point for the anisotropic scaling model with the "in between" dimension $D_{el} = 23/9 = 2.5555 D$ proposed by Schertzer and Lovejoy (1985a), in which the horizontal and vertical have power law spectra but with different exponents β_h and β_v. When—as observed—$\beta_v > \beta_h$ (and it is not obvious), this implies that structures with horizontal extent L have vertical extents L^{H_z} with $H_z = (\beta_v - 1)/(\beta_h - 1)$ so that they become flatter and flatter at larger and larger scales (see Figure 3.7b for an illustration). In addition, the volume of typical structures is $LLL^{H_z} = L^{D_{el}}$, where $D_{el} = 2 + H_z$ is the "elliptical" dimension characterizing the effective dimension of the (nonintermittent, i.e., space-filling) structures.

Although the existence of different scaling exponents in the horizontal and vertical was confirmed by *in situ* measurements of many different atmospheric variables (see notably Lilley et al. 2004, 2008; Lovejoy et al. 2007a, 2007b, 2008; Radkevitch et al. 2008; Lovejoy and Schertzer, 2010a; and especially Chapter 6 in Lovejoy and Schertzer 2013), the overall situation took a long time to clarify. It turned out that there was a problem with the interpretation of aircraft data in the horizontal: spurious breaks in the scaling were caused by aircraft following isobars rather than isoheights (Lovejoy et al. 2009a). This had the effect of leading to a break in the horizontal scaling where the vertical displacement of the aircraft became large enough that the fluctuations in wind were dominated by the much stronger vertical shears (and larger vertical exponents). This was recently directly demonstrated by the first joint vertical–horizontal structure function analyses of the horizontal wind from 14,500 aircraft flights (Pinel et al. 2012).

Although it was important to clarify the status of the *in situ* wind measurements, the basic result—anisotropic scaling—had in fact already been very clearly established through the use of remotely sensed aerosol backscatter from airborne lidar (Lilley et al. 2004, 2008; Lovejoy et al. 2007a, 2007b, 2008; Radkevitch et al. 2008; see box 6.1 in Lovejoy and Schertzer 2013, for analogous results using CloudSat data and cloud liquid water radar reflectivities). The key differential stratification can be observed directly by eye in Figure 3.6a and b, which shows vertical cross sections of lidar aerosol backscatter fields with resolutions down to 3 m in the vertical. Starting at the low-resolution image (Figure 3.6a), we can see that structures are generally highly stratified. However, zooming in (Figure 3.6b) we can already make out waves and other vertically (rather than horizontally) oriented structures. In Figure 3.6c, we confirm—by direct spectral analysis—that the fields are scaling in both the horizontal and vertical directions and that the exponents are indeed different in both directions: the critical exponent ratio $(\beta_h - 1)/(\beta_v - 1) = H_z$

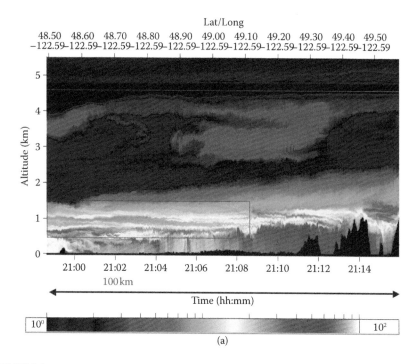

(a)

FIGURE 3.6

(a) Typical vertical–horizontal lidar backscatter cross-section acquired on August 14, 2001. The scale (bottom) is logarithmic: darker is for smaller backscatter (aerosol density surrogate), lighter is for larger backscatter. The black shapes along the bottom are mountains in the British Columbia region. The line at 4.6 km altitude shows the aircraft trajectory. *(Continued)*

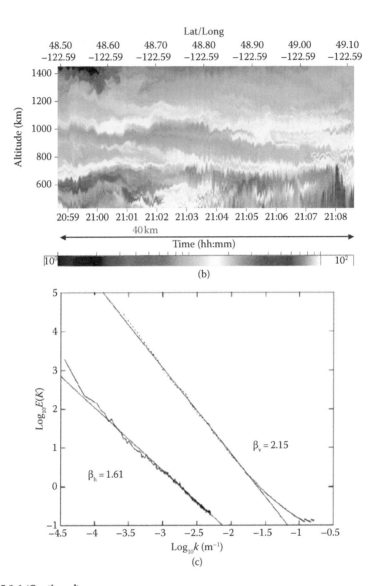

(b)

(c)

FIGURE 3.6 (Continued)

(b) Enlarged content of the (700–1,600 m) box in (a). Note that small structures become more vertically aligned, whereas large structures are fairly flat. The aspect ratio is 1:96. Zoom of the previous image, showing that at the small scales, structures are beginning to show vertical (rather than horizontal) "stratification" (even though the visual impression is magnified by the 1:40 aspect ratio, the change in stratification at smaller and smaller scales is visually obvious). (Reproduced from Lilley, M., et al., *Phys. Rev. E*, 70, 036307, 2004. With permission.)

(c) The lower curve is the power spectrum for the fluctuations in the lidar backscatter ratio, a surrogate for the aerosol density (B) as a function of horizontal wave number k (in m^{-1}) with a line of best fit with slope $\beta_h = 1.61$. The upper trace is the power spectrum for the fluctuations in B as a function of vertical number k with a line of best fit with slope $\beta_v = 2.15$; hence $H_z = 0.61/1.15 \approx 0.53$. (Adapted from Lilley, M., et al., *Phys. Rev. E*, 70, 036307, 2004.)

is quite near the theoretical value 5/9, which is the ratio of the (horizontal) Kolmogorov value 1/3 and the (vertical) Bolgiano–Obukhov value 3/5. The wave number at which the spectra in Figure 3.6c typically cross is in the range from $(10 \text{ cm})^{-1}$ to $(100 \text{ cm})^{-1}$; this is the scale at which structures are roughly "roundish," called the "sphero-scale."

Anisotropic Scaling, the Phenomenological Fallacy, and the Missing Quadrillion

We have presented a series of striking wide-range scaling spectra covering many significant atmospheric and surface fields. In this "tour" of the scaling, we have exclusively used a common statistical analysis technique (the power spectrum). The conclusion that scaling is a fundamental symmetry principle of wide applicability is hard to escape, although it is not universally embraced. The main difficulty seems to be that we are inculcated with the "scale bound" (Mandelbrot 1981) idea that every time we zoom into structures (by magnifying them), we expect to see qualitatively different structures and different processes. When trying to understand and model such phenomena, we are taught to search for mechanistic explanations involving processes acting over only narrow ranges of scale, and when we find them, we typically assume that they are incompatible with statistical theories and explanations. In atmospheric science, a particularly extreme form of this was recently highlighted by Lovejoy (2014), who compared a (still frequently cited) 1970s "mental picture" of atmospheric processes that proposed a spectrum dominated by narrow range spikes (such as the diurnal and annual cycles), with the rest being essentially uninteresting white-noise "background" processes (Mitchell 1976). By comparing this speculation with modern data, it was found that the mental picture was in error by a factor of $\approx 10^{15}$ and that almost all of the variance was in the nontrivial scaling background. Similarly, publication of the nearly perfect space–time scaling of thermal infrared data over the Pacific (figure 3c, Pinel et al. 2014), was delayed for over a year due to strenuous referee objections based on a perceived incompatibility between the scaling statistics and the usual mechanistic phenomenology of tropical meteorology (e.g., the dominance of waves). There was a similar reaction to the scaling statistics of satellite-based Martian reanalyses in Lovejoy et al. (2014) and Chen et al. (2016).

Such reactions illustrate the "phenomenological fallacy" (Lovejoy and Schertzer 2007a), which arises when phenomenological approaches are only based on morphologies rather than underlying dynamics. The fallacy is to identify phenomenologically defined forms, structures, or morphologies with distinct dynamical mechanisms when in actual fact a unique dynamical mechanism acting over a wide range of scales can also lead to structures that change with scale. The fallacy thus has two aspects. In the first, form and mechanism are confounded so that different morphologies are taken as *prima facie* evidence for the existence of different dynamical mechanisms. In the second, scaling is reduced to its special isotropic "self-similar" special case in which small and large scales are statistically related by an isotropic "zoom" or blowup (Figure 3.7a). In fact, scaling is a much more general symmetry: it suffices for small and large scales to be related in a way that does not introduce a characteristic scale. Moreover, the relation between scales can involve differential squashing, rotation, and so on, so that small and large scales can share the same dynamical mechanism yet nevertheless have quite different appearances. Compare Figure 3.7a with b, which is the same

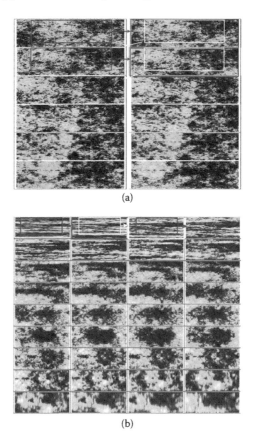

(a)

(b)

FIGURE 3.7
(a) A self-similar (isotropic) multifractal cloud simulation. Each image is enlarged by a factor of 1.7 (the areas enlarged are shown in yellow and red rectangles for the first few enlargements, top rows). (b) A sequence "zooming" into a vertical cross-section of an anisotropic multifractal cloud with $H_z = 5/9$. Starting at the upper left corner, moving from left to right and from top to bottom, we progressively zoom in by factors of 1.21 (total factor ≈ 1,000). Note that while at large scales, the clouds are strongly horizontally stratified, when viewed close up they show structures in the opposite direction. The sphero-scale is equal to the vertical scale in the leftmost simulation on the bottom row. The film version of this (and other anisotropic space–time multifractal simulations) can be found at http://www.physics.mcgill.ca/~gang/multifrac/ index.htm. (Adapted from Lovejoy, S., and D. Schertzer, *The weather and climate: Emergent laws and multifractal cascades*, Cambridge University Press, Cambridge, 2013.)

process but with different horizontal and vertical exponents corresponding to differential (scale-dependent) "squashing."

In order to illustrate how morphologies can change with scale when the scaling is anisotropic, consider Figure 3.8. This is a multifractal simulation of a rough surface with the parameters estimated for the topography; its anisotropy is in fact rather simple in the framework of the generalized scale invariance (GSI) (Schertzer and Lovejoy 1985b; for a review, see Chapter 7 of Lovejoy and Schertzer 2013). More precisely, it is an example of linear GSI (with a diagonal generator) or "self-affine" scaling. The technical complexity with respect to self-similarity is that the exponents are different in orthogonal directions, which are the eigenspaces of the generator, so that structures are systematically "squashed" (stratified) at larger and larger scales. The underlying epistemological difficulty, which was not simple to overcome and which still puzzles phenomenologists, corresponds to a deep change in the underlying symmetries.

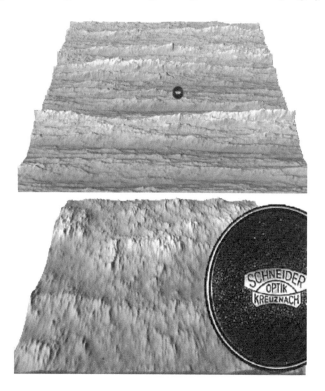

FIGURE 3.8
This self-affine simulation illustrates the "phenomenological fallacy," because both the top and bottom look quite different while having the same anisotropic mechanism at scales differing by a factor of 64 (top and bottom blowup). The figure shows the proverbial geologist's lens cap at two resolutions differing by a factor of 64. Seen from afar (top), the structures seem to be composed of left-to-right ridges; however, closer inspection (bottom) shows that in fact this is not the case at the smaller scales. (Reproduced from Lovejoy, S., and D. Schertzer, *Nonlin. Processes Geophys.*, 14, 1–38, 2007b. With permission.)

The top figure illustrates the morphology at a "geologist's scale" as indicated by the traditional lens cap reference. If this were the only data available, one might invoke a mechanism capable of producing strong left–right striations. However, if one only had the bottom image available (at a scale 64 times larger), then the explanation (even "model") of this would probably be rather different. In actual fact, we know by construction that there is a unique mechanism responsible for the morphology over the entire range.

Figure 3.9 gives another example of the phenomenological fallacy, this time with the help of multifractal simulations of clouds. Again (roughly) the observed cascade parameters were used, yet each with a vertical "sphero-scale" (this is the scale where structures have roundish vertical cross sections) decreasing by factors of 4, corresponding to zooming out at random locations. It is evident from the vertical cross-section (bottom row) that the degree of vertical stratification increases from left to right. These passive scalar cloud simulations (liquid water density, bottom two rows; single scattering radiative transfer, top row) show that by zooming out (left to right) diverse morphologies appear. Although a phenomenologist might be tempted to introduce more than one mechanism to explain the morphologies

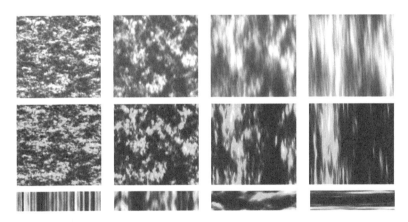

FIGURE 3.9
Examples of continuous in scale anisotropic multifractals in 3D ($256 \times 256 \times 64$), showing the effect of changing the sphero-scale (l_s) on multifractal models of clouds with $H_z = 5/9$. The cloud statistical parameters are as follows: $\alpha = 1.8$, $C_1 = 0.1$, and $H = 1/3$ (similar to CloudSat and aerosols; see Table 3.1). From left to right, we decrease l_s (corresponding to zooming out by factors of 4) so that we see the initially vertically aligned structures (bottom left) becoming quite flat at scales 64 times larger (right). At the same time, the horizontal structures have scaling anisotropies so that they too change orientation and elongation (the horizontal sphero-scale starts at 1 pixel, far left; for a review, see Lovejoy and Schertzer [2013, chapters 6 and 7] for this generalized scale invariance). The middle row is a false-color rendition of the liquid water density field; the bottom row is the corresponding vertical sections (side view); the top row is the corresponding single scatter visible radiation; the mean optical thickness is 2; isotropic scattering phase function; sun incident at 45° to the right. (Reproduced from Lovejoy, S., et al., *Atmos. Chem. Phys.*, 9, 1–19, 2009a. With permission.)

at different scales, in the figure we are simply seeing the consequence of single underlying mechanism repeating scale after scale. The phenomenological fallacy can undermine many classical ideas. For example, Lovejoy and Schertzer (2013, Box 6.1) argued with the help of CloudSat analyses that the classical two-scale theories of convection are incompatible with data that are scaling—that division into qualitatively distinct small and large regimes is unwarranted.

Radiative Transfer in Fractal and Multifractal Clouds

Most wavelengths used for remote sensing are sensitive to the radiative transfer properties of clouds, and we have seen (Figure 3.1a through d) that the associated radiances are scaling over a wide range of scales. It is thus reasonable to simulate clouds using scaling models for the distribution of scatterers and absorbers and then to model the radiative transfer through them at different wavelengths. Such models will be needed either to understand the clouds themselves—of fundamental importance in atmospheric science—or to understand the role of clouds in modulating the transmission/absorption characteristics of the underlying surface fields. An understanding of cloud and radiation variability and their interrelations over wide is also a challenging problem in the physics of disordered media. Figure 3.10 shows some examples for a purely scattering atmosphere (albeit with single scattering only). Figure 3.10a and b shows the external "faces" of the 3D liquid water concentration field for four realization of an anisotropic (stratified) multifractal cloud with exponents near to those observed; the difference in the figures is due to the different degrees of stratification, as can be seen best from the side views (Figure 3.10b). Figure 3.10c and d shows the corresponding visible radiation fields above and below the cloud using single scattering only and an isotropic phase function (no absorption; numerical details on the multifractal simulations can be found in Lovejoy and Schertzer [2010c] and Lovejoy and Schertzer [2010b]; for some theory and numerics for multiple scattered radiative transfer in multifractal clouds, see Lovejoy et al. [2009b] and Watson et al. [2009]).

The classical theory of radiative transfer is elegant (Chandrasekhar 1950) but is only relevant in 1D ("plane parallel," horizontally homogeneous) media, yet the use of 1D models has long dominated the field. This is because when we turn to horizontally inhomogeneous media, there is no consensus on the appropriate model of heterogeneity, nor is the transport problem analytically tractable. As a consequence, the effect of horizontal variability was underestimated and usually reduced to the problem of inhomogeneity of the external cloud/medium boundaries (e.g., cubes, spheres, and cylinders;

Busygin et al. 1973; McKee and Cox 1976; Preisendorfer and Stephens 1984) with the internal cloud and radiance fields still being considered smoothly varying if not completely homogeneous. When stronger internal horizontal inhomogeneity was considered, it was typically confined to narrow ranges of scale so that various transfer approximations could be justified (Weinman and Swartzrauber 1968; Welch et al. 1980).

When the problem of transfer in inhomogeneous media finally came to the fore, the mainstream approaches were heavily technical (see Gabriel [1993] for a review), with emphasis on intercomparisons of general purpose

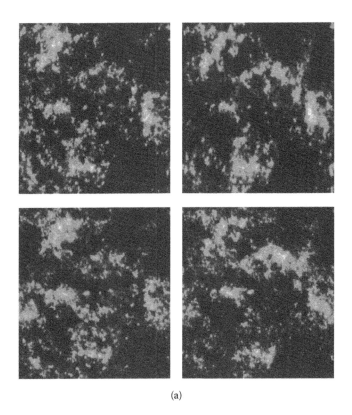

(a)

FIGURE 3.10
(a) The top layers of three-dimensional cloud liquid water density simulations (false colors); all have anisotropic scaling with horizontal exponents $d = 1$, $c = 0.05$, $e = 0.02$, $f = 0$ (see Lovejoy and Schertzer, 2013, for a review and definition of these parameters); stratification exponent $H_z = 0.555$; and multifractal statistical exponents $\alpha = 1.8$, $C_1 = 0.1$, and $H = 0.333$. They are simulated on a $256 \times 256 \times 128$ point grid. The simulations in the top row have horizontal sphero-scales of $l_s = 8$ pixels (left column), 64 pixels (right column), with additional (trivial) anisotropy exponents $\xi = 0$ (top row), $\xi = 3/4$ (bottom row); they all have $\xi_z = 0.25$ (the ξ parameters—not to be confused with the structure function exponent $\xi(q)$—are explained in Lovejoy and Schertzer [2013], from which all these figures were reproduced). Note that in these simulations, $l_s = 8$ and 64 applies to both the vertical and horizontal cross-sections (i.e., $l_s = l_{sz}$).   *(Continued)*

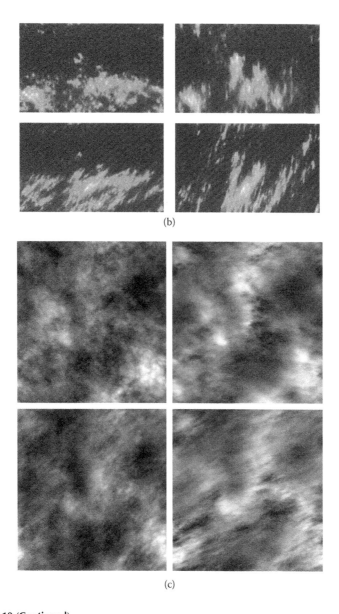

(b)

(c)

FIGURE 3.10 (Continued)
(b) A side view. (c) The top view with single scattering radiative transfer; incident solar radiation at 45° from the right; mean vertical optical thickness = 50. *(Continued)*

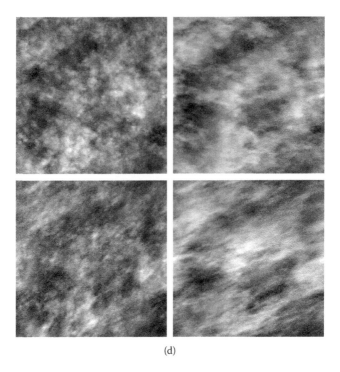

(d)

FIGURE 3.10 (Continued)
(d) The same as Figure 3.10c except viewed from the bottom.

numerical radiative transfer codes (see the C^3 initiative; Cahalan et al. 2005) and the application to large eddy simulation cloud models (Mechem 2002). At a more theoretical level, the general problem of the consequences of small-scale cloud variability on the large-scale radiation field has been considered using wavelets (Ferlay and Isaka 2006) but has only been applied to numerical modelling. As a consequence, these "3D radiative transfer approaches" have generally shed little light on the scale-by-scale statistical relations between cloud and radiation fields in realistic scaling clouds (e.g., see the collection in Marshak 2005). Overall, there has been far too much emphasis on techniques and applications with little regard for understanding the basic scientific issues.

The simplest interesting transport model is diffusion (on fractals see the reviews by Bouchaud and Georges [1990] and Havlin and Ben-Avraham [1987] and on multifractals see Meakin [1987], Weissman [1988], Lovejoy et al. [1993a], Lovejoy et al. [1998], Marguerite et al. [1997]). However—except in 1D (Lovejoy et al. 1993a, 1995)—diffusion is not in the same universality class as radiative transport (Lovejoy et al. 1990).

The first studies of radiative transport on fractal clouds (with a constant density on the support) were by Barker and Davies (1992), Cahalan (1994),

Cahalan (1989), Cahalan et al. (1994), Davis et al. (1990), Gabriel et al. (1990), Gabriel (1986), Lovejoy et al. (1990), and Lovejoy (1989). These works used various essentially academic fractal models and focused on the (spatial) mean (i.e., bulk) transmission and reflectance. They clearly showed that (1) fractality generally leads to nonclassical ("anomalous") thick cloud scaling exponents; (2) the latter were strongly dependent on the type of scaling of the medium; and (3) the exponents are generally independent of the phase function (Lovejoy et al. 1990).

Some general theoretical results exist for conservative cascades ($H = 0$; see the section "Multifractals, Structure Functions" for a definition of H and α and the section "Multifractals and Trace Moment Analysis" for cascades), single scattering for log-normal clouds (Lovejoy et al. 1995) ($\alpha = 2$), and for more general (i.e., $\alpha < 2$), "universal" multifractal clouds dominated by low density "Lévy holes" (frequent low-density regions where most of the transport occurs; Watson et al. 2009). The latter shows how to "renormalize" cloud density, that is, to relate the mean transmission statistics to those of an equivalent homogeneous cloud. Lovejoy et al. (2009b) extended these (numerically) to $H > 0$ and with multiple scattering including the case of very thick clouds. By considering the (fractal) path of the multiply scattered photons, it was found that due to long-range correlations in the cloud, the photon paths are "subdiffusive," and that the corresponding fractal dimensions of the paths tend to increase slowly with mean optical thickness. Reasonably accurate statistical relations between N scatter statistics in thick clouds and single scatter statistics in thin clouds were developed, showing that the renormalized single scatter result is remarkably effective. This is because of two complicating effects acting in contrary directions: the "holes," which lead to increased single scatter transmission, and the tendency for multiply scattered photons to become "trapped" in optically dense regions, thus decreasing the overall transmission.

All results to date are for statistically isotropic media; for more realism, future work must consider scaling stratification as well as the statistical properties of the radiation fields and their (scaling) interrelations with cloud density fluctuations.

Multifractals: Structure Functions

Quantifying the Variability over Scales: Fluctuations and Structure Functions

Spectral analysis is often convenient, but it is not always easy to interpret. In addition, it is only a second-order statistic so that—unless the process is quasi-Gaussian—it characterizes neither the intermittency nor the extremes. The real space alternatives are based on fluctuations of various sorts. Wavelets provide a general formalism for defining and handling fluctuations;

indeed, it is so general that the choice of wavelets is often made on the basis of mathematical convenience or elegance. Although such wavelets may be useful in localizing singularities (if they are indeed localized!) or in estimating statistical scaling exponents, the interpretation of the fluctuations is often opaque. An exception is the Haar fluctuation, which was in fact the first wavelet (Haar 1910; before the full theory). Haar fluctuations are simple to understand; the wavelet formalism is not needed in order to be able to easily apply them in remote sensing and geophysical applications. We could also mention the detrended fluctuation analysis (Peng et al. 1994; Kantelhardt et al. 2002) method, which is nearly a wavelet method but, unfortunately with fluctuations that are not simple to interpret.

The Haar fluctuation $\Delta I(\Delta x)$ at timescale Δx of the intensity field I is simply the difference of the mean of I over the first and second halves of the interval Δx:

$$\left(\Delta I(\Delta x)\right)_{\text{Haar}} = \left| \frac{2}{\Delta x} \int_{x-\Delta x/2}^{x} I(x')dx' - \frac{2}{\Delta x} \int_{\Delta x}^{x-\Delta x/2} I(x')dx' \right| \tag{3.10}$$

where the subscript Haar was added to distinguish it from other common definitions of fluctuation, and the x dependence was suppressed because we assume that the fluctuations are statistically homogeneous. With an appropriate "calibration" constant (a factor of 2 is used below and is canonical), in scale regions where $H > 0$, the Haar fluctuations are nearly equal to the differences; in scale regions where $H < 0$, they are nearly equal to the anomalies:

$$\left(\Delta I(\Delta x)\right)_{\text{dif}} = \left| I(x + \Delta x) - I(x) \right|$$

$$\left(\Delta I(\Delta x)\right)_{\text{anom}} = \left| \frac{1}{\Delta x} \int_{x}^{x+\Delta x} I'(x')dx' \right| ; I' = I - \overline{I} \tag{3.11}$$

where $\overline{I}$ is the mean over the entire series. The Haar fluctuation is essentially the difference fluctuation of the anomaly; it is also equal to the anomaly fluctuation of the difference.

Now that we have defined the fluctuations, we need to characterize them; the simplest way is through (generalized) structure functions (generalized to fluctuations other than the usual differences, and generalized to moments of order other than the usual value of 2): $\left\langle \Delta I(\Delta x)^{q} \right\rangle$ where "⟨·⟩" indicates statistical (ensemble) averaging; this is the qth order structure function.

Physically, if the system is scaling, then the fluctuations are related to the driving flux φ by the following:

$$\Delta I(\Delta x) = \varphi_{\Delta x} \Delta x^{H} \tag{3.12}$$

where the subscript Δx on φ indicates that it is the flux at resolution Δx. The structure function is

$$\left\langle \Delta I(\Delta x)^q \right\rangle = \left\langle \varphi_{\Delta x}^q \right\rangle \Delta x^{qH} \tag{3.13}$$

Turbulent fluxes φ are conserved from scale to scale so that $\left\langle \varphi_{\Delta x} \right\rangle$ = constant (independent of scale), implying that $\left\langle \Delta I(\Delta x) \right\rangle \propto \Delta x^H$, so that H is the mean fluctuation exponent. Beyond the simplicity of interpretation, the Haar fluctuations give a good characterization of the variability for stochastic processes with $-1 < H < 1$. In contrast, fluctuations defined as differences or as anomalies are only valid over the narrower ranges $0 < H < 1$ and $-1 < H < 0$, respectively (see Lovejoy and Schertzer 2012; Lovejoy et al. 2013). Outside these ranges in H, the fluctuation at scale Δx is no longer dominated by wave numbers $\approx \Delta x^{-1}$, so that the fluctuations depend spuriously on details of the finite data sample, typically either the highest or the lowest frequencies that happen to be present in the sample. For H outside these ranges, other definitions of fluctuations (wavelets) must be used. However, empirically, almost all geophysical fields fall into this range.

In the spatial domain, remotely sensed fields typically have $0 < H < 1$ (see Table 3.1), so that there is not much to be gained by using Haar fluctuations instead of differences. However, in the time domain, $-1 < H < 0$ is quite general for atmospheric fields with resolutions of 5–10 days or more; for oceanic fields, of resolutions ≈ 1 year or more (see chapter 10 of Lovejoy and Schertzer 2013), so that Haar fluctuations have the advantage of being able to handle both temporal and spatial analyses, whereas differences are often not appropriate in the corresponding temporal domains.

The generic scaling process is multifractal so that, in general, φ has the following statistics:

$$\left\langle \varphi_{\Delta x}^q \right\rangle \propto \Delta x^{-K(q)} \tag{3.14}$$

where $K(q)$ is a convex function (the constant of proportionality depends on the external scale of the the process; see Equation 3.20). Substituting this into Equation 3.13, we obtain

$$\left\langle \Delta I(\Delta x)^q \right\rangle \propto \Delta x^{\xi(q)}; \ \xi(q) = qH - K(q) \tag{3.15}$$

where $\xi(q)$ is the "structure function exponent." Although we return to this in more detail below, for the moment note that the mean ($q = 1$) flux $<\varphi_{\Delta x}>$ is independent of Δx, so that $K(1) = 0$ and hence $\xi(1) = H$. Note also that for quasi-Gaussian processes, none of the moments of $\varphi_{\Delta x}$ have any scale dependence, so that $K(q) = 0$ and $\xi(q) = qH$ (all the scale dependencies are characterized by H). Finally, the RMS fluctuation $\left(\left\langle \Delta I(\Delta x)^2 \right\rangle^{1/2} \right)$ has the exponent $\xi(2)/2$ so that the error in using the quasi-Gaussian approximation for the variance (i.e., $\xi(2) = 2H$) is $\xi(2)/2 - H = K(2)/2$.

Soil Moisture and Vegetation Indices

In the section titled "The Horizontal," we displayed the excellent spectral scaling of the MODIS bands used to determine soil moisture and vegetation indices, noting that these surrogates are defined (Equation 3.9) in an *ad hoc* way using the (subjective and finest available) resolution. Figure 3.11a shows the corresponding (difference fluctuation based) structure functions for the MODIS bands contributing to the vegetation and soil moisture indices. As expected from the spectra, except for the smallest scales where there are sampling and sensor smoothing artefacts, the scaling is excellent. The slopes (estimates of $\xi(q)$) are shown in Figure 3.11b.

The fact that the bands have nontrivial scaling implies that if the surrogates are defined at different resolutions, their statistical properties will be different; in other words, the single-scale surrogate (sss) can be (at most) correct at a single resolution. To see this more clearly, these sss (at scale ratio $\lambda = L/\Delta x$, where L is the large [image scale] and Δx is the resolution) can be contrasted with the corresponding multiscale surrogates

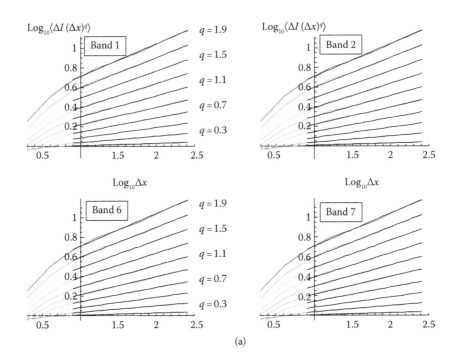

(a)

FIGURE 3.11

(a) Structure functions based on difference fluctuations for the MODIS bands discussed in the text, plotted as a function of lags $\log_{10}\Delta x$ (with distance Δx measured in pixels = 512 m). The regression lines (slope $\xi(q)$, fit for the range Δx = 8–256 pixels) are also shown; the linearity is an indication of the quality with which the scaling is respected. Adapted from Lovejoy et al. (2007a).   *(Continued)*

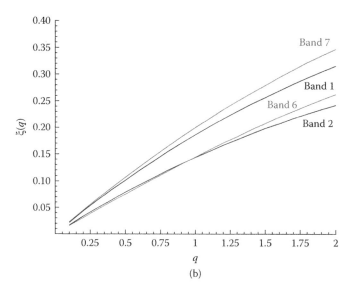

FIGURE 3.11 (Continued)
(b) The structure function exponent $\xi(q)$ for MODIS bands 7, 1, 6, and 2 (top to bottom, the slopes from Figure 3.11a). The figure is derived from Lovejoy et al. (2007a).

(mss, at scale ratio λ). Mathematically, the difference can be expressed as follows:

$$\sigma_\lambda^{(s)} = \left[\sigma_\Lambda^{(s)}\right]_\lambda; \quad \sigma_\Lambda^{(s)} = \frac{I_{i,\Lambda} - I_{j,\Lambda}}{I_{i,\Lambda} + I_{j,\Lambda}}$$

$$\sigma_\lambda^{(m)} = \frac{I_{i,\lambda} - I_{j,\lambda}}{I_{i,\lambda} + I_{j,\lambda}}; \quad I_{i,\lambda} = \left[I_{i,\Lambda}\right]_\lambda$$

(3.16)

where $\Lambda = L/l$ is the maximum scale ratio (satellite image scale L)/(single pixel scale l) and the notation $\left[I_{i,\Lambda}\right]_\lambda$ and $\left[\sigma_\Lambda^{(s)}\right]_\lambda$ denotes averaging from this finest resolution up to the intermediate ratio $\lambda < \Lambda$ (i.e., $\Delta x > l$). The mss is the surrogate that would be obtained by applying an identical algorithm (Equation 3.9) to satellite data at the lower resolution, whereas the sss is the surrogate at the same resolution but based instead on the finest scale available.

From the single- and multiple-scale definitions of the surrogates, we can define the difference fluctuations:

$$\Delta\sigma^{(s)}(\Delta x) = \left|\sigma_\Lambda^{(s)}(x + \Delta x) - \sigma_\Lambda^{(s)}(x)\right|$$

$$\Delta\sigma^{(m)}(\Delta x) = \left|\sigma_\lambda^{(m)}(x + \Delta x) - \sigma_\lambda^{(m)}(x)\right|$$

(3.17)

where for the multiple-scale definition (second line), we have $\Delta x = L/\lambda$. In Figure 3.12a (bottom), we show that the resulting mss obtained across the

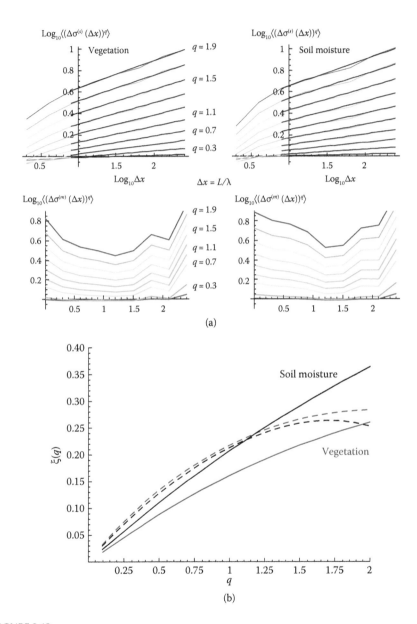

FIGURE 3.12
(a) A comparison of the single- and multiple-scale indices (top and bottom rows, respectively) defined by Equation 3.17, that is, by degrading indices defined at the finest resolution to an intermediate resolution (single scale index) and by first degrading the resolution of the bands and then determining the indices (multiple-scale index). The moments from top to bottom are in order $q = 1.9, 1.7, 1.5, 1.3, 1.1, 0.9, 0.7, 0.5, 0.3$, and 0.2. Adapted from Lovejoy et al. (2007a). (b) Comparison of the structure function exponents for the vegetation and soil moisture indices (red and black, respectively), estimated using the single-scale and multiple-scale index definitions (solid and dashed lines, respectively; the slopes in Figure 3.12a). Derived from Lovejoy et al. (2007a).

same range of scales as the sss (Figure 3.12a, top) is quite different (it is statistically significantly larger). The comparison of the corresponding exponents ($\xi(q)$) is given in Figure 3.12b.

Haar Fluctuations and Martian Topography

An impressive application of the structure function method with Haar fluctuations was recently published by Landais (2014) and Landais et al. (2015), using data from the Mars Orbiter Laser Altimeter (MOLA). Figure 3.13a shows an analysis over more than four orders of magnitude of scale, displaying two scaling regimes with a break at ≈10 km. In this case, the altimeter has numerous holes (some large) caused notably by obstruction of the signal by dust clouds. The usual method of filling such holes is by interpolation, but if there is a significant amount of interpolation, then the results can be badly corrupted. The reason is that even the lowest order interpolation (linear interpolation) is once differentiable it has $H = 1$, whereas the topography is only differentiable of order $H \approx 0.52$ (see Figure 3.13b). Therefore, linear interpolation will mix segments of $H = 1$ with segments of $H = 0.52$, generally

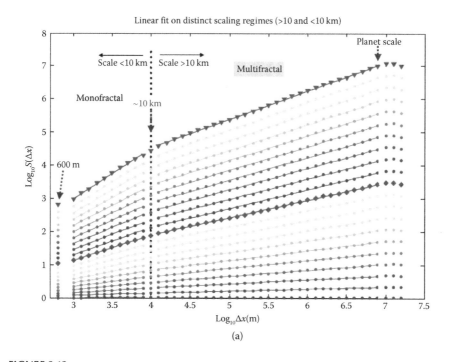

(a)

FIGURE 3.13

(a) The Haar fluctuations from 5×10^5 Mars Orbiter Laser Altimeter altitude estimates for moments of order 0.1, 0.2, ..., 1.9, and 2 (bottom to top). Δx is in meters; $S(\Delta x)$ is the corresponding structure function for the $q = 1$ moment (the second red line from the bottom); the units are also in meters; and $S(\Delta x)$ gives the mean vertical change over a distance Δx. *(Continued)*

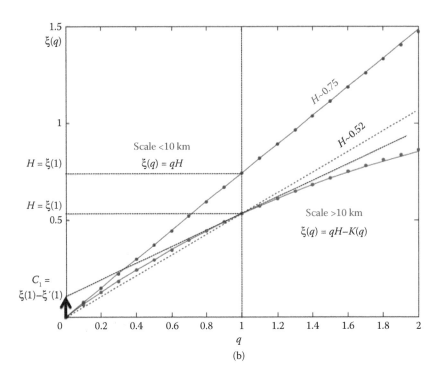

FIGURE 3.13 (Continued)

(b) The structure function exponent $\xi(q)$ estimated from the regression slopes in Figure 3.13a for the Martian topography (the blue curve). The bottom red curve is for the region >10 km and the top blue curve is for scales <10 km. While the latter is nearly linear and hence monofractal (with $\xi(1) = H \approx 0.75$), the former is curved and multifractal. For this curve, we also show the graphical estimates of the mean fluctuation exponent H using $H = \xi(1) \approx 0.52$ and (from the tangent at $q = 1$) the intermittency correction near the mean, $C_1 = \xi(1) - \xi'(1) \approx 0.1$. (Adapted from Landais, F., et al., *Nonlin. Processes Geophys. Discuss.*, 2, 1007–1031, 2015.)

yielding a spurious result. Thus, in order to estimate the Haar fluctuations it is best to use an algorithm that does not require interpolation. One such simple algorithm is described in the appendix of Lovejoy (2014) and is freely available at: http://www.physics.mcgill.ca/~gang/software/index.html; it was used to generate the figure.

Multifractals and Trace Moment Analysis

Characterizing $\xi(q)$, $K(q)$: Universal Multifractals

The previous examples estimating $\xi(q)$ from MOLA and MODIS data underline a basic problem: that the scaling generally involves an entire nonlinear convex function $\xi(q)$ (equivalently the concave function $K(q) = qH - \xi(q)$) for its characterization, and empirically this is the equivalent of determining an infinite number of parameters. In order to make the problem manageable (to reduce the parameters to a finite number), we can make use of a multiplicative analogue of the usual (additive) central limit theorem for random variables. This shows that under fairly general circumstances, $K(q)$ is determined by only two parameters that define multifractal "universality classes":

$$K(q) = \frac{C_1}{\alpha - 1}(q^\alpha - q); 0 \leq \alpha \leq 2 \qquad (3.18)$$

where α is the Levy index and C_1 is the codimension of the mean (Schertzer and Lovejoy 1987). From Equation 3.18, we see that $C_1 = K'(1)$; this provides a convenient way of estimating the parameters. Figure 3.13b shows how it can be graphically estimated from $\xi(q)$; for α, it is also possible to use $\alpha = K''(1)/K'(1)$. Table 3.1 shows the resulting parameter estimates for many geophysical fields, including several that were remotely sensed. In addition, note that the difference between the exponents of the RMS and mean (important for interpreting the slopes in the RMS Haar graphs) $\xi(2)/2 - H = K(2)/2 = A_\alpha C_1$, where $A_\alpha = (2^{\alpha-1} - 1)/(\alpha - 1)$. Because empirically $1.5 > \alpha > 2$ (Table 3.1), we find $0.83 > A_\alpha > 1$ (near 1) so that often C_1 provides a good estimate of the error in using the RMS exponent $\xi(2)/2$ in place of H. From the table it is evident that, in space, the difference $\xi(2)/2 - H$ can readily be ≈ 0.1, which is significant. The intermittency quantified by C_1, α leads to much larger variability than would be expected from classical (quasi-Gaussian) processes. Additional difficulties in the interpretation of data analyses arise when the scaling is anisotropic (see Lovejoy and Schertzer [2007b] or section 7.1.6 of Lovejoy and Schertzer [2013]).

Note that with the help of remote sensing, many solid earth fields have also been shown to be scaling over various ranges. These include the rock density, magnetic susceptibilities, surface magnetic fields, and surface gravity fields; see the review by Lovejoy and Schertzer (2007b) for examples, references, and theory for these geopotential fields. Visible and infrared emissions from geothermal regions have been studied by Laferrière and Gaonac'h (1999), Harvey et al. (2002), Gaonac'h et al. (2003), and Beaulieu et al. (2007).

Cascades and Trace Moment Analysis

We investigated the nonlinearity of $\xi(q)$, $K(q)$ (intermittency) directly from the structure function, estimating $\xi(q)$ from the exponents of the qth order moments and then estimating $K(q)$ as $\xi(q) + q\xi(1)$ (see Equation 3.15, recalling that $K(1) = 0$). However, frequently, the H value is quite a bit larger than the C_1 value, so that the linear contribution to $\xi(q)$ is much larger than the nonlinear contribution, making C_1, α difficult to accurately estimate. However, by estimating φ and hence $K(q)$ directly, we can obtain improved estimates of C_1, α. We are also able to estimate the outer scale of the underlying cascade process.

To see how this works on gridded data, we can conveniently estimate ΔI as the absolute finite difference along the transect. The normalized high-resolution flux is then obtained by dividing Equation 3.12 by its ensemble average to obtain the following:

$$\frac{\varphi}{\langle \varphi \rangle} = \frac{\Delta I}{\langle \Delta I \rangle} \tag{3.19}$$

Finally, this high-resolution flux can be systematically degraded to lower resolution by averaging. The generic multifractal process is a multiplicative cascade; if such a process starts at outer scale L_{eff}, then the statistics follow:

$$M = \left\langle \varphi_{\lambda'}^q \right\rangle = \lambda'^{K(q)}; \ \lambda' = L_{eff}/\Delta x \tag{3.20}$$

where λ' is the scale ratio of the outer scale to the resolution scale of the degraded flux Δx and $\varphi_{\lambda'}$ is the normalized flux at scale ratio λ'. When $\left\langle \varphi_{\lambda'}^q \right\rangle$ is estimated in this way (by taking q powers of the successively degraded flux), it is called a *trace moment*. In empirical analyses, L_{eff} is not known *a priori*; it has to be estimated from the data. Here, we replace it by a convenient reference scale, $L_{ref} = 20{,}000$ km, which is the largest distance on Earth (half the circumference), and use $\lambda = L_{ref}/\Delta x$. If Equation 3.20 holds, then for all q the lines of $\log \left\langle \varphi_{\lambda}^q \right\rangle$ against $\log \lambda$ will cross at a scale corresponding to $\lambda = \lambda_{eff} = L_{ref}/L_{eff}$.

Figure 3.14a shows the results of estimating the various moments of order $2 \geq q \geq 0$ for the TRMM orbiting weather radar data analyzed along the orbit direction. It is evident from the log–log linearity that the scaling is excellent up to near-planetary scales of 10,000 km, and the convergence of the lines to a common outer scale, λ_{eff}, is a direct confirmation of the multiplicative ("cascade") nature of the statistics (Equation 3.20). The lines plausibly cross at a scale of the order of the size of the planet (see Table 3.1). The fact that $L_{eff} = L_{ref}/\lambda_{eff}$ can be a little larger than the size of Earth is because, even at planetary scales (20,000 km), there is some residual variability due to the interaction of the precipitation field with other atmospheric fields—L_{eff} is simply the "effective" scale at which the cascade would have had to start in order to explain the statistics over the observed range.

Precipitation is particularly interesting because, according to Table 3.1, it is the most intermittent of the common geophysical fields (the value of the intermittency parameter C_1 is the highest, ≈ 0.4) and therefore is presumably the most difficult to measure. Figure 3.14b shows the corresponding results for other precipitation products (both gauge- and reanalysis-based) as well

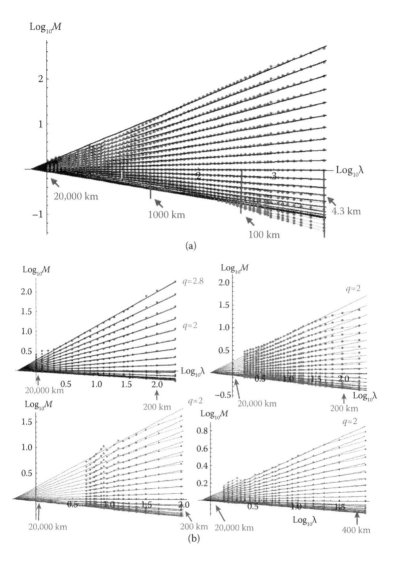

FIGURE 3.14

(a) The TRMM reflectivities (4.3-km resolution). The moments M (see Equation 3.20) are for $q = 0$, 0.1, 0.2, ..., 2, taken along the satellite track. The poor scaling (curvature) for the low q values (bottom with negative slopes) can be explained as artefacts of the fairly high minimum detectable signal. $L_{ref} = 20,000$ km so that $\lambda = 1$ corresponds to 20,000 km; the lines cross at the effective outer scale $\approx 32,000$ km, $C_1 \approx 0.63$. (Reproduced from Lovejoy, S., et al., *Geophys. Resear. Lett.*, 36, L01801, 2009c. With permission.) (b) East–west analyses of the gridded precipitation products discussed in the text. Upper left: The TRMM 100×100 km, 4-day averaged product. Upper right: The ECMWF interim stratiform rain product (all latitudes were used). Note that the data were degraded in constant angle bins so that the outer scale is 180°. To compare with the other analyses, a mean map factor of 0.69 has been applied (the mean east–west outer scale was $\approx 14,000$ km). Lower left: The CPC hourly gridded rainfall product (US only). (Reproduced from Lovejoy, S., et al., *Adv. Water Res.*, 45, 37–50, 2012. With permission.) *(Continued)*

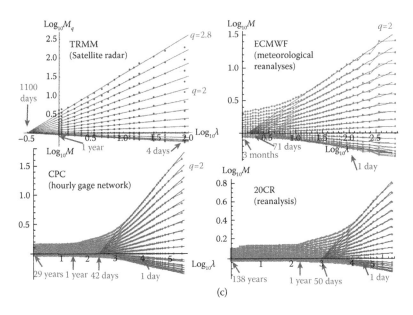

FIGURE 3.14 (Continued)
(c) The temporal analyses of the various precipitation products analyzed spatially in
Figure 3.14b (for the TRMM satellite radar at 4 days to 1 year; the ECMWF stratiform rain
product, 3 hours to 3 months; and the CPC network, 1 hour to 29 years; the 20CR reanalysis
was for 45°N, every 6 hours for 138 years). (Reproduced from Lovejoy, S., et al., *Adv. Water Res.*,
45, 37–50, 2012. With permission.)

as for a lower-resolution TRMM product, all analyzed in the east–west
direction. It is evident that the qualitative behaviors are very similar. More
detailed analysis (Lovejoy et al. 2012) shows that there are nevertheless significant differences in parameters, notably C_1. Similarly, in the macroweather
regime (for precipitation, ≈2–4 days up to 40 years in the industrial epoch),
this is also true (de Lima and Lovejoy 2015; Lovejoy and de Lima 2015), leading to the conclusion that the problem of accurately estimating areal precipitation is still unsolved.

Interestingly, the same data can be analyzed in the temporal domain.
The reason for expecting scaling in time as well as space is that the wind
field that connects the two is scaling (see, e.g., the spectrum in Figure 3.1c).
Figure 3.14c shows that the straight lines on the log–log plot converge—
the behavior expected for multiplicative cascades (Equation 3.20) and
the signature of multifractality—but at timescales a bit longer than the
outer scale of the weather regime (the point timescale at which the scaling becomes poor, closer to ≈10 days). The main exception is the TRMM
data, which seem to have excellent scaling out to the limit of the 5,900
orbits (about 1 year). It is possible that the reason for this significantly
larger outer timescale is that the TRMM data are mostly captured over

the tropical oceans and the oceans have a significantly longer outer scale (about 1 year) due to the much lower energy rate densities in the ocean (see chapter 8 of Lovejoy and Schertzer, 2013, for a review of the theory and data).

Other significant examples of the use of trace moments are for characterizing the Earth's energy budget as a function of scale. Figure 3.15a and b shows the results for the TRMM satellite (from two of the channels whose spectra were analyzed in Figure 3.1a and b) at visible and thermal infrared wavelengths, respectively. Upon inspection, it is evident that the multiplicative cascade Equation 3.20 holds very accurately

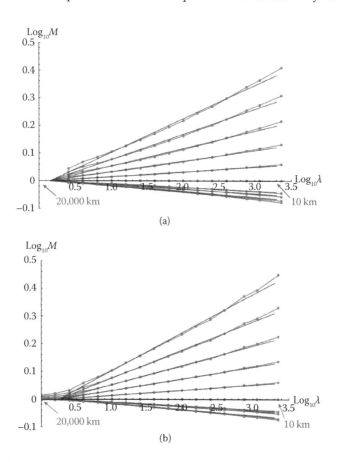

(a)

(b)

FIGURE 3.15
(a) TRMM visible data (0.63 mm) from the VIRS instrument, Channel 1, with fluxes estimated at 8.8 km. Only the well-lit 15,000-km orbit sections were used. L_{ref} = 20,000 km, so that λ = 1 corresponds to 20,000 km, the lines cross at $L_{eff} \approx 9,800$ km. (Reproduced from Lovejoy, S., et al., *Geophys. Resear. Lett.*, 36, L01801, 2009c. With permission.) (b) Same as Figure 3.15a except for VIRS thermal IR (Channel 5, 12.0 μm), $L_{eff} \approx 15,800$ km (Reproduced from Lovejoy, S., et al., *Geophys. Resear. Lett.*, 36, L01801, 2009c. With permission.) *(Continued)*

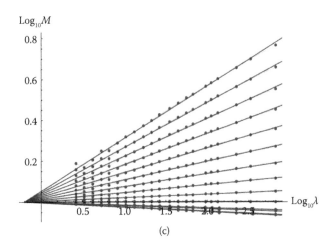

(c)

FIGURE 3.15 (Continued)
(c) Logs of normalized moments M (Equation 3.20) *vs.* $\log_{10}\lambda$ for 2 months (1,440 images) of MTSAT, thermal IR, 30-km resolution over the region 40°N to 30°S, 130° east–west over the western Pacific, the average of east–west and north–south analyses. L_{ref} = 20,000 km so that $\lambda = 1$ corresponds to 20,000 km; the lines cross at the effective outer scale ≈ 32,000 km (from Pinel 2012) and $C_1 \approx 0.074$ (close to the TRMM thermal IR results, Table 4.7a, VIRS 4, 5).

(to about ±0.5%) up to about 5,000 km. Figure 3.15c shows the analogous result for the (geostationary) MTSAT thermal IR radiances over the Pacific (same as the data in Figure 3.1c). Finally, Figure 3.16 shows the results of trace moment analysis on meteorological state variables estimated through a reanalysis (see Figure 3.2). It is therefore not surprising that numerical weather prediction models also show excellent scaling and similar multiplicative trace moments (see Stolle et al. [2009] and Stolle et al. [2012] for analyses of several forecast models). *In situ* aircraft data of meteorological variables of state also show very similar cascade characteristics (Lovejoy et al. 2010).

Finally, we can return to the topography data whose spectra were analyzed in Figure 3.3. Figure 3.17 shows the cascade structure of the topographic gradients obtained by combining the four different data sets used in Figure 3.3 spanning the range 20,000 km down to submetric scales. As for the spectrum, the scaling holds quite well until around 40 m. Gagnon et al. (2006) argued that this break is due to the presence of trees (for the high resolution data set used over Germany, 40 m is roughly the horizontal scale at which typical vertical fluctuations in the topography are of the order of the height of a tree). Over the range of planetary scales down to ≈40 m, it was estimated that the mean residue of the universal scaling form with parameters C_1 = 0.12, α = 1.79 (for all moments $q \leq 2$) was ±45% over this range of nearly 10^5 in scale (this error estimate was for the "reduced" moments $\langle \varphi^q \rangle^{1/q}$, e.g., for $q = 2$, root mean square moments).

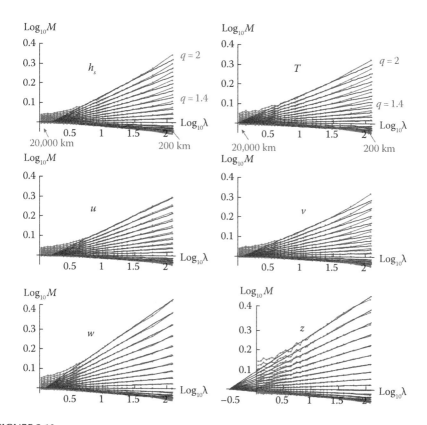

FIGURE 3.16

The trace moment analysis of the 700-mb ECMWF interim reanalysis products analyzed in the east–west direction, data at 700 mb (about 3 km altitude) between ±45° latitude. The fields are specific humidity (h_s), temperature (T), east–west wind (u), north–south wind (v), vertical wind (w), and geopotential height (z); fields at 24-hour intervals for the year 2006 were used. The moments are for orders q = 2, 1.9, and 1.8 (descending from top to bottom); λ = 1 corresponds to 20,000 km. See Table 3.1 for some parameter estimates. (Reproduced from Lovejoy, S., and D. Schertzer, *The weather and climate: Emergent laws and multifractal cascades*, Cambridge University Press, Cambridge, 2013. With permission.)

Space–Time Scaling: Example of MTSAT

Remotely sensed products remapped into convenient coordinate systems are increasingly available at regular time intervals. This allows for the systematic study and assessments of trends. It also requires us to understand the full (joint) space–time variability. As an example, in this section we study global scale thermal infrared data from MTSAT (see Figure 3.1c for the 1D spectra and Figure 3.18a through c for three 2D subspaces) and show how knowledge of its spectral density in (k_x, k_y, ω) space (horizontal wave number

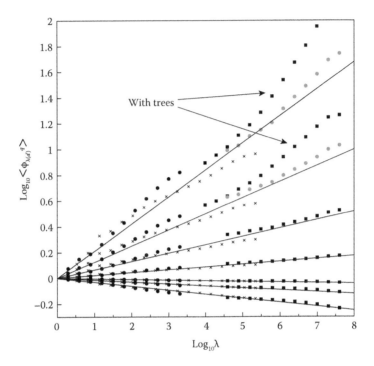

FIGURE 3.17
The trace moments of the data sets whose spectra were analyzed in Figure 3.4: ETOPO5 (circles), continental United States (Xs), and lower Saxony (squares). For the latter, a subsample was also analyzed (light circles) that was (mostly) free of trees, the difference indicating the effect of trees. The regression lines distinguish between the q values of the associated points: $q = 2.18$, 1.77, 1.44, 1.17, 0.04, 0.12, and 0.51 (top to bottom). (Reproduced from Gagnon, J. S., et al., *Nonlin. Proc. Geophys.*, 13, 541–570, 2006. With permission.)

and frequency) can be used to determine AMVs, which are used as surrogates for winds.

The MTSAT data set used for this study is comprised of nearly 1,300 hours of hourly geostationary MTSAT data (at 30-km resolution) from an $8,000 \times 13,000$ km² region centered on the equatorial Pacific (see Pinel et al. 2014, for details). It is convenient to use Fourier techniques. Introduce the nondimensional space–time lag $\underline{\Delta R}$ and nondimensional wave vector $\underline{K}$:

$$\underline{\Delta R} = \left(\underline{\Delta r}, \Delta t\right); \ \underline{\Delta r} = \left(\Delta x, \Delta y\right)$$
$$\underline{K} = \left(\underline{k}, \omega\right); \ \underline{k} = \left(k_x, k_y\right)$$

(3.21)

The nondimensionalization can be achieved by using the pixel and sampling periods or the planet scale (space) and the corresponding lifetime of planetary structures (time, roughly 10 days). We can now estimate the space–time ("st")

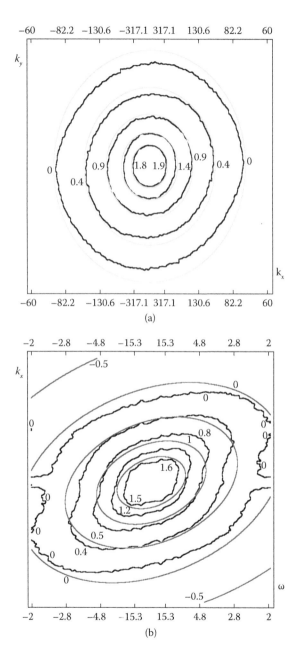

FIGURE 3.18
(a) Contours of log $P(k_x, k_y)$, the spatial spectral density. Black represents empirical contours; colored lines are from the fit using Equation 3.32 with $a = 1.2 \pm 0.1$. (b) Contours of log $P(k_x, \omega)$, the zonal wave number/frequency subspace. Black represents empirical contours; colored lines are from the fit using Equation 3.32. The orientation is a consequence of the mean zonal wind, -3.4 m/s. *(Continued)*

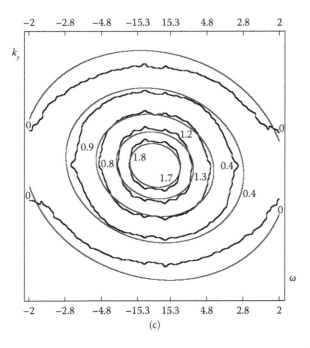

(c)

FIGURE 3.18 (Continued)
(c) Contours of log $P(k_y, \omega)$, the meridional wave number/frequency subspace. There is very little if any "tilting" of structures because the mean meridional wind was small: 1.1 m/s. Black represents empirical contours; colored lines are from the fit using Equation 3.32. (Reproduced from Pinel, J., et al. *Atmos. Resear.*, 140–141, 95–114, 2014. With permission.)

spectral density $P_{st}(\underline{K}) \propto \left\langle \left| \tilde{I}(\underline{K}) \right|^2 \right\rangle$ (see Equation 3.2). We can determine the three 1D spectra by successively integrating out the conjugate variables:

$$E_x(k_x) = \int P_{st}(k_x, k_y, \omega) dk_x \, d\omega \propto k_x^{-\beta_x}$$

$$E_y(k_y) = \int P_{st}(k_x, k_y, \omega) dk_y \, d\omega \propto k_y^{-\beta_y} \qquad (3.22)$$

$$E_t(\omega) = \int P_{st}(k_x, k_y, \omega) dk_x \, dk_y \propto \omega^{-\beta_t}$$

where the right-hand side equalities are consequences of assuming space–time scaling; the exponents can in principle all be different. Figure 3.1c shows the result: it is evident that the east–west, north–south, and temporal spectra are nearly identical up to ≈(5–10 days)$^{-1}$ and are nearly perfect power laws (most of the deviations from linearity in the figure can be accounted for by the finite resolution and finite data "window"; see the black line that theoretically takes these limits into account). The main exceptions are the two small spectral "bumps" at (1 day)$^{-1}$, (12 hours)$^{-1}$. The fact that the 1D spectral densities are nearly identical shows that $\beta_x = \beta_y = \beta_t = \beta$ and that the spectrum satisfies the isotropic scaling symmetry:

$$P_{st}(\lambda \underline{K}) = \lambda^{-s} P_{st}(\underline{K}); \ s = \beta + 2 \tag{3.23}$$

where empirically, $s = \beta + 2 \approx 3.4$. Note that at any given scale P_s will generally display anisotropy; Equation 3.22 simply implies that the anisotropy doesn't change with scale (see Pinel et al. 2014, for the full analysis).

By necessity, the scale symmetry (Equation 3.22) can only hold up to planetary scales ($L_e = L_{ref} = 20{,}000$ km); this implies a breakdown in the time domain at scales τ, which is interpreted as the lifetime of planetary scale structures. In Figure 3.1c, we find $\tau \approx 5{-}10$ days; Lovejoy and Schertzer (2010a) describe how this is determined by the turbulent energy flux ε (power/mass): $\tau = \varepsilon^{-1/3} L_e^{2/3}$, where ε itself is determined by the solar flux. This theory well describes the spectrum of the many atmospheric variables—ocean temperatures as well as the Martian weather and macroweather (Lovejoy et al. 2014), see also Chen et al. 2016.

In order to solve the functional Equation 3.22 expressing the scaling symmetry on P, it is convenient to introduce real and Fourier space–time scale functions $[\![\Delta R]\!], [\![K]\!]_F$. The scale functions are generalizations of the usual notion of vector norm (distance). They generally satisfy a scale function equation (for this GSI, see Chapter 7 in Lovejoy and Schertzer 2013). In the general case (needed, for example, to deal with vertical stratification), we have the following:

$$\left[\!\left[\lambda^G \underline{K} \right]\!\right]_F = \lambda \left[\!\left[\underline{K} \right]\!\right]_F \tag{3.24}$$

where G is the generator of the anisotropy (in the basic linear case, G is a matrix so that the anisotropy depends on scale, but not position). In the special isotropic case in Equation 3.23, we simply have $G = 1 =$ the identity we have:

$$\left[\!\left[\lambda \underline{K} \right]\!\right]_F = \lambda \left[\!\left[\underline{K} \right]\!\right]_F \tag{3.25}$$

With this, the empirical spectra (Figure 3.18, Equation 3.22) can be written as follows:

$$P_{st}(\underline{K}) \approx P_0 \left[\!\left[\underline{K} \right]\!\right]^{-s} \tag{3.26}$$

where P_0 is a constant. A basic result (a corollary of the Wiener–Khinchin theorem) relates the second-order structure function and the spectral density: it states that $\langle \Delta I^2 \rangle$ and $2(1{-}P)$ are Fourier transform pairs. With this, we obtain the following:

$$\left\langle \Delta I (\Delta \underline{R})^2 \right\rangle = \left[\!\left[\underline{\Delta R} \right]\!\right]^{\xi(2)}; \ P_{st}(\underline{K}) \approx \left[\!\left[\underline{K} \right]\!\right]_F^{-s} \tag{3.27}$$

with $s = 2 + \beta$ as before.

Theoretical considerations based on turbulence theory (Lovejoy and Schertzer 2010a; Pinel et al. 2014) lead to the convenient result for the nondimensional scale functions:

$$\left\lVert \Delta \underline{R} \right\rVert^2 = \Delta \underline{R}^T \underline{B} \Delta \underline{R}; \quad \left\lVert \underline{K} \right\rVert_F^2 = \underline{K}^T \underline{B}^{-1} \underline{K} \tag{3.28}$$

with the matrices $\underline{B}, \underline{B}^{-1}$ given by

$$\underline{B} = \begin{pmatrix} 1 & 0 & -\mu_x \\ 0 & a^2 & -a^2\mu_y \\ -\mu_x & -a^2\mu_y & 1 \end{pmatrix};$$

$$\underline{B}^{-1} = \frac{1}{1-\mu_x^2-a^2\mu_y^2} \begin{pmatrix} 1-a^2\mu_y^2 & \mu_y\mu_x & \mu_x \\ \mu_y\mu_x & \left(1-\mu_x^2\right)/a^2 & \mu_y \\ \mu_x & \mu_y & 1 \end{pmatrix} \tag{3.29}$$

μ_x and μ_y are the components of the nondimensional average advection velocity:

$$\underline{\mu} = \left(\overline{v_x}, \overline{v_y}\right)/V_w; \quad V_w = \left(\overline{v_x^2} + a^2\overline{v_y^2}\right) \tag{3.30}$$

where V_w is the large-scale turbulent velocity and a is the north–south/east–west aspect ratio (taking into account the fact that the large-scale gradients are typically a times larger in the north–south direction than in the east–west direction). If we now introduce the modified nondimensional speeds and frequency:

$$\mu_x' = \frac{\mu_x}{\sqrt{1-\left(\mu_x^2+a^2\mu_y^2\right)}}; \quad \mu_y' = \frac{\mu_x}{\sqrt{1-\left(\mu_x^2+a^2\mu_y^2\right)}}; \quad \omega' = \frac{\omega'}{\sqrt{1-\left(\mu_x^2+a^2\mu_y^2\right)}} \tag{3.31}$$

we find:

$$\left\lVert \underline{K} \right\rVert_F^2 = \underline{K}^T \underline{B}^{-1} \underline{K} = \left(\omega' + \underline{k} \cdot \underline{\mu}\right)^2 + k_x^2 + k_y^2/a^2 \tag{3.32}$$

The corresponding spectral density is as follows:

$$P(k_x, k_y, \omega) = P_0\left(\left(\omega' + \underline{k} \cdot \underline{\mu}\right)^2 + k_x^2 + k_y^2/a^2\right)^{-s/2} \tag{3.33}$$

Equation 3.32 has the interpretation as a mean advection by $\underline{\mu}'$ with a spatial squashing by factor a in the east–west (x) direction.

In order to test this in detail, Pinel et al. (2014) considered the spectral densities in the three 2D subspaces (k_x, ω), (k_y, ω), and (k_x, k_y) obtained by integrating out the (single) conjugate variables from $P_{st}(k_x, k_y, \omega)$ (Figure 3.18a through c). This enabled the theory to be tested and the parameters to be estimated. A further analysis (Pinel and Lovejoy 2014) showed that the (small) residuals could be interpreted in terms of multiplicative wave-like perturbations to P_{st}. This analysis is theoretically motived by improvement of the well-known "turbulence background" of space–time IR radiances analyzed in the traditional wave context by Wheeler and Kiladis (1999) and Hendon and Wheeler (2008).

Finally, Pinel et al. (2014) showed how to use the real space structure function (based on the quadratic form determined by the matrix $\underline{B}$) as a theoretical justification for a series of essentially *ad hoc* techniques starting with Hubert and Whitney (1971) for measuring "satellite winds," now more accurately called *AMVs* (Atmospheric Motion Vectors). Using the real space parts of Equations 3.27 through 3.29, we obtain an equation for the (second-order) structure function (Equation 3.26) in terms of the winds. The structure function is linearly related to the autocorrelation function and hence gives useful formulae for relating the maximum correlation (which turns out to be the minimum structure function) and the winds. Although there are variants, this maximum cross-correlation method is currently used operationally with MTSAT and GOES geostationary IR, visible satellite imagery (see, e.g., Szantai and Sèze [2008] for an overview and comparison).

Conclusions

As resolutions improve, remote sensing allows us to take a veritable voyage through scales, effectively zooming further and further into complex geofields. The journey potentially takes us over 10 orders of magnitude in scale—from planetary to submillimetric. In order to understand, characterize, and model such geocomplexity, we need an appropriate theoretical framework. The traditional framework is scale bound; it assumes *a priori* that as we change scales we uncover qualitatively different processes and phenomena. It postulates simple, usually deterministic models, at best explaining the variability over a narrow range of space–time scales. It is usually justified by an appeal to phenomenology, to the fact that structures at different scales often look different.

In this paper, we argued that on the contrary most of the geovariability (for example, most of the spectral variance) is in the wide-range scaling background processes. With the help of numerical multifractal simulations, we pointed out the phenomenological fallacy: that scaling is generally anisotropic, in which case structures *do* change appearance with scale, even though the underlying mechanism simply repeats scale after scale—it is scale invariant. If in these cases, the morphologies (rather than statistics) were used to infer the dynamics, then several distinct mechanisms might be invoked, whereas there was in fact only one—hence the fallacy. Although it is true that we have discussed natural geosystems, which often respect a scale-invariance symmetry, it is possible that the same considerations might also apply to at least some human-created morphologies such as the spatial distribution of urban areas.

This paper is an update of Pecknold et al. (1997), which concentrated on developing the basic framework of multifractal processes and GSI needed for handling extremely intermittent scaling variability and anisotropic scaling, respectively. We discussed some striking new global scale examples: visible, infrared, passive, and active microwave sensing of the atmosphere; meteorological reanalyses (meteorological state variables) of ocean color, soil moisture, and vegetation indices; topography (Earth and Mars). We focused on improvements in (real space) fluctuation analysis methods (especially the Haar fluctuations), as well as on trace moment analysis, which can now be routinely used to estimate the external scale of the process and to directly test the multiplicative nature of the intermittency. Finally—reflecting the increasing availability of regular temporal series of images—we reviewed some recent work on space–time analyses from geostationary infrared satellite data. It showed, remarkably, that the (horizontal) space–time scaling has an isotropic exponent, while at the same time having "trivial" anisotropy that reflects the advection and turbulent velocities and can be used as theoretical bases for AMV determinations.

References

Balmino, G. (1993), The spectra of the topography of the Earth, Venus and Mars, *Geophys. Res. Lett.*, 20(11), 1063–1066.

Balmino, G., K. Lambeck, and W. Kaula (1973), A spherical harmonic analysis of the Earth's topography, *J. Geophys. Res.*, 78(2), 478–481.

Barker, H. W., and J. A. Davies. (1992), Cumulus cloud radiative properties and the characteristics of satellite radiance wavenumber spectra, *Remote Sens. Environ.*, 42, 51–64.

Beaulieu, A., G. Gaonac'h, and S. Lovejoy. (2007), Anisotropic scaling of remotely sensed drainage basins: The differential anisotropy scaling method, *Nonlin. Proc. Geophys.*

Bell, T. H. (1975), Statistical features of sea floor topography, *Deep Sea Res.*, 22, 883–891.

Berkson, J. M., and J. E. Matthews. (1983), Statistical properties of seafloor roughness, in *Acoustics and the Sea-Bed*, edited by N. G. Pace, pp. 215–223, Bath University Press, Bath, England.

Bouchaud, J. P., and A. Georges. (1990), Anomalous diffusion in disordered media: Statistical mechanisms, models and physics applications, *Phys. Rep.*, 195, 127–293.

Burgert, R., and W. W. Hsieh. (1989), Spectral analysis of the AVHRR sea surface temperature variability off the west coast of vancouver Island, *Atmos.-Ocean.*, 27, 577–587.

Busygin, V. P., N. A. Yevstatov, and E. M. Feigelson. (1973), Optical propeorties of cumulus clouds and radiant fluxes for cumulus cloud cover, *Izv. Acad. Sci. USSR Atmos. Oceanic Phys.*, 9, 1142–1151.

Cahalan, R. (1994), Bounded cascade clouds: Albedo and effective thickness, *Nonlin. Proc. Geophys.*, 1, 156–167.

Cahalan, R. F. (1989), Overview of fractal clouds, in *Advances in Remote Sensing Retrieval Methods*, 17–34, edited by A. Deepak, H. Flemming and J. Theon, A. Deepak, Hampton Va.

Cahalan, R. F., L. Oreopoulos, A. Marshak, K. F. Evans, A. B. Davis, R. M. Pincus, K. H. Yetzer, et al. (2005), The I3RC: Bringing together the most advanced radiative transfer tools for cloudy atmospheres, *Bull. Am. Meteorol. Soc.*, 86, 1275–1293.

Cahalan, R. F., W. Ridgeway, W. J. Wiscoombe, T. L. Bell, and J. B. Snider. (1994), The albedo of fractal stratocumulus clouds, *J. Atmos. Sci.*, 51, 2434–2455.

Chandrasekhar, S. (1950), *Radiative Transfer*, Clarendon Press, Oxford.

Davis, A., S. Lovejoy, P. Gabriel, D. Schertzer, and G. L. Austin. (1990), Discrete angle radiative transfer. Part III: Numerical results on homogeneous and fractal clouds, *J. Geophys. Res.*, 95, 11729–11742.

de Lima, M. I. P., and S. Lovejoy. (2015), Macroweather precipitation variability up to global and centennial scales, *Wat. Resour. Res.*, 51, 9490–9513.

Deschamps, P. Y., R. Frouin, and M. Crepon. (1984), Sea surface temperature of the coastal zones of France observed by the HCMM-Satellite, *J. Geophys. Res.*, 89, 8123–8149.

Deschamps, P. Y., R. Frouin, and L. Wald. (1981), Satellite determination of the mesoscale variability of the sea surface temperature, *J. Phys. Oceanogr.*, 11, 864–870.

Ferlay, N., and H. Isaka. (2006), Multiresolution analysis of radiative transfer through inhomogeneous media: Part I: Theoretical development, *J. Atmos. Sci.*, 63, 1200–1212.

Fox, C. G., and D. E. Hayes. (1985), Quantitative methods for analyzing the roughness of the seafloor, *Rev. Geophys.*, 23, 1–48.

Gabriel, P., S. Lovejoy, A. Davis, D. Schertzer, and G. L. Austin. (1990), Discrete angle radiative transfer. Part II: Renormalization approach to scaling clouds, *J. Geophys. Res.*, 95, 11717–11728.

Gabriel, P., S. Lovejoy, D. Schertzer, and G.L. Austin. (1986), Radiative transfer in extremely variable fractal clouds, Paper presented at 6th conference on atmospheric radiation, *Amer. Meteor. Soc.*, Boston, Williamsburg, VA.

Gabriel, P. M., S.-C Tsay, and G. L. Stephens. (1993), A fourier-riccati approach to radiative transfer Part I: Foundations, *J. Atmos. Sci.*, 50, 3125–3147.

Gagnon, J. S., S. Lovejoy, and D. Schertzer. (2006), Multifractal earth topography, *Nonlin. Proc. Geophys.*, 13, 541–570.

Gaonac'h, H., S. Lovejoy, and D. Schertzer. (2003), Resolution dependence of infrared imagery of active thermal features at Kilauea volcano, *Int. J. Remote Sensing*, 24, 2323–2344.

Gilbert, L. E. (1989), Are topographic data sets fractal?, *Pageoph*, 131, 241–254.

Haar, A. (1910), Zur Theorie des orthogonalen Funktionsysteme, *Mathematische Annalen*, 69, 331–371.

Harvey, D. A., H. Gaonac'h, S. Lovejoy, and D. Schertzer. (2002), Multifractal characterization of remotely sensed volcanic features: A case study from Kiluaea volcano, Hawaii, *Fractals*, 10, 265–274.

Havlin, S., and D. Ben-Avraham. (1987), Diffusion in disordered media, *Adv. Phys.*, 36, 695–798.

Hendon, H. H., and M. Wheeler. (2008), Some space-time spectral analyses of tropical convection and planetary waves, *J. Atmos. Sciences*, 65, 2936–2948.

Hubert, L. F., and L. F. J. Whitney. (1971), Wind estimation from geostationary-satellite pictures, *Monthly Wea. Rev.*, 9, 665–672.

Kantelhardt, J. W., S. A. Zscchegner, K. Koscielny-Bunde, S. Havlin, A. Bunde, and H. E. Stanley. (2002), Multifractal detrended fluctuation analysis of nonstationary time series, *Physica A*, 316, 87–114.

Laferrière, A., and H. Gaonac'h (1999), Multifractal properties of visible reflectance fields from basaltic volcanoes, *J. Geophys. Res.*, 104, 5115–5126.

Lampkin, D. J., and S. R. Yool. (2004), Monitoring mountain snowpack evolution using near-surface optical and thermal properties, *Hydrol. Process.*, 18, 3527–3542.

Landais, F. (2014), *Description statistique de la topographie martienne*, (in French), U. Paris Sud (Orsay), Paris, p. 41.

Landais, F., F. Schmidt, and S. Lovejoy. (2015), Universal multifractal Martian topography, *Nonlin. Process. Geophys. Discuss.*, 2, 1007–1031. doi: 10.5194/npgd-2-1007-2015.

Lavallée, D., S. Lovejoy, D. Schertzer, and P. Ladoy. (1993), Nonlinear variability and landscape topography: Analysis and simulation, in *Fractals in geography*, edited by L. De Cola and N. Lam, pp. 171–205, Prentice-Hall, Englewood, NJ.

Lesieur, M. (1987), *Turbulence in fluids*, Martinus Nijhoff Publishers, Dordrecht.

Lilley, M., S. Lovejoy, D. Schertzer, K. B. Strawbridge, and A. Radkevitch. (2008), Scaling turbulent atmospheric stratification, Part II: Empirical study of the the stratification of the intermittency, *Quart. J. Roy. Meteor. Soc.*, 134, 301–331. doi: 10.1002/qj.1202.

Lilley, M., S. Lovejoy, K. Strawbridge, and D. Schertzer. (2004), 23/9 dimensional anisotropic scaling of passive admixtures using lidar aerosol data, *Phys. Rev. E*, 70, 036307-1-7.

Lovejoy, S. (2013), What is climate?, *EOS*, 94(1), 1–2.

Lovejoy, S. (2015), A voyage through scales, a missing quadrillion and why the climate is not what ou expect, *Climate Dyn.* 44, 3187–3210. doi: 10.1007/s00382-014-2324-0.

Lovejoy, S., W. Brian, S. Daniel, and B. Gerd. (1995), *Scattering in multifractal media, Paper presented at Particle Transport in Stochastic Media*, American Nuclear Society, Portland, OR.

Lovejoy, S., W. J. C. Currie, Y. Tessier, M. Claeredeboudt, J. Roff, E. Bourget, and D. Schertzer. (2000), Universal multfractals and ocean patchiness phytoplankton, physical fields and coastal heterogeneity, *J. Plankton Res.*, 23, 117–141.

Lovejoy, S., and M. I. P. de Lima. (2015), The joint space-time statistics of macroweather precipitation, space-time statistical factorization and macroweather models *Chaos* 25, 075410. doi: 10.1063/1.4927223.

Lovejoy, S., P. Gabriel, A. Davis, D. Schertzer, and G. L. Austin. (1990), Discrete angle radiative transfer. Part I: Scaling and similarity, universality and diffusion, *J. Geophys. Res.*, 95, 11699–11715.

Lovejoy, S., J. P. Muller, and J. P. Boisvert. (2014), On mars too, expect macroweather, *Geophys. Res. Lett.* 41, 7694–7700. doi: 10.1002/2014GL061861.

Lovejoy, S., S. Pecknold, and D. Schertzer. (2001b), Stratified multifractal magnetization and surface geomagnetic fields, part 1: Spectral analysis and modelling, *Geophys. J. Inter.*, 145, 112–126.

Lovejoy, S., J. Pinel, and D. Schertzer. (2012), The global space-time cascade structure of precipitation: Satellites, gridded gauges and reanalyses, *Adv. Water Res.*, 45, 37–50. doi:10.1016/j.advwatres.2012.03.024.

Lovejoy, S., and D. Schertzer. (1989), Fractal clouds with discrete angle radiative transfer, in *International Radiation Symposium 88*, edited by J. Lenoble and J. F. Geleyen, pp. 99–102, Deepak, Hampton, VA.

Lovejoy, S., and D. Schertzer. (2007a), Scaling and multifractal fields in the solid earth and topography, *Nonlin. Processes Geophys.*, 14, 1–38.

Lovejoy, S., and D. Schertzer. (2007b), Scale, scaling and multifractals in geophysics: Twenty years on, in *Nonlinear dynamics in geophysics*, 311–337 edited by J. E. A. A. Tsonis, Elsevier.

Lovejoy, S., and D. Schertzer. (2010a), Towards a new synthesis for atmospheric dynamics: Space-time cascades, *Atmos. Res.*, 96, 1–52. doi: 10.1016/j.atmosres.2010.01.004.

Lovejoy, S., and D. Schertzer. (2010b), On the simulation of continuous in scale universal multifractals, Part II: Space-time processes and finite size corrections, *Comput. Geosci.*, 36, 1404–1413. doi: 10.1016/j.cageo.2010.07.001.

Lovejoy, S., and D. Schertzer. (2010c), On the simulation of continuous in scale universal multifractals, Part I: Spatially continuous processes, *Comput. Geosci.*, 36, 1393–1403 doi: 10.1016/j.cageo.2010.04.010.

Lovejoy, S., and D. Schertzer. (2011), Space-time cascades and the scaling of ECMWF reanalyses: Fluxes and fields, *J. Geophys. Res.*, 116, D14117, doi: 10.1029/2011JD015654.

Lovejoy, S., and D. Schertzer. (2012), Haar wavelets, fluctuations and structure functions: Convenient choices for geophysics, *Nonlinear Proc. Geophys.*, 19, 1–14 doi: 10.5194/npg-19-1-2012.

Lovejoy, S., and D. Schertzer. (2013), *The weather and climate: Emergent laws and multifractal cascades*, Cambridge University Press, Cambridge, p. 496.

Lovejoy, S., D. Schertzer, V. Allaire, T. Bourgeois, S. King, J. Pinel, and J. Stolle. (2009c), Atmospheric complexity or scale by scale simplicity?, *Geophys. Resear. Lett.*, 36, L01801. doi:01810.01029/02008GL035863.

Lovejoy, S., D. Schertzer, M. Lilley, K. B. Strawbridge, and A. Radkevitch. (2008), Scaling turbulent atmospheric stratification, Part I: Turbulence and waves, *Quart. J. Roy. Meteor. Soc.*, 134, 277–300. doi: DOI: 10.1002/qj.201.

Lovejoy, S., D. Schertzer, and P. Silas. (1998), Diffusion on one dimensional multifractals, *Water Res. Res.*, 34, 3283–3291.

Lovejoy, S., D. Schertzer, P. Silas, Y. Tessier, and D. Lavallée. (1993b), The unified scaling model of atmospheric dynamics and systematic analysis in cloud radiances, *Annal. Geophys.*, 11, 119–127.

Lovejoy, S., D. Schertzer, and J. D. Stanway. (2001c), Direct evidence of planetary scale atmospheric cascade dynamics, *Phys. Rev. Lett.*, 86(22), 5200–5203.

Lovejoy, S., D. Schertzer, and I. Tchiguirinskaia. (2013), Further (monofractal) limitations of climactograms, *Hydrol. Earth Syst. Sci. Discuss.*, 10, C3086–C3090 doi: 10/C3181/2013/.

Lovejoy, S., D. Schertzer, and Y. Tessier. (2001a), Multifractals and Resolution independent remote sensing algorithms: The example of ocean colour, *Inter. J. Remote Sensing*, 22, 1191–1234.

Lovejoy, S., D. Schertzer, and B. Watson. (1993a), Radiative transfer and multifractal clouds: Theory and applications, in *I.R.S. 92*, edited by A. Arkin. et al., pp. 108–111.

Lovejoy, S., A. Tarquis, H. Gaonac'h, and D. Schertzer. (2007a), Single and multiscale remote sensing techniques, multifractals and MODIS derived vegetation and soil moisture, *Vadose Zone J.*, 7, 533–546. doi: 10.2136/vzj2007.0173.

Lovejoy, S., A. F. Tuck, S. J. Hovde, and D. Schertzer. (2007b), Is isotropic turbulence relevant in the atmosphere?, *Geophys. Res. Lett.*, doi: 10.1029/2007GL029359, L14802.

Lovejoy, S., A. F. Tuck, and D. Schertzer. (2010), The Horizontal cascade structure of atmospheric fields determined from aircraft data, *J. Geophys. Res.*, 115, D13105. doi: 10.1029/2009JD013353.

Lovejoy, S., A. F. Tuck, D. Schertzer, and S. J. Hovde. (2009a), Reinterpreting aircraft measurements in anisotropic scaling turbulence, *Atmos. Chem. Phys.*, 9, 1–19.

Lovejoy, S., B. Watson, Y. Grosdidier, and D. Schertzer. (2009b), Scattering in thick multifractal clouds, Part II: Multiple scattering, 388, 3711–3727. doi:10.1016/j.physa.2009.05.037.

Mandelbrot, B. (1981), Scalebound or scaling shapes: A useful distinction in the visual arts and in the natural sciences, *Leonardo*, 14, 43–47.

Mandelbrot, B. B. (1967), How long is the coastline of Britain? Statistical self-similarity and fractional dimension, *Science*, 155, 636–638.

Marguerite, C., D. Schertzer, F. Schmitt, and S. Lovejoy. (1998), Copepod diffusion within multifractal phytoplankton fields, *J. Marine Biol.*, 16, 69–83.

Marshak, A., and A.B. Davis (Ed.). (2005), *3D Radiative transfer in cloudy atmospheres*, Springer, Berlin, p. 686.

McKee, T., and S. K. Cox. (1976), Simulated radiance patterns for finite cubic clouds, *J. Atmos. Sci.*, 33, 2014–2020.

McLeish, W. (1970), Spatial spectra of ocean surface temperature, *J. Geophys. Res.*, 75, 6872–6877.

Meakin, P. (1987), Random walks on multifractal lattices, *J. Phys. A*, 20, L771–L777.

Mechem, D. B., Y. L. Kogan, M. Ovtchinnikov, A. B. Davis, R. R. Cahalan, E. E. Takara, and R. G. Ellingson. (2002), Large-eddy simulation of PBL stratocumulus: Comparison of multi-dimensional and IPA longwave radiative forcing, Paper presented at 12th ARM Science Team Meeting Proceedings, St. Petersburg, Florida.

Mitchell, J. M. (1976), An overview of climatic variability and its causal mechanisms, *Quaternary Res.*, 6, 481–493.

Park, K.-A., and J. Y. Chung. (1999), Spatial and temporal scale variations of sea surface temperature in the east sea using Noaa/Avhrr data, *J. Phys. Oceanogr.*, 55, 271–288.

Pecknold, S., S. Lovejoy, and D. Schertzer. (2001), Stratified multifractal magnetization and surface geomagnetic fields, part 2: Multifractal analysis and simulation, *Geophys. Inter. J.*, 145, 127–144.

Pecknold, S., S. Lovejoy, D. Schertzer, and C. Hooge. (1997), Multifractals and the resolution dependence of remotely sensed data: Generalized Scale Invariance and Geographical Information Systems, in *Scaling in Remote Sensing and*

Geographical Information Systems, edited by M. G. D. Quattrochi, pp. 361–394, Lewis, Boca Raton, FL.

Peng, C.-K., S. V. Buldyrev, S. Havlin, M. Simons, H. E. Stanley, and A. L. Goldberger. (1994), Mosaic organisation of DNA nucleotides, *Phys. Rev. E.*, 49, 1685–1689.

Perrin, J. (1913), *Les Atomes*, NRF-Gallimard, Paris.

Pinel, J., and S. Lovejoy. (2014), Atmospheric waves as scaling, turbulent phenomena, *Atmos. Chem. Phys.*, 14, 3195–3210. doi: 10.5194/acp-14-3195-2014.

Pinel, J., S. Lovejoy, and D. Schertzer. (2014), The horizontal space-time scaling and cascade structure of the atmosphere and satellite radiances, *Atmos. Resear.*, 140–141, 95–114. doi: 10.1016/j.atmosres.2013.11.022.

Pinel, J., S. Lovejoy, D. Schertzer, and A. F. Tuck. (2012), Joint horizontal - vertical anisotropic scaling, isobaric and isoheight wind statistics from aircraft data, *Geophys. Res. Lett.*, 39, L11803. doi: 10.1029/2012GL051698.

Preisendorfer, R. W., and G. I. Stephens. (1984), Multimode radiative transfer in finite optical media, I fundamentals, *J. Atmos. Sci.*, 41, 709–724.

Radkevitch, A., S. Lovejoy, K. B. Strawbridge, D. Schertzer, and M. Lilley. (2008), Scaling turbulent atmospheric stratification, Part III: Emplrical study of Space-time stratification of passive scalars using lidar data, *Quart. J. Roy. Meteor. Soc.*, doi: 10.1002/qj.1203.

Richardson, L. F. (1961), The problem of contiguity: An appendix of statistics of deadly quarrels, *Gen. Syst. Yearbook*, 6, 139–187.

Rodriguez-Iturbe, I., and A. Rinaldo. (1997), *Fractal river basins*, Cambridge University Press, Cambridge, p. 547.

Rouse, J. W., Jr., R. H. Haas, J. A. Schell, and D. W. Deering. (1973), *Monitoring the vernal advancement and retrogradation (green wave effect) of natural vegetation*, Rep. Prog. Rep. RSC 1978-1, Remote Sensing Cent., Texas A&M University, College Station, TX.

Sachs, D., S. Lovejoy, and D. Schertzer. (2002), The multifractal scaling of cloud radiances from 1m to 1km, *Fractals*, 10, 253–265.

Saunders, P. M. (1972), Space and time variability of temperature in the upper Ocean, *Deep-Sea Res.*, 19, 467–480.

Schertzer, D., and S. Lovejoy. (1985b), Generalised scale invariance in turbulent phenomena, *Physico Chem. Hydrodyn. J.*, 6, 623–635.

Schertzer, D., and S. Lovejoy. (1985a), The dimension and intermittency of atmospheric dynamics, in *Turbulent Shear Flow*, edited by L. J. S. Bradbury. et al., pp. 7–33, Springer-Verlag, Berlin.

Schertzer, D., and S. Lovejoy. (1987), Physical modeling and analysis of rain and clouds by anisotropic scaling of multiplicative processes, *J. Geophys. Res.*, 92, 9693–9714.

Seuront, L., F. Schmitt, D. Schertzer, Y. Lagadeuc, and S. Lovejoy. (1996), Multifractal analysis of eulerian and lagrangian variability of physical and biological fields in the Ocean, *Nonlinear Proc. Geophys.*, 3, 236–246.

Steinhaus, H. (1954), Length, Shape and Area, *Colloq. Math.*, III, 1–13.

Stolle, J., S. Lovejoy, and D. Schertzer. (2009), The stochastic cascade structure of deterministic numerical models of the atmosphere, *Nonlin. Proc. Geophys.*, 16, 1–15.

Stolle, J., S. Lovejoy, and D. Schertzer. (2012), The temporal cascade structure and space-time relations for reanalyses and Global Circulation models, *Quart. J. Royal Meteor. Soc.*, 38, 1895-1913.

Szantai, A., and G. Sèze. (2008), Improved extraction of low-level atmospheric motion vectors over West-Africa from MSG images, in *9th International Winds Workshop, 14-18 April 2008, Amer. Meteor. Soc.*, pp. 1–8, Annapolis, MD.

Tchiguirinskaia, I., S. Lu, F. J. Molz, T. M. Williams, and D. Lavallée. (2000), Multifractal versus monofractal analysis of wetland topography, *SERRA*, 14(1), 8–32.

Venig-Meinesz, F. A. (1951), A remarkable feature of the Earth's topography, *Proc. K. Ned. Akad. Wet. Ser. B Phys. Sci.*, 54, 212–228.

Watson, B. P., S. Lovejoy, Y. Grosdidier, and D. Schertzer. (2009), Multiple scattering in thick multifractal clouds, Part I: Overview and single scattering scattering in thick multifractal clouds, *Physica. A.*, 388, 3695–3710. doi: 10.1016/j.physa.2009.05.038.

Weinman, J. A., and P. N. Swartzrauber. (1968), Albedo of a striated medium of isotrpically scattering particles, *J. Atmos. Sci.*, 25, 497–501.

Weissman, H., S. Havlin. (1988), Dynamics in multiplicative processes, *Phys. Rev. B.*, 37, 5994–5996.

Welch, R. M., S. Cox, and J. Davis. (1980), *Solar radiation and clouds*, Meteorological Monographs, *Amer. Meteor. Soc.* 39.

Wheeler, M., and G. N. Kiladis. (1999), Convectively coupled equatorial waves: Analysis of clouds and temperature in the wavenumber-frequency domain, *J. Atmos. Sci.*, 56, 374–399.

4

Fusion of Multiscaled Spatial and Temporal Data: Techniques and Issues

Bandana Kar and Edwin Chow

CONTENTS

Introduction

Whereas spatial data sets correspond to a specific location on the Earth, temporal data capture the temporal dynamics of processes and systems. Because physical and social processes occur at a location at a specific time, spatial and temporal data are extensively used to represent and manage features corresponding to physical and social environments and to understand the interaction between these environments. Traditionally, spatial data sets are collected *via* remote sensing (imagery and data are collected from air and space using airborne cameras, satellites, and sensors), Global Positioning System (GPS) (a network of satellites that provide precise coordinate locations), and field-based methods (e.g., total station instruments are used to collect spatial data and questionnaire surveys are used to collect attribute data). However, in the twenty-first century, the growth and advancements in geospatial technologies, such as the launch of commercially operated satellites, have enabled the generation of large volumes of remotely sensed data for military and civilian purposes at high spatial, temporal, spectral, and radiometric resolutions. For instance, the WorldView-3 satellite sensor provides multispectral images at a spatial resolution of 0.31 m and a temporal resolution of 1–4.5 days (DigitalGlobe 2015).

The advancements in information and communication technologies (ICT) and the rise of Web 2.0 in the twenty-first century have also contributed to the generation and sharing of large volumes of spatial and temporal data. Internet mapping technologies, such as Google Earth and Microsoft Virtual Earth, have enabled public sharing of spatial data and aided with visualization of these data. Social media (e.g., Twitter), social networking sites (e.g., Facebook and Foursquare—a location-based networking site), and location-based services (e.g., Ushahidi—an open-source multimedia mapping platform) have empowered citizens to generate and share spatial data (i.e., images and pictures) in near real time (Heinzelman and Waters 2010; Pu and Kitsuregawa 2014). Open Street Map (OSM) is another example of a citizen-driven open data source that harnesses citizen knowledge in creating spatial data about road networks (OSM 2015). Public institutions have also undertaken efforts to create spatial data in near real time for different activities. In 2006, the United Nations Office for Outer Space Affairs (UNOOSA) established the Space-Based Information for Disaster Management and Emergency Response (UN-SPIDER) to help with data collection, disaster risk reduction, and response through stakeholder participation and knowledge sharing (UN 2006). Overall, both citizens and experts, public and private institutions are generating and sharing data either voluntarily (a.k.a. Volunteered geographic information [VGI]) or involuntarily at varying spatial scales and resolutions in near real time to over a certain duration.

Obviously, there is no paucity of spatial and temporal data. Whereas spatial data obtained *via* traditional technologies (e.g., remote sensing, GPS) are available in a structured format at specific times for specific locations, these data are not always available at high spatial and temporal resolutions, can contain erroneous information due to resolution, and can be too expensive to procure to be useful in near real time (Cova 1999; Gao et al. 2011; Pu and Kitsuregawa 2014). Crowdsourced data available *via* social media and social networking sites, in contrast, are unstructured, available randomly and sporadically, and 1%–3% of these data contain geospatial information, of which a majority is fudged or spoofed to protect user privacy (Leetaru et al. 2013; Weidemann and Swift 2013). These data, nonetheless, provide access to high quality and inexpensive data in near real time (Elwood 2008; Goodchild and Li 2012). Despite the obvious usability of these data sets, structured and unstructured spatial and temporal data need to be integrated to increase their usage for different applications. This chapter discusses (1) fusion techniques used to integrate multisourced and multiscaled data; (2) analytical issues associated with data fusion and multiscaling analysis; and (3) geovisualization techniques used to visualize multiscaled data. In the following section, a discussion of techniques used to integrate and analyze multiscaled data is presented. Following this discussion, data fusion techniques are examined. In the later part of this chapter, issues existing with data fusion and geovisualization techniques used to visualize multiscaled data are considered. Finally, potential research gaps in data fusion and multiscaling are discussed.

Multiscale Modeling

Geography as a discipline focuses on the study of physical and social environments and their interactions over space and time. Therefore, considerable research has been undertaken to study the role of spatial and temporal scales of analysis in geospatial research and applications (Fotheringham and Wong 1991; Pereira 2002). With the increasing availability of multiscaled data, multiscale modeling has become a prominent research area. However, prior to discussing multiscale modeling, an overview of scale as a concept in geospatial science is presented.

The concept of *scale*, though ubiquitous, has different connotations, but scale in general represents the relative or absolute size of an object or a system (Mason 2001). *Spatial scale* primarily focuses on space in geography and has at least four connotations: geographical scale, cartographic scale, operational scale, and spatial resolution (Lam and Quattrochi 1992; Goodchild and Quattrochi 1997; Mason 2001). *Geographical scale* is the spatial extent of a study (e.g., a study conducted for a county or a state). *Cartographic scale* or *map scale* is the representation of the relationship between "map distance" and "ground distance." It is represented as a ratio or a representative fraction (RF) in which the numerator corresponds to the map distance and the denominator refers to the ground distance. A map scale of 1:10,000 implies that one unit of map distance is equivalent to 10,000 units of ground distance. Expressed in relative terms, a large denominator (i.e., 1:20,000) refers to a small-scale map (resulting in loss of information) and a small denominator refers to a large-scale map (with information that is more detailed). *Operational scale*, otherwise known as *phenomenon scale*, signifies the geographic extent at which certain processes operate and the scale at which any change in these processes is more pronounced. For instance, atmospheric processes can occur at micro-, meso-, and macroscales, and climate change processes can be explored at local, regional, and global scales. *Spatial resolution*, also known as *analysis scale*, *granularity*, or *observational scale*, is the unit of analysis at which measurements are made and data are analyzed; it is the dimension of the smallest distinguishable part of a study. The smaller the unit of analysis, the finer is the resolution. For instance, a remote sensing image with a pixel size of 10×10 m has a finer resolution than an image with a pixel size of 100×100 m (Lam and Quattrochi 1992; Goodchild and Quattrochi 1997; Mason 2001; Sayre 2005).

Temporal scale generally corresponds to duration (the total time period) and granularity, or the time interval at which certain processes occur or the unit at which certain analysis is conducted (McMichael 2010). For instance, every year tropical cyclones occur along the Gulf Coast between June 1 and November 1. A study investigating tropical cyclones would have a temporal scale of 5 months but could use a granularity of 1 month to study the seasonal dynamics of tropical cyclones over the 5-month period.

Likewise, a study examining population distribution at a certain location could use a temporal scale of 1 hour to determine the hourly variability of a population. When dealing with temporal scale, attention must also be paid to the fact that time could be considered a discrete or a continuous element (Pelekis et al. 2014). For example, because tropical cyclones occur every year during the same time period, studies researching tropical cyclones can represent time as a discrete element. By contrast, processes that change constantly, such as population growth and mobility, require using time as either a uniform value or an irregular value. If population dynamics are examined during the day and night at a specific place, then a uniform interval of 8 hours due to workplace movement can be used as the temporal scale. However, if population dynamics are observed across a city or a metropolitan area that experiences constant mobility due to the presence of offices, commercial, and residential complexes, then the temporal scale cannot be a discrete number, but rather must be a continuous number that changes according to the time of study (Pelekis et al. 2014). Although spatial and temporal scales have always been the focus of geospatial research, *social scale* must be considered in studies focusing on humans and social processes (Marston 2000). Social scale can be measured through observable demographic or cultural differences (e.g., income or racial composition) or the interactions between societal structures and human agencies (Marston 2000). Therefore, social scale in a study can represent an individual, a household, a community, or another unit appropriate to the purpose of the study.

Because physical and social systems and processes vary across scales, a model implemented at a specific scale may not be accurate or efficient enough to be implemented at another scale (Weinan and Lu 2011). Thus, in multiscale modeling either (1) multiple models are used concurrently at different scales to describe a relationship or a process or a system or (2) one model is used sequentially at different scales to describe a process or a system (Wu et al. 2000; Weinan and Lu 2011). Needless to say, multiscale modeling consists of multiscale analysis (understanding the relationship between models at different scales of analysis), multiscale models (developing a model that integrates models implemented at different scales), and multiscale algorithms (computational algorithms used for multiscale ideas) (Weinan and Lu 2011). In geospatial science, sequential and hierarchical multiscale models are often used such that a process and its spatial variation are examined at a specific scale by ignoring how the process might occur at other scales at the same time by using one model (Sun and Li 2002; Weinan and Lu 2011). Multiscale models depicting the relationship between a scale and a process across different spatial scales at a given time have also been developed (Levine 1992; Chow and Hodgson 2009; Kar and Hodgson 2012). A number of spatial and statistical techniques have also been developed for multiscale modeling over space and time.

Moellering and Tobler (1972) first introduced *scale variance analysis*. This method determines the spatial variability of a pattern or a relationship at each scale of analysis and the contribution of variance at each scale to the total variance. Built on the principle of spatial autocorrelation, *semi-variance analysis and geostatistics* provide a framework to quantify the variability of an attribute or a process in a spatial or temporal domain where each observation is separated by a certain distance interval (Robertson and Gross 1994; Burrough 1995; Xu et al. 2013). *Entropy analysis,* which originates from information theory, is used to model the multiscale variation of spatial properties at different hierarchical spatial scales (Batty 1976). Spatial entropy is a derivation of the original entropy function, which is used to measure the probabilistic distribution of variance of a process across different spatial scales of analysis (Batty 1976). *Spectral analysis,* built on the concept of autocorrelation, quantifies the variation of a process across multiple spatial scales and temporal granularity and determines the mathematical function depicting the underlying pattern at each scale of analysis (Ripley 1978).

Fourier transformation is a type of spectral analysis, specifically used on data or processes that vary across space or time. Fourier analysis separates physical space or time series data into frequency components—a series of sinusoidal functions (Fourier 1878). The frequency domain can correspond to a discrete space or time of analysis. Figure 4.1 demonstrates the Fourier transformation of two images—an image with a centered black box (upper left) and an image with a black box in the lower right corner (lower left). It is apparent that in both transformations (upper right for the upper left image and lower right for the lower left image), the frequencies along the vertical and horizontal axes are same. Although both transformed images have the same frequencies, the transformations fail to identify the location where the frequencies occur despite obvious difference in location of the targets in the original images. Fourier transformation is extensively used in image processing, particularly in high- and low-pass filtering to remove noise from an image. In high-pass filtering, an image is transformed from the space domain to the frequency domain, and then the high frequencies generally associated with noise are removed. Finally, the transformed frequency domain is reversed by using inverse Fourier transformation to an image containing no noise (Jensen 2009). This is the principle used for filtering in most image-processing software including ERDAS Imagine (Smith et al. 1995). Although these filtering techniques help with noise reduction and edge detection, Fourier transformation fails to detect high-frequency target signals that might have the same frequency as the noise, thereby leading to the removal of the target from the image; it also fails to provide spatial information for a frequency domain.

Wavelet analysis is a multiscale modeling technique that is similar to Fourier transform. This technique enables study of the variability of a process or data sets across multiple spatial and temporal resolutions to identify the dominant scale of analysis that will provide high accuracy (Bradshaw and Spies 1992; Guo et al. 2007). Wavelet transformation is an advancement over Fourier transform because it allows extraction of image information

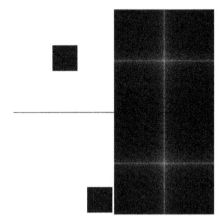

FIGURE 4.1
The Fourier transform (upper right) of an image with a centered black box (upper left) is the same as the Fourier transform (lower right) of an image with a black box at the lower right corner (lower left).

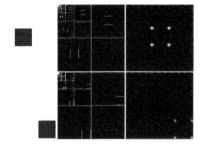

FIGURE 4.2
Original centered black box image.

at different positions and scales both in the spatial and frequency domains (Watson 1999). Figure 4.2 depicts the five-level discrete Meyer wavelet transformation (center column) and third-level discrete Meyer wavelet transform of the images used in Figure 4.1 for the Fourier transform. The images on the right display both the frequency and the location of targets corresponding to the frequency within the images.

Myint (2001, 2006, 2010) and Myint et al. (2004) used wavelet analysis with imagery of varying spatial and temporal resolutions to classify and map urban environments. They found that wavelet transformation was more effective and accurate in multiscale analysis in comparison to fractal analysis, spatial autocorrelation, and spatial co-occurrence approaches. In their study of spatial patterns of ecological features, Mi et al. (2005) and Keitt and Urban (2005) found wavelet transformation to be more effective in spatial and temporal partitioning of ecological patterns in comparison to quadrat analysis. Falkowski et al. (2006) used wavelet analysis to autodetect the location, height, and crown diameter of individual trees within a mixed conifer open canopy stand. Crowsey (2012) used wavelet transformation to assess tropical cyclone- and tornado-induced wind damage to residential structures instead of change detection techniques. Using wavelet analysis along with discriminant analysis for a number of tropical cyclones and tornadoes, including Hurricanes Ike and Katrina, Crowsey found that wavelet analysis allowed automatic damage classification of residential structures and provided high accuracy in classification outcomes. However, the discriminant analysis model used a large number of independent variables derived from wavelet transformation of pre- and postevent imagery to increase the accuracy of classified image products. Overall, wavelet analysis has been found to be more effective in undertaking multiscale analysis and detecting the spatiotemporal variability of processes, which is generally not possible using Fourier transformation.

The shape of an object is the visual attribute used in pattern recognition. Although changing the spatial scale of analysis impacts the size of a spatial unit used to study the variability of a process or a relationship, its shape generally remains the same. A *fractal* as defined by Mandelbrot (1982) describes the fragmented nature of a geometric shape or an object that can be split into parts that are similar in shape to the original object. Fractal dimension measures *self-similarity* and quantifies the complexity (i.e., change in details with change in scale) that cannot be described by Euclidean geometry (Mandelbrot 1983).

Whereas in Euclidean geometry a point has zero dimensions, a line has two dimensions, and a volume has three dimensions, in reality that is not the case (Lam et al. 2002). A curvilinear feature, for instance, a coast boundary, can have a complex shape. Although a coast boundary's topological dimension is one, its fractal dimension is not one, as this feature can be broken into many self-similar parts. Given the complex shape of geographic features, Mandelbrot (1967) used fractal analysis to study the change in coastline length due to change in measurement unit. Because fractal analysis provides information contained within geospatial data and remote sensing images at varying spatial, temporal, and spectral resolutions while accounting for spatial complexities, it is used in multiscale modeling of processes that vary across spatial and temporal scales but contain similar parameters (Mesev and Longley 2001; Quattrochi et al. 2001; Tate and Wood 2001). Although fractal analysis has been used with remote sensing images, its use with vector-based data is still in its infancy. *Lacunarity analysis,* an extension of fractal analysis, measures the variation of heterogeneity or texture of an object and the spatial pattern of a process at multiple spatial and temporal scales (Plotnick et al. 1996; Quan et al. 2014).

Block quadrat variance analysis is a pattern analysis that implements a gridded approach to observe the variation in the patterns of point data in each grid and compares the variation with grid sizes (Dale 1999). The grid can represent spatial resolution or temporal granularity. The *Bayesian model*, founded upon Bayes' theorem, expresses the conditional probabilities of an event's occurrence at a specific spatial and/or temporal scale (Kolacayk and Huang 2001; Raghavan et al. 2014). Because this model accounts for the uncertainties associated with an event being measured at different scales of analysis, it provides high accuracy.

While the techniques discussed above have successfully been used in multi-scale modeling, all techniques fail to account for spatial autocorrelation—a measure of the extent to which spatial features and their values are clustered together in space (Tobler 1979). Spatial autocorrelation being a given in spatial analysis, using traditional statistical techniques (such as ordinary least-squares regression) that assume the observations scattered across space are independent of each other with spatial data is not very effective. Furthermore, a regression technique implements a global model to determine the parameters without accounting for the presence of spatial autocorrelation among variables. Getis–Ord Gi* and local Moran's *I* are two local cluster analysis techniques that account for positive spatial autocorrelation existing between observations distributed in a neighborhood of certain distance (Griffith 2003). These techniques, however, do not account for the variability present in a relationship between predictor and response variables due to change in the spatial scale of analysis, spatial nonstationarity, or the variation of a process across space (Huang et al. 2010).

Regression techniques, especially spatial autoregressive models (simultaneous and conditional), are used to examine the impact of a location on the relationship between dependent and independent variables (ASA 2014). Geographically weighted regression (GWR) is a spatial regression model that implements local models, such that a set of local parameters are estimated for each observation based on a neighborhood of observations and/or based on observations present within a spatial unit of analysis (Fotheringham et al. 1998). Although the selection of neighborhood distance is similar to Getis–Ord Gi*, GWR accounts for the impact of spatial nonstationarity on the relationship and the impact of spatial scale. GWR, however, has been criticized for failing to address the impact of temporal scale—temporal variation of a process—on the relationship between predictor and response variables. Temporally weighted regression (TWR) is a special case of GWR that explains the impact of temporal clustering of observations and their temporal variability (nonstationarity) on the relationship between variables (Huang et al. 2010). Because physical and social processes vary across space and time simultaneously, Huang et al. (2010) developed geographically and temporally weighted regression (GTWR), which is an extension of the GWR model but incorporates a temporal dimension to justify both spatial and temporal nonstationarity. Instead of incorporating a separate weight matrix for time, the model uses a weight matrix in which distance of an observation at a specific location and time to all other observations is computed. The authors reported that the GTWR reduced the absolute error and performed better than GWR and TWR. Fotheringham et al. (2015) also tested GTWR using house price data for London during 1980–1998 and found that the model performed better than GWR. Wu et al. (2014) developed a geographically and temporally weighted autoregressive model (GTWAR) that explained both spatiotemporal autocorrelation and nonstationarity. The authors implemented a two-stage least-squares model to analyze housing price variation in Shenzhen, China, and reported that GTWAR outperformed the other spatiotemporal models discussed above in increasing analytical accuracy.

Clearly, a number of statistical and spatial techniques are available to model multiscaled data. The major challenge in implementing these techniques, however, is the difficulty in establishing multiscale linkages among spatial and temporal units and in accounting for the presence of correlation and dependence among these units. The lack of knowledge about the best technique depending on the scales of analysis also makes multiscale modeling a challenge. In the era of big data, reducing data dimensionality through scalable methods also needs to be investigated further in order to explore multiscale pattern distribution within hierarchical scales.

Data Fusion

Data fusion is a process that *allows combination, correlation, and integration of data and information from a single and/or multiple sources at multiple spatial and temporal scales to improve the quality of data and analytical outcomes* (Lou and Kay 1989; Hall 1992; Castanedo 2013). Data fusion is used to increase the reliability of large data sets and reduce data ambiguity due to variable scales, resolutions, and data structures (Khaleghi et al. 2011). Information fusion is a type of data fusion that allows integration of conceptual, contextual, and textual information from different sources (Torra 2003).

Data fusion is a multidisciplinary research area that addresses the following challenges and issues associated with different data sets (Khaleghi et al. 2011): *data imperfection*, the presence of uncertainty and impreciseness in data sets; *outliers and spurious data*, the inability of sensors and data collection tools/techniques to distinguish between inconsistencies present in the environment; *conflicting data*, data sets that provide incomplete information or different information due to variation in scale of analysis; *data modality*, data and information are available from sources that are similar (redundant) or different (complementary); *data correlation*, bias in measurements due to the presence of noise and/or spatial autocorrelation, as is the case with geospatial data; *data alignment/registration*, an essential step when dealing with spatial data obtained from sensors so that the data correspond to a location on Earth rather than an abstract space and help with alignment of information from multiple sources corresponding to the same location before analysis; *data association*, difficulty in identifying the source or target from which measurements were made (e.g., in optical remote sensing, it is difficult to distinguish features that are impacted by water); *processing framework*, the approach used to process data sets—centralized versus decentralized; *operational timing*, incorporating the temporal variability of a process or a relationship in an analysis to account for the impact of temporal scale on analysis; *static* versus *dynamic phenomena*, if a process or a relationship is constantly changing (e.g., human mobility), then the analysis must account for this change instead of using a discrete time interval; *data dimensionality*, reducing data dimension to reduce processing time while extracting pertinent information from large data sets (e.g., hyperspectral images provide data in more than 200 spectral channels) without losing a lot of data.

Every data set suffers from one or more of the issues identified above. A number of algorithms have been developed to fuse input data sets while reducing the impact of some of these data-related challenges. While dealing with data containing uncertainties or that is incomplete or impacted by granularity of analysis, the following techniques are used for fusion (Khaleghi et al. 2011): *probabilistic fusion*: this method builds on the Bayesian framework and relies on probability distribution or probability density functions to express data uncertainty. It is possible to incorporate new data in the Bayes estimator to increase the accuracy of the final product. The Kalman filter is a special type of Bayes filter that is used extensively in data fusion; *Bayesian inference*: Bayesian inference is built on Bayes' theorem, which determines the *a posteriori* probability of a hypothesis based on the *a priori* probability of the hypothesis and conditional probability of observational evidence available for the hypothesis (Klein 2004; Zeng et al. 2006); *the Dempster–Shafer evidence theory*: this is a generalization of Bayesian theory that computes the likelihood of a hypothesis being true by using probability and uncertainty intervals on observational data (Hall and Llinas 1997; Klein 2004). The algorithm computes the probability of an event or object being captured by a sensor and determines the uncertainty associated with the probability of object identification. In the case of multiple sensors, the total probability should be 100%. However, Dempster's rule may create a situation where the total conflicting probability is larger than 100%, which is a limitation of this algorithm (Klein 2004); *fuzzy logic*: this is used to distinguish processes or phenomena that are difficult to separate into distinct categories due to the absence of hard boundaries (e.g., a house and its surrounding physical space captured within a 30 × 30 m pixel of a Landsat image will be difficult to separate into hard classes) (Klein 2004); *possibility fusion*: this theory is based on fuzzy set theory and is used to address data incompleteness. The possibility distribution determines the possibility (the degree to which an object is plausible) and necessity (the degree to which it is certain the object belongs to the class) measures that can be considered as a special case of upper and lower probabilities of a probability theory to help determine the possibility of an object belonging to a specific class; *rough set fusion*: used to fuse imprecise data obtained at varying granularities. The method uses precise upper and lower bounds using set theory operators and does not require any additional information; *hybrid fusion*: this technique is computationally intensive and uses a combination of fusion techniques to address different data-related issues.

In the case of correlated data, principal component analysis (PCA) and other similar techniques are used to remove correlated data before conducting other analyses. Likewise, the techniques discussed above used to account for spatiotemporal autocorrelation can also be used with data sets subjected to correlation. If the presence of correlation is unknown, then covariance intersection can be used to determine the variability between data sets (Khaleghi et al. 2011). When dealing with data sets containing missing information (e.g., Landsat 7 Enhanced Thematic Mapper Plus [ETM+] had missing data due to the scan line corrector [SLC] being switched off) and/or conflicting information, either an attempt is made to identify and predict missing information and remove outliers by using prior data sets, or the Dempster–Shafer evidence theory framework and Bayesian probabilistic framework are used (Khaleghi et al. 2011). Machine-learning algorithms are also used in information fusion, especially in the big data era (Vidhate and Kulkarni 2012). For instance, *artificial neural networks* (ANNs) are commonly used with large data sets containing erroneous information to study the relationship between input and output variables. An ANN can be trained using supervised or unsupervised training data to optimize the error associated with input–output relation. ANN is also used to integrate data obtained from multiple sensors (Zeng et al. 2006).

Given the availability of images at multiple spatial, temporal, spectral, and radiometric resolutions, information extracted from one image source might not always be sufficient for an analysis. However, information available in different images can be combined by using feature and classifier fusion algorithms. While feature fusion algorithms aggregate extracted features from different sources into a single set, which is then used for classification, classifier algorithms combine decisions derived from individual classifiers using separate sources (Bahrampour et al. 2015). The two image fusion techniques that have become popular recently are dictionary learning and sparse representation matrix.

In dictionary learning, a dictionary of training data sets is created, which represents an approximation of signals in a *sparse representation matrix* (Bahrampour et al. 2015). Dictionary learning algorithms, a.k.a. dictionaries and sparse coding, can be supervised and unsupervised. Because these algorithms minimize misclassification error, they are used for signal denoising, image classification, and image reconstruction (Bahrampour et al. 2015). Huang (2012) developed the sparse representation–based spatiotemporal reflectance fusion model (SPSTFM) that used a sparse representation matrix to integrate information obtained from high spatial resolution–low temporal resolution imagery with low spatial resolution–high temporal resolution imagery. The SPSTFM model used two dictionaries created from high and low spatial resolution difference image patches for two different time periods (*t1* and *t3*), which were then used in sparse coding to reconstruct the high-resolution difference image patch for a time period *t2* (*t1* < *t2* < *t3*) and its corresponding reflectance values. Following a similar dictionary learning and sparse coding approach, Huang et al. (2013a) developed the spatial and spectral fusion model (SASFM) to fuse images of low spatial–high spectral resolution with images of high spatial–low spectral resolution. Both SASFM and SPSTFM were found to perform better than other algorithms used for image fusion.

Image registration, a precursor to image fusion, focuses on aligning two images into a common coordinate system so that changes between the images can be identified and information obtained from the images can be integrated for analysis. A number of techniques are used to increase the accuracy of image registration (Shah and Mistry 2014). These can be classified as follows:

1. *Intensity-based method*: In this method, the scalar values of an image pixel in two images are aligned by quantifying the similarity in intensity of the two images. Generally, the sum of the squared difference in intensity between the images and correlation coefficients is used to measure similarity.

2. *Feature-based registration*: In this method, feature points are identified in both images, which are then matched using the corresponding points and image intensity values. Finally, image transformation techniques are used to map the target image with the reference image for each set of features.

3. *Geometric transformation*: Affine transformation is a type of geometric transformation technique in which linear transformation parameters are computed by using the coordinates of control points, which are used to geometrically transform the images for registration. In the next step, pixel registration is done to align pixel values in both images for the same object.

4. *Surface methods*: In this method, corresponding surfaces in different images for a specific object are identified and the transformation that best aligns the surfaces is implemented to register the images.

5. *Curve method*: In this method, first, a set of points and an open curve corresponding to the points in each image are identified. The two open curves are then used to register the two images.

6. *Correlation method*: This method is used in the case of monomodal images, in which a correlation between several images of the same object is determined to identify features for registration.

   A discussion of other methods in addition to these image registration methods can be found in Fitzpatrick et al. (2000) and Shah and Mistry (2014).

Although it is assumed that two satellite sensors will record the same information when viewing the same target at the same time, it is not always the case, which may be because of the variation in spatial and spectral resolutions of the sensors. Image intercalibration is an issue that results because two instruments do not make identical measurements (Chander et al. 2013). Therefore, image intercalibration must be addressed by transforming collocated data to a comparable scale before integrating remotely sensed data. However, before transformation, it is crucial to (1) quantify the relative bias or the difference between the two instruments so that the results can be generalized between satellite pairs, (2) correct the bias, and (3) identify sources of bias and eliminate them. Some of the most commonly used effective intercalibration methods are as follows: *Simultaneous nadir overpass*: in this method, the calibration difference for spectrally matched channels of a pair of instrument is determined by observing the same object at nearly the same time. By viewing a large number of objects over varying time periods, the uncertainties are reduced; *Statistical intercalibration*: statistical techniques are used to homogenize the statistical mean of observations by using a large set of samples collected by the pair of instruments over an extended period of time for many targets; *Double differencing method*: if it is not possible to compare two instruments, then a reference instrument is used and the difference of each instrument to the intermediate reference instrument is determined for comparison. The assumption in this method is that the intermediate reference instrument is unbiased. A detailed discussion of other image intercalibration techniques is available (see Chander et al. 2013).

In addition to the techniques discussed above, exploratory data analysis (EDA) techniques are used to determine the characteristics of a population based on observed samples from multiple sensors and data sets. This knowledge helps identify data sets that can be fused and the transformations that need to be made before integrating data sets. Nonstatistical techniques and spatial techniques such as wavelet transformation, high-pass filter, and geostatistics are also used for data fusion (Xu et al. 2004; Gangkofner et al. 2008). Because each fusion technique has its advantages and limitations, the selection of a fusion technique depends on the purpose of a study and the available data sets. However, to achieve high accuracy, a combination of fusion techniques and ancillary data/information must be used. Use of a fusion technique is also dependent on expert knowledge in developing rules; determining acceptable probabilities and confidence levels; and identifying data sets, information, and sources that will provide high accuracy.

Practical and Analytical Issues of Fusing Multiscaled Data

Based on data acquisition technology and specifications (e.g., platform, spatial and temporal resolutions, accuracy, and precision), data fusion approaches can be classified into the following taxonomy (Table 4.1).

Regardless of the type of data fusion, the ultimate goal is to enhance spatial analysis, interpretation, and decision-making. It is, therefore, crucial to

TABLE 4.1

Taxonomy of Data Fusion

Data Fusion		Example	
Type	Description	Vector	Raster
Multispatial	Fusion of data acquired by the same acquisition method but resulting in varying spatial resolutions	Data comparison across census blocks, block groups, and tracts	IKONOS multispectral and panchromatic bands
Multitemporal	Fusion of data acquired by the same acquisition method at multiple times	Track data of wildlife species	Landsat time series
Multi-instrument	Fusion of data acquired by different acquisition methods but the same data model. There may be implicit change(s) in spatio temporal dimension	Incorporation of surveyed hard breaklines into triangulated irregular network	MODIS + AVHRR
Multidata model	Fusion of data acquired by different acquisition methods and data models. There may be implicit change(s) in spatio-temporal as well as attribute dimensions	Lidar and hyperspectral, tweets, and remote sensing imageries	

preprocess the data to be used individually prior to data fusion. In image fusion, for example, single image preprocessing procedures, such as atmospheric correction, geometric and radiometric rectification, contrast stretching, and filtering, among others, are commonly used to enhance each image. Some indiscriminative data sources, such as lidar, may require a series of complex procedures (e.g., ground/nonground separation, hydrologic enforcement) prior to data fusion (Renslow 2013).

For multispatial data fusion, a resampling or [dis]aggregation approach should be used to ensure a consistent unit of analysis for overlay. In remote sensing, resampling involves the splitting, merging, and interpolation of pixels from one spatial resolution to another (Jensen 2009). Resampling, however, might cause image misalignment for pan-sharpening, leading to unreliable quality (Jing et al. 2012). Therefore, instead of resampling original data sets to a coarser or finer spatial resolution, it is a common practice to match the coarser-scale imagery with the finer-scale imagery to preserve critical spatial information (e.g., heterogeneity) and pattern. In vector data, aggregation or disaggregation can be used in a similar fashion. This procedure is especially common in studies using census data to match the demographic and socioeconomic characteristics available at various census aggregation levels (e.g., population count at blocks, median household income at block groups, and commute pattern at census tracts). The resampling or [dis]aggregation procedure tends to induce the modifiable areal unit problem (MAUP), which results in uncertainties due to scaling or zoning of the geospatial data and affects subsequent use of the resulting data sets (Fotheringham and Wong 1991).

Multitemporal data fusion typically involves change detection, longitudinal comparison, or time-series analysis of geospatial data. In remote sensing, a common procedure is to adopt co-registration (i.e., image-to-image rectification) to align bitemporal imageries for subsequent analysis. Du et al. (2013) proposed a two-stage framework that conducts pan-sharpening of individual images followed by fusion techniques to improve the performance of change detection. Longitudinal studies using census data may involve redistricting due to changing administrative boundaries. One possible solution is to reinterpolate the attribute based on areal changes assuming a specific distribution of that entity in space (e.g., spatial uniformity or weighted by another factor). The quality of areal interpolation, however, is dependent on the degree of spatial heterogeneity. In general, higher spatial heterogeneity would challenge the underlying assumption and introduce higher uncertainties in the resulting temporal analysis (Brindley et al. 2005).

In the context of this chapter, multi-instrument data fusion integrates the geospatial data of similar geographic phenomena measured by various data capture methodologies using the same geospatial data model. Theoretically, geospatial data could span various geographic phenomena and data models, which could be classified as a multidata model for ease of discussion in terms of the analytical procedures and challenges involved. Due to the varying specifications of data acquisition instruments, the resulting geospatial data sets generated would have different spatial and temporal discrepancies, and all the aforementioned procedures (e.g., resampling and co-registration) would be applicable to multi-instrument data fusion. Depending on the data sets to be fused, this typically requires a specialized operation to handle data integration. For example, a breakline representing a surveyed linear feature can be incorporated into a triangulated irregular network using Delaunay triangulation (e.g., constrained *vs* conforming) based on the insertion of Steiner points (Cohen-Steiner et al. 2004). Bhaskar et al. (2014) combined traffic count point data collected by loop detectors and the travel time of discoverable Bluetooth devices within a zone collected by a Bluetooth media access control scanner to estimate traffic density. In image fusion, multi-instrument measurements (e.g., imaging spectroscopy and radargrammetry) tend to provide more spectral and textural details for feature classification (Reiche et al. 2015). However, careful calibration must be exercised prior to image fusion. An alternative solution is to use ratio indices, such as the normalized difference vegetation index (NDVI), that "cancel out a large portion of the multiplicative noise attributed to illumination differences, cloud shadows, topographic variations, and atmospheric conditions" (Hwang et al. 2011). Data reduction techniques like factor analysis or PCA may also be used to reduce the dimensionality of multivariate data (e.g., hyperspectral imagery) and enhance the interpretation of fused products (Pal et al. 2007).

Image-based fusion applications can be classified into pixel-based, feature-level, and decision-level (Pohl and Van Genderen 1998). In general, the pixel-based solutions are raster-based and involve spatial interpolation techniques to transform discrete vector data to a continuous raster surface. A common procedure in lidar image fusion is to create raster surfaces (e.g., elevation, density, and intensity) by interpolating lidar point cloud that can be overlaid with multispectral imagery for subsequent analysis (Torabzadeh et al. 2014). In contrast, feature-level approaches are vector-based and can leverage techniques like dynamic segmentation to enhance region growing and feature extraction from raster imagery (Zhang et al. 2014). For example, optical imagery enables the detection of building edges, whereas lidar data can be used to extract roof patches for three-dimensional building reconstruction (Li et al. 2013). Unlike pixel-based and feature-level image fusion, the decision approach does not require prior data transformation to a common structure for synthesis. The latter method generalizes a set of decision-level rules from individual data sources and then applies those rules to derive a synthesized product (Ghosh et al. 2014). Examples of decision-level fusion include expert system, majority voting, Dempster–Shafer evidence theory, and fuzzy integral (Du et al. 2013; Ghosh et al. 2014). The main advantage of decision-level fusion is the generation of human-interpretable rules to verify and advance our understanding of the phenomenon under investigation. Depending on the scope of application, it may be desirable to use any of these fusion techniques.

Geovisualization and Multiscaled Data

Although techniques exist to fuse multiscaled data, it is often difficult to interpret the results of fused data sets. Geovisualization techniques (e.g., kaleido-maps and dynamic prism) are generally used to visually represent multiscaled data by projecting each data set to a dimension. Geographic visualization (or geovisualization) originates from cartography—the subdiscipline of geography that focuses on creating and communicating information *via* maps. With increasing computer usage, geovisualization has become an interdisciplinary research field that "integrates approaches from visualization in scientific computing, cartography, image analysis, information visualization, exploratory data analysis (EDA), and geographic information systems (GIS) to provide theory, methods and tools for visual exploration, analysis, synthesis, and presentation of geospatial data" (MacEachren and Kraak 2001; Kerren et al. 2007).

Maps are used to represent spatial and nonspatial data in a two-dimensional form. The map designed by Minard (Napoleon's mapmaker)

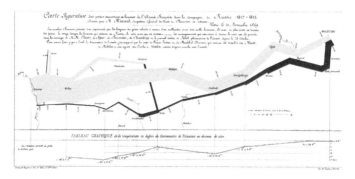

FIGURE 4.3
Minard's visual depiction of Napoleon's invasion of Russia.

in 1869 (Figure 4.3), which depicted the march of Napoleon's army to and its retreat from Moscow (Spence 2014), has been the inspiration for visualization, as it provided information about the number of soldiers at different times and locations, and the change in numbers of soldiers with the progression of time due to change in temperature. Since then, a number of two-dimensional visualization techniques have been developed to represent both spatial and nonspatial data (Nöllenburg 2007). A *cartogram* is commonly used to represent spatial data such that the area of a spatial unit on the map is proportional to the number of a single variable rather than the true area of the unit. A *choropleth map* is also widely used to depict a single or multiple variables in a combination of color, texture, and pattern for nonoverlapping areas. Despite their wide usability and effectiveness in displaying single or multiple variables, these two visualization techniques suffer from loss of information due to generalization of attributes across geographic space. Glyph-based techniques and geometric techniques are used to mitigate this limitation. *Glyph-based* techniques include star plots and Chernoff faces, where each dimension of the star plot or the face corresponds to an attribute. The glyph-based techniques can be combined with two-dimensional maps to reduce information loss. Scatter plots and parallel coordinate plots are the most commonly used geometric techniques for visualizing multidimensional data, where each dimension represents an attribute. Graph-drawing techniques are also widely used to display the connectivity between space (road networks and trade connections) and social network analysis (Figure 4.4). However, selecting the appropriate number of nodes and edges to display connections in an informative and legible manner is a challenge.

Most of these two-dimensional techniques are useful for displaying spatial and nonspatial data and multiple attributes. Spatial data are always associated with a time component, and therefore displaying spatial and temporal data using a two-dimensional technique is a challenge. An alternative approach is to use three-dimensional techniques, such as the space–time cube, which is used to depict multiscale data (spatial and temporal). It build on

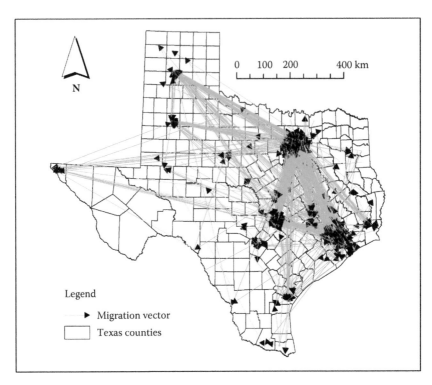

FIGURE 4.4
An example of the graph-based visualization technique.

Hägerstrand's (1970) conceptualization of depicting social behavior across time in a space–time model (Figure 4.5). In a space–time cube, the two dimensions represent geographical space, while the third dimension represents time. Instead of time, the third dimension of the cube can represent another data set, which is the case when fusing multiple data sets or the same data set in multiple scales. The reference map in the cube is displayed in the coordinate plane at time 0, while other changes that happened at different times are displayed vertically or horizontally at their corresponding spatial locations (MacEachren 1995). The space–time path is a specific representation of the space–time cube, in which the location of an object is considered as a three-dimensional point in space and time (Kraak 2003) (Figure 4.6). Time-flattening is a two-dimensional representation of the space–time cube where each map is two-dimensional and depicts the information at a specific time slice (Bach et al. 2014). Animation of these two-dimensional maps can be used to depict the multitemporal distribution of spatial data, but it is not useful for depicting fused data sets (Andrienko and Andrienko 2008). The space–time cube is also useful for depicting trajectory movements like the movement patterns of Napoleon's soldiers, which is difficult in a two-dimensional map. A kaleidomap is a two-dimensional representation of time series data that build on

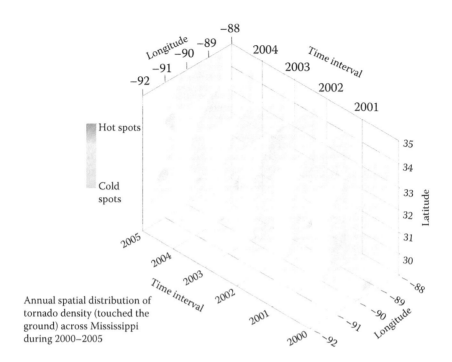

Annual spatial distribution of tornado density (touched the ground) across Mississippi during 2000–2005

FIGURE 4.5
An example of a space–time cube, in which time is represented as the third dimension.

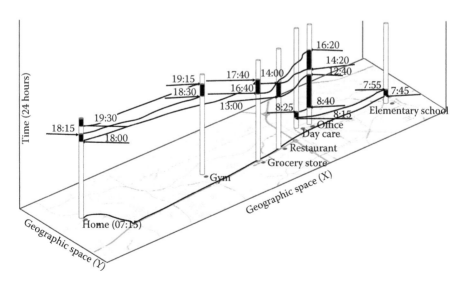

FIGURE 4.6
An example of a space–time path.

the principle of a cascade plot and a child's kaleidoscope. The circular form is another two-dimensional technique used to represent multiple variables and their temporal distribution such that the radial axis and the arc segment correspond to two different types of time intervals (Bale et al. 2007).

Shanbhag et al. (2005) developed three techniques to add a time dimension to two-dimensional choropleth maps. Instead of coloring a spatial unit, in the first technique the area of the unit was partitioned into sections where each section's area represented a specific time interval. In the second technique, time intervals were represented as concentric rings, such that the inner ring corresponded to earliest time for the temporal data. In the third technique, the polygonal area was divided into vertical slices where each slice corresponded to a specific time interval. These techniques depict both spatial and temporal data and can very well be manipulated to display multiscale and fused data sets. However, from an effective visualization perspective, the number of divisions (corresponding to a time interval or an attribute or a data set) must be limited to avoid cluttering. Hewagamage et al. (1999) created a spiral depiction of time in three-dimensional format that is similar to a space–time cube, except that the linear timeline is coiled to a spiral and the events occurring at the specific time and location are displayed as icons along the spiral. This technique, though informative, can be cluttered in case of a lot of time data and can occlude certain parts of the spiral depending upon its orientation. A detailed discussion of the different techniques and existing tools used to visualize time and time-oriented data can be found in Aigner et al. (2011).

Emerging Issues and Future Research

The quality of an individual data set is paramount in data fusion. The failure of the SLC in the Landsat 7 ETM+, for example, resulted in about 22% data loss, particularly along the edge of each scene after May 31, 2003 (US Geological Survey [USGS], 2013). Many scientists have proposed remediation methods to fill the data gap of Landsat 7 ETM+ images (Chen et al. 2011). While the remediated SLC-off data were reported to be still useful for change detection (Wijedasa et al. 2012), Wulder et al. (2008) observed that the selection of a remediation technique can result in varying degrees of trade-off in commission and omission errors. For error assessment purposes, these uncertainties would possibly contribute to false positives or negatives (e.g., under- or overestimation of change) in specific applications.

Similarly, atmospheric conditions, such as Rayleigh scattering, ozone, aerosols, and water vapors, affect the sensitivity of AVHRR, MODIS, VIIRS, and SPOT in varying multispectral bands (e.g., red and near-infrared) and subsequently the products generated from these data sets, such as the NDVI (van Leeuwen et al. 2006; Scheftic et al. 2014). Despite using cross-sensor translation equations to develop comparable products, insufficient knowledge about the sensor-specific atmospheric conditions, processing strategies, algorithm preferences, and their interactions will introduce uncertainty that will impact the use of remote sensing for different applications including multisensor monitoring applications (van Leeuwen et al. 2006).

Beside data quality, another major challenge in data fusion is redundancy. The availability of geospatial data at very high spatial resolutions inevitably introduces more noise into the underlying spatiotemporal pattern or distribution to be discovered. It is desirable to retain only relevant and complementary information from multiple data sources prior to or during data fusion. On the other hand, there is a need to include data available from multiple views or perspectives. Conventional remote sensing imagery is often limited to providing a complete geometric representation of geographic objects due to cloud cover, shadow, and its [off-] nadir view. Certain characteristic of geographic features can only be measured from the field or low-altitude aerial platforms. For example, in the aftermath of the 2014 South Napa earthquake, approximately 200 buildings were yellow- or red-tagged with minor to moderate levels of damage that could not be detected from conventional remote sensing imagery (Chen 2014). With the emergence of low-altitude oblique angle images acquired by unmanned aerial vehicle and crowdsourced images from social media (e.g., Flickr), there is a tremendous need to extend data fusion to incorporate and enable multiview data sources (Gamba et al. 2005). In general, future research should focus on data simplification and/or data reduction across various data models and platforms while retaining the pertinent information available from multiple sources.

Data fusion typically involves geospatial data of varying scales (e.g., spatial and temporal), and scale remains an outstanding research challenge to be addressed. For example, the post density of lidar point clouds could affect the topographic representation of a raster digital elevation model in terms of average slope as well as local slope in a 3 × 3 window (Chow and Hodgson 2009). Common approaches acknowledging MAUP in multiscale data fusion should (1) include sensitivity analysis using various spatial resolutions and zoning configurations at the same spatial resolution; (2) avoid the use of aggregated data, which may be made possible with big data, such as SMS (Corbane et al. 2012), tweets (Tsou et al. 2013), and Web demographics (Chow et al. 2011); and (3) use an "optimal" spatial resolution appropriate to a specific application (Hengl 2006).

Indeed, multiscale analysis has been proposed to explore possible enhancement of data fusion. For example, Huang et al. (2013) proposed Bayesian data fusion to unify the spatial, spectral, and temporal fusions based on the degree of belief in resolving and preserving the characteristics of multisource imagery. Another interesting approach is to utilize multiscale analysis as a way to quantify the uncertainty associated with the recurring information revealed at each scale. Fisher et al. (2005) explored the ambiguity of landscape morphometry (e.g., peaks and dunes) among various kernel filter sizes to be used as fuzzy sets. Using this fuzzy multiscale analysis framework, they identified static topographic landmarks (e.g., peaks and ridges) to the named toponym database, as well as the changing pattern of dynamic coastal dune fields, using multitemporal data. While their approach has shown to be effective for multispatial and multitemporal data fusion, future research can look into possible extension and implications into multisource data fusion across instruments and data models.

While data fusion is endorsed to improve quality of spatial analysis and modeling, its necessity as a prerequisite for some applications is occasionally challenged. In the context of rapid mapping for humanitarian relief, Witharana et al. (2014) examined 12 techniques of pan-sharpening fusion and applied multiresolution segmentation and a support vector machine classifier to extract refugee camps. The authors reported that there was negligible difference in the accuracy of classified image objects between the nonfused images and fused images. While their work did not discount the value of data fusion, it is clear that the current understanding about the synergies, best practices, and subsequent analysis appropriate to various types of data fusion has remained limited. The big data obtained from crowdsourced sites (e.g., Twitter and Facebook) tend to be unstructured and contain a substantial amount of nonspatial data but a limited amount of geospatial information (about 1%–3% of social media data is geotagged) that is generally fudged or spoofed to protect user privacy (Leetaru et al. 2013; Weidemann and Swift 2013) and are available within a limited time frame. With the increasing availability of big data in a variety of data formats to support near real-time analysis, their integration with structured geospatial data and other data sets necessitates the development of techniques and tools appropriate to fusing these data while ensuring that the big data quality is fit for purpose.

Visualization is not a new concept, as seen from Minard's map of 1869. Over the years, geovisualization has grown to be an interdisciplinary field focusing on the visualization of spatial and temporal data. A number of visualization tools and techniques (two- and three-dimensional) implementing statistical and cartographic methods have been developed to combine data and information from multiple sources and at multiple scales and to increase the informativeness and effectiveness of multiscaled and multivariate data sets. Although the inclusion of a number of techniques in one visualization tool is conceptually informative and interactive (e.g., GeoVista), these tools remain ineffective from a user's perspective (Nöllenburg 2007). Despite advancements in the field of geovisualization, inclusion of spatial and temporal data in these techniques remains a daunting challenge (Kaya et al. 2014). The two-dimensional visualization techniques (parallel coordinate plots, scatter plot matrix, data cube built on a scatter plot matrix, and a hierarchical tree structure) can still be used for visualizing fused and multiscaled data, but there is no specific technique that can be used to visualize multiscaled and fused data. Extensive research has been undertaken in the fields of multiscaling, geovisualization, and data fusion, but research combining these subdisciplines is still a long way from maturity.

References

Advanced Spatial Analysis (ASA). 2014. *Bayesian spatial modeling*. http://gispopsci. org/bayesian-spatial-modeling/ (accessed August 25, 2015).

Aigner, W., S. Miksch, H. Schumann, and C. Tominski. 2011. *Visualization of time-oriented data*. Springer. London: Dordrecht Heidelberg, New York.

Andrienko, G., and N. Andrienko. 2008. Spatio-temporal aggregation for visual analysis of movements. In *Proceedings of IEEE Symposium on Visual Analytics Science and Technology*, IEEE organization, OH.

Bach, B., P. Dragicevic, D. Archambault, C. Hurter, and S. Carpendale. 2014. A review of temporal data visualizations based on space-time cube operations. *Eurographics conference on visualization*, United Kingdom. https://hal.inria.fr/ file/index/docid/1006140/filename/spacetime-authorversion.pdf (accessed February 20, 2016)

Bahrampour, S., N. M. Nasrabadi, A. Ray, and W. K. Jenkins. 2015. Multimodal task-driven dictionary learning for image classification. *IEEE* arXiv:1502.01094v1

Bale, K., P. Chapman, N. Barraclough, J. Purdy, N. Aydin, and P. Dark. 2007. Kaleidomaps: A new technique for the visualization of multivariate time-series data. *Information Visualization* 6: 155–167.

Batty, M. 1976. Entropy in spatial aggregation. *Geographical Analysis* 8: 1–21.

Bhaskar, A., T. Tsubota, L. M. Kieu, and E. Chung. 2014. Urban traffic state estimation: Fusing point and zone based data. *Transportation Research Part C* 48: 102–142.

Bradshaw, G. A., and T. A. Spies. 1992. Characterizing canopy gap structure in forests using wavelet analysis. *Journal of Ecology* 80: 205–215.

Brindley, P., S. M. Wise, R. Maheswaran, and R. P. Haining. 2005. The effect of alternative representations of population location on the areal interpolation of air pollution exposure. *Computers, Environment and Urban Systems* 29(4): 455–469.

Burrough, P. A. 1995. Spatial aspects of ecological data. In *Data analysis in community and landscape ecology*, eds. R. H. G. Jongman, C. J. F. Ter Braak and O. F. R. Van Tongeren, 213–265. Cambridge: Cambridge University Press.

Castanedo, F. 2013. A review of data fusion techniques. *The Scientific World Journal*. http://www.hindawi.com/journals/tswj/2013/704504/ (accessed August 25, 2015).

Chander, G., T. J. Hewison, N. Fox, X. Wu, and X. Xiong. 2013. Overview of inter-calibration of satellite instruments. *USGS Staff—Published Research Paper 702*. http://digitalcommons.unl.edu/usgsstaffpub/702 (accessed August 25, 2015).

Chen, J., X. Zhu, J. E. Vogelmann, F. Gao, and S. Jin. 2011. A simple and effective method for filling gaps in Landsat ETM+ SLC-off images. *Remote Sensing of the Environment* 115: 1053–1064.

Chen, Z. 2014. *Disaster-scene crowdsourcing—A short tutorial to DS-crowd*. http:// info.umkc.edu/digitlab/smart-app-disaster-scene-crowdsourcing (accessed August 25, 2015).

Chow, T. E., and M.E. Hodgson. 2009. Effects of Lidar post-spacing and DEM resolution to mean slope estimation. *International Journal of Geographic Information Science* 23(10): 1277–1295.

Chow, T. E., Y. Lin, and W. D. Chan. 2011. The development of a web-based demographic data extraction tool for population monitoring. *Transactions in GIS* 15 (4): 479–494.

Cohen-Steiner, D., E. C. de Verdière, and M. Yvinec. 2004. Conforming Delaunay triangulations in 3D. *Computational Geometry* 28(2–3): 217–233.

Corbane, C., G. Lemoine, and M. Kauffmann. 2012. Relationship between the spatial distribution of SMS messages reporting needs and building damage in 2010 Haiti disaster. *Natural Hazards and Earth System Sciences* 12: 255–265.

Cova, T. J. 1999. GIS in emergency management. In *Geographical information systems: Principles, techniques, applications, and managements*, eds. P. A. Longley, M. F. Goodchild, D. J. Maguire, and D. W. Rhind, 845–858. Wiley Hoboken, NJ, USA.

Crowsey, R. C. 2012. *Coastal hurricane damage assessment via wavelet transform of remotely sensed imagery*. PhD dissertation, University of Southern Mississippi.

Dale, M. R. T. 1999. *Spatial pattern analysis in plant ecology*. Cambridge: Cambridge University Press.

DigitalGlobe. 2015. *WorldView-3*. http://content.satimagingcorp.com.s3.amazonaws.com/static/satellite-sensor-specification/WorldView-3-PDF-Download.pdf (accessed August 25, 2015).

Du, P., S. Liu, J. Xia, and Y. Zhao. 2013. Information fusion techniques for change detection from multi-temporal remote sensing images. *Information Fusion* 14(1): 19–27.

Elwood, S. 2008. Volunteered geographic information: Future research directions motivated by critical, participatory, and feminist GIS. *GeoJournal* 72(3–4): 173–183. doi: 10.1007/s10708-008-9186-0.

Falkowski, M. J., A. M. S. Smith, A. T. Hudak, P. E. Gessler, L. A. Vlerling, and N. L. Crookston. 2006. Automated estimation of individual conifer tree height and crown diameter via two-dimensional spatial wavelet analysis of lidar data. *Canadian Journal of Remote Sensing* 32: 153–161.

Fisher, P., J. Wood, and T. Cheng, 2005. Fuzziness and ambiguity in multi-scale analysis of landscape morphometry. In *Fuzzy modeling with spatial information for geographic problems*, eds. F. E. Petry, V. B. Robinson, and M. A. Cobb, 209–232. Springer, Germany.

Fitzpatrick, J. M., D. L. G. Hill, and C. R. Maurer, Jr. 2000. Image registration. In *Medical image processing and analysis*, eds. M. Sonka and J. M. Fitzpatrick, 488–496. Bellingham, WA: SPIE Press.

Fotheringham, A. S., M. E. Charlton, and C. Brunsdon. 1998. Geographically weighted regression: A natural evolution of the expansion method for spatial data analysis. *Environment and Planning A* 30: 1905–1927.

Fotheringham, A. S., R. Crespo, and J. Yao. 2015. Geographical and temporal weighted regression (GTWR). *Geographical Analysis* 47(4): 431–452. doi: 10.1111/gean.12071.

Fotheringham, A. S., and D. W. S. Wong. 1991. The modifiable areal unit problem in multivariate statistical analysis. *Environment and Planning A* 23: 1025–1044.

Fourier, J. B. J. 1878. *The analytical theory of heat*, translated with notes by Alexander Freeman. London: University Press.

Gamba, P., F. Dell'Acqua, and B. V. Dasarathy. 2005. Urban remote sensing using multiple data sets: Past, present, and future. *Information Fusion* 6: 319–326.

Gangkofner, U. G., P. S. Pradhan, and D. W. Holcomb. 2008. Optimizing the high-pass filter addition technique for image fusion. *Photogrammetric Engineering & Remote Sensing* 74(9): 1107–1118.

Gao, H., X. Wang, G. Barbier, and H. Liu. 2011. Promoting coordination for disaster relief—From crowdsourcing to coordination. *Social Computing, Behavioral-Cultural Modeling and Prediction* 6589: 197–204. doi: 10.1007/978-3-642-19656-0_29.

Ghosh, A., R. Sharma, and P. K. Joshi. 2014. Random forest classification of urban landscape using Landsat archive and ancillary data: Combining seasonal maps with decision level fusion. *Applied Geography* 48: 31–41.

Goodchild, M. F., and L. Li. 2012. Assuring the quality of volunteered geographic information. *Spatial Statistics* 1:110–120. doi:10.1016/j.spasta.2012.03.002.

Goodchild, M. F., and D. A. Quattrochi. 1997. Scale, multiscaling, remote sensing, and GIS. In *Scale in remote sensing and GIS*, eds. D. A. Quattrochi and M. F. Goodchild, 1–12. Boca Raton, FL: CRC Press.

Griffith, D. 2003. *Spatial autocorrelation and spatial filtering: Gaining understanding through theory and scientific visualization*. Springer Science & Business Media, Germany.

Guo, L. Z., S. A. Billings, and D. Coca. 2007. *Multiscale identification of spatio-temporal dynamical systems using a wavelet multiresolution analysis*. ACSE research report # 947. http://eprints.whiterose.ac.uk/74606/1/947.pdf (accessed August 25, 2015).

Hägerstrand, T. 1970. What about people in regional science? *Regional Science Association* 24(1): 6–21. doi:10.1007/BF01936872.

Hall, D. L. 1992. *Mathematical techniques in multisensor data fusion*. Norwood, MA: Artech House.

Hall, D.L., and J. Llinas. 1997. An introduction to multisensor fusion, *Proceedings of the IEEE*, 85(1): 6–23.

Heinzelman, J., and C. Waters. 2010. *Crowdsourcing crisis information in disaster-affected Haiti, United States Institute of Peace*. http://www.usip.org/sites/default/files/SR252%20-%20Crowdsourcing%20Crisis%20Information%20in%20Disaster-Affected%20Haiti.pdf (accessed August 25, 2015).

Hengl, T. 2006. Finding the right pixel size. *Computers and Geosciences* 32: 1283–1298.

Hewagamage, K. P., M. Hirakawa, and T. Ichikawa. 1999. Interactive visualization of spatiotemporal patterns using spirals on a geographical map. In *Proceedings of the IEEE Symposium on Visual Languages*, Washington, DC: IEEE Computer Society.

Huang, B., and H. Song. 2012. Spatiotemporal reflectance fusion via sparse representation. *IEEE Transactions on Geoscience and Remote Sensing* 50(10): 3707–3716. doi:10.1109/TGRS.2012.2186638.

Huang, B., H. Song, H. Cui, J. Peng, and Z. Xu. 2013a. Spatial and spectral image fusion using sparse matrix factorization. *IEEE Transactions on Geoscience and Remote Sensing* 52(3): 1693–1704. doi:10.1109/TGRS.2013.2253612

Huang, B., B. Wu, and M. Barry. 2010. Geographically and temporally weighted regression for modeling spatio-temporal variation in house prices. *International Journal of Geographical Information Science* 24(3): 383–401.

Huang, B., H. Zhang, H. Song, J. Wang, and C. Song, 2013b. Unified fusion of remote-sensing imagery: Generating simultaneously high-resolution synthetic spatial-temporal-spectral earth observations. *Remote Sensing Letters* 4(6): 561–569.

Hwang, T., C. Song, P. V. Bolstad, and L. E. Band. 2011. Downscaling real-time vegetation dynamics by fusing multi-temporal MODIS and Landsat NDVI in topographically complex terrain. *Remote Sensing of Environment* 115: 2499–2512.

Jensen, J. R. 2009. *Introductory digital image processing: A remote sensing perspective.* Upper Saddle River, NJ: Prentice Hall.

Jing, L., Q. Cheng, H. Guo, and Q. Lin. 2012. Image misalignment caused by decimation in image fusion evaluation. *International Journal of Remote Sensing* 33(16): 4967–4981.

Kar, B., and M. E. Hodgson. 2012. Relationship between observational scale and modeled potential residential loss from a storm surge. *GIScience & Remote Sensing* 49(2): 202–227.

Kaya, E., M. T. Eren, C. Doger, and S. Balcisoy. 2014. *Do 3D visualizations fail? An empirical discussion on 2D and 3D representations of the spatio-temporal data.* http://www.eurasiagraphics.hacettepe.edu.tr/papers/05-Kaya.pdf (accessed August 25, 2015).

Keitt, T. H., and D. L. Urban. 2005. Scale-specific inference using wavelets. *Ecology* 86: 2497–2504.

Kerren, A., A. Ebert, and J. Meyer. 2007. Human-centered visualization environments. *LNCS Tutorial.* Springer, Germany.

Khaleghi, B., A. Khamis, and F. O. Karray. 2011. Multisensor data fusion: A review of the state-of-the-art. *Information Fusion.* 14(1): 28–44. doi:10.1016/ j.inffus.201 1doi:.08.001.

Klein, L. A. 2004. Sensor and data fusion: A tool for information assessment and decision making. SPIE Press Monograph Vol. PM138SC, SPIE Press, USA, doi: 10.1117/3.563340.

Kolacayk, E. D., and H. Huang. 2001. Multiscale statistical models for hierarchical spatial aggregation. *Geographical Analysis* 33(2): 95–118.

Kraak, M.-J. 2003. The space-time cube revisited from a geovisualization perspective. *Proceedings of the 21st international cartographic conference (ICC),* Durban, South Africa, August 10–16, 2003.

Lam, N. S.-N., H. L. Qiu, D. A. Quattrochi, and C. W. Emerson. 2002. An evaluation of fractal methods for characterizing image complexity. *Cartography and Geographic Information Science* 29: 25–35.

Lam, N. S.-N., and D. A. Quattrochi. 1992. On the issues of scale, resolution and fractal analysis in the mapping sciences. *The Professional Geographer* 44(1): 88–98.

Leetaru, K., S. Wang, G. Cao, and A. Padmanabhan. 2013. *Mapping the global Twitter heartbeat: The geography of Twitter.* First Monday. http://firstmonday.org/ojs/index.php/fm/article/view/4366/3654?utm_source=buffer&utm_medium=twitter&utm_campaign=Buffer&utm_content=buffer506bd (accessed August 25, 2015).

Levine, S. A. 1992. The problem of pattern and scale in ecology. *Ecology* 73(6): 1943–1967.

Li, Y., H. Wu, R. An, H. Xu, Q. He, and J. Xu. 2013. An improved building boundary extraction algorithm based on fusion of optical imagery and LIDAR data. *Optik* 124: 5357–5362.

Lou, R. C., and M. G. Kay. 1989. Multisensor integration and fusion in intelligent systems. *IEEE Transactions in Systems, Man and Cybernetics* 19(5): 901–931.

MacEachren, A. M. 1995. *How maps work: Representation, visualization, and design.* New York: The Guilford Press.

MacEachren, A. M., and M. Kraak. 2001. Research challenges in geovisualization. *Cartography and Geographic Information Science* 28(1): 3–12.

Mandelbrot, B. 1967. How long is the coast of Great Britain? Statistical self-similarity and fractional dimensions. *Science* 156: 636–638.

Mandelbrot, B. 1982. *The fractal geometry of nature.* W. H. Freeman and Company, USA.

Marston, S. A. 2000. The social construction of scale. *Progress in Human Geography* 24(2): 219–242.

Mason, A. 2001. Scale in geography. In *Interrnational encyclopedia of the social and behavioral sciences,* eds. N. J. Smelser and P. B. Baltes, 13501–13504. Oxford: Pergamon Press.

McMichael, C. E. 2010. Physical geography. *The Encyclopedia of Earth.* http://www.eoearth.org/ (accessed February 20, 2016).

Mesev, V., and P. A. Longley. 2001. Measuring the 'urban': Measuring and modeling a regional settlement hierarchy. In *Modeling scale in geographical information science,* eds. N. J. Tate and P. M. Atkinson, 53–68. New York: Wiley.

Mi, X., H. Ren, Z. Ouyang, W. Wei, and K. Ma. 2005. The use of the Mexican Hat and the Morlet wavelets for detection of ecological patterns. *Plant Ecology* 179: 1–19.

Moellering, H., and W. Tobler. 1972. Geographical variances. *Geographical Analysis* 4: 34–64.

Myint, S. W. 2001. *Wavelet analysis and classification of urban environment using high-resolution multispectral image data.* Ph.D. dissertation, Louisiana State University.

Myint, S. W. 2006. A new framework for effective urban land use and land cover classification: A wavelet approach. *GIScience & Remote Sensing* 43: 155–178.

Myint, S. W. 2010. Multi-resolution decomposition in relation to characteristic scales and local window sizes using an operational wavelet algorithm. *International Journal of Remote Sensing* 31: 2551–2572.

Myint, S. W., N. S.-N. Lam, and J. M. Tyler. 2004. Wavelets for urban spatial feature discrimination: Comparisons with fractal, spatial autocorrelation, and spatial co-cccurrence approaches. *Photogrammetric Engineering & Remote Sensing* 70: 10.

Nöllenburg, M. 2007. Geographic visualization. *Human-Centered Visualization Environments: Lecture Notes in Computer Science* 4417: 257–294.

OpenStreetMap (OSM). 2015. *Open Street Map.* https://www.openstreetmap.org/about

Pal, S. K., T. J. Majumdar, and A. K. Bhattacharya. 2007. ERS-2 SAR and IRS-1C LISS III data fusion: A PCA approach to improve remote sensing based geological interpretation. *Photogrammetry and Remote Sensing* 61: 281–297.

Pelekis, N., B. Theodoulidis, I. Kopanakis, and Y. Theodoridis. 2014. *Literature review of spatio-temporal database models.* http://pdf.aminer.org/000/339/016/database_conceptual_modeling_for_integrating_e_a_model_and_e.pdf (accessed August 25, 2015).

Pereira, G. M. 2002. A typology of spatial and temporal scale relations. *Geographical Analysis* 34(1): 21–33.

Plotnick, R. E., R. H. Gardner, W. W. Hargrove, and K. Prestegaard. 1996. Lacunarity analysis: A general technique for the analysis of spatial patterns. *Physical Review E* 53(5): 5461–5468.

Pohl, C., and J. L. Van Genderen. 1998. Multisensor image fusion in remote sensing: Concepts, methods and applications. *International Journal of Remote Sensing* 19(5): 823–854.

Pu, C., and M. Kitsuregawa. 2014. *Big data and disaster management: A report from the JST/ NSF joint workshop.* https://grait-dm.gatech.edu/wp-content/uploads/2014/03/ Big DataAndDisaster-v34.pdf (accessed August 25, 2015).

Quan, Y., Y. Xu, Y. Sun, and Y. Luo. 2014. Lacunarity analysis on image patterns for texture classification. *IEEE Xplore.* http://www.cv-foundation.org/openaccess/content_cvpr_2014/papers/Quan_Lacunarity_Analysis_on_2014_CVPR_paper.pdf (accessed August 25, 2015).

Quattrochi, D. A., C. W. Emerson, N. Lam, and H. Qiu. 2001. Fractal characterization of multitemporal remote sensing data. In *Modeling scale in geographical information science,* eds. N. J. Tate and P. M. Atkinson, 13–34. New York: Wiley.

Raghavan, R. K., D. Neises, D. G. Goodin, D. A. Andresen, and R. R. Ganta. 2014. Bayesian spatio-temporal analysis and geospatial risk factors of human monocytic ehrlichiosis. *PLoS One* 9(7). doi: 10.1371/journal.pone.0100850.

Reiche, J., J. Verbesselt, D. Hoekman, and M. Herold. 2015. Fusing landsat and SAR time series to detect deforestation in the tropics. *Remote Sensing of Environment* 156: 276–293.

Renslow, M. 2013. *Airborne Topographic Lidar Manual.* American Society of Photogrammetry and Remote Sensing (ASPRS), USA.

Ripley, B. D. 1978. Spectral analysis and the analysis of pattern in plant communities. *Journal of Ecology* 66: 965–981.

Robertson, G. P., and K. L. Gross. 1994. Assessing the heterogeneity of belowground resources: Quantifying pattern and scale. In *Exploitation of environmental heterogeneity by plants: Ecophysiological processes above- and belowground,* eds. M. M. Caldwell and R. W. Pearcy, 237–253. San Diego, CA: Academic Press.

Sayre, N. F. 2005. Ecological and geographical scale: Parallels and potential for integration. *Progress in Human Geography* 29(3): 276–290.

Scheftic, W., X. Zeng, P. Broxton, and M. Brunke 2014. Intercomparison of seven NDVI products over the United States and Mexico. *Remote Sensing* 6(2): 1057–1084.

Shah U. S., and D. Mistry. 2014. Survey of image registration techniques for satellite images. *International Journal for Scientific Research & Development* 1(11). ISSN (online): 2321-0613

Shanbhag, P., P. Rheingans, and M. desJardins. 2005. Temporal visualization of planning polygons for efficient partitioning of geo-spatial data. In *IEEE Symposium on Information Visualization,* October 23–25, Minneapolis, MN, USA, pp. 211–218.

Smith, C., N. Pyden, and P. Cole. 1995. *ERDAS field guide. 628.* Atlanta, GA: ERDAS, Inc.

Spence, R. 2014. *Information visualization: An introduction.* Reading: Addison-Wesley.

Sun, M. L., and Y. S. Li. 2002. Multi-scale character of spatial data in GIS environment and discussion of ITS key problems. *SiChuan Surveying* 25(4): 154–157.

Tate, N. J., and J. Wood. 2001. Fractals and scale dependencies in topography. In *Modeling scale in geographical information science,* eds. N. J. Tate and P. M. Atkinson, 35–52. New York: Wiley.

Tobler, W. R. 1979. Smooth pycnophylactic interpolation for geographical regions. *Journal of the American Statistical Association* 74: 519–536.

Torabzadeh, H., F. Morsdorf, and M. E. Schaepman. 2014. Fusion of imaging spectroscopy and airbornes laser scanning data for characterization of forest ecosystems—A review. *ISPRS Journal of Photogrammetry and Remote Sensing* 97: 25–35.

Torra, V. (ed). 2003. *Information fusion in data mining.* Secaucus, NJ: Springer-Verlag New York, Inc.

Tsou, M. H., J. A. Yang, D. Lusher, S. Han, B. Spitzberg, D. Gupta, J. M. Gawron, and L. An. 2013. Mapping social activities and concepts with social media Twitter and web search engines Yahoo and Bing: A case study in 2012 US Presidential Election. *Cartography and Geographic Information Science* 40(4): 337–348.

United Nations (UN). 2006. *United Nations Platform for Spacebased Information for Disaster Management and Emergency Response (UN-SPIDER).* http://www.unoosa.org/pdf/publications/st_space_65E.pdf (accessed August 25, 2015).

USGS. 2013. *SLC-off Products: Background.* http://landsat.usgs.gov/products_slcoffbackground.php (accessed August 25, 2015).

Van Leeuwen, W. J. D., B. J. Orr, S. E. Marsh, and S. M. Hermann. 2006. Multi-sensor NDVI data continuity: Uncertainties and implications for vegetation monitoring applications. *Remote Sensing of Environment* 100(1): 67–81.

Vidhate, D., and Kulkarni, P. 2012. Cooperative machine learning with information fusion for dynamic decision making in diagnostic applications. In *Proceedings of the 2012 International Conference on Advances in Mobile Network, Communication and Its Applications*, 70–74. Washington, DC: IEEE Computer Society.

Watson, G. H. 1999. Introduction to wavelet analysis. In *Application of mathematical signal processing techniques to mission systems*, 12. Koln, Germany: RTO EN-7.

Weidemann, C., and J. Swift. 2013. Social media location intelligence: The next privacy battle—An ArcGIS add-in and analysis of geospatial data collected from Twitter.com. *International Journal of GeoInformatics* 9(2): 21–27.

Weinan, E., and J. Lu. 2011. Multiscale moeling. *Scholarpedia* 6(10): 11527. http://www.scholarpedia.org/article/Multiscale_modeling (accessed September 5, 2015).

Wijedasa, L. S., S. Sloan, D. G. Michelakis, and G. R. Clements. 2012. Overcoming limitations with Landsat imagery for mapping of peat swamp forest in Sundaland. *Remote Sensing* 4: 2595–2618.

Witharana, C., D. L. Civco, and T. H. Meyer. 2014. Evaluation of data fusion and image segmentation in earth observation based rapid mapping workflows. *ISPRS Journal of Photogrammetry and Remote Sensing* 87: 1–18.

Wu, B., R. R. Li, and B. Huang. 2014. A Geographically and temporally weighted autoregressive model with application to housing prices. *International Journal of Geographical Information Science* 28(5): 1186–1204.

Wu, J., D. E. Jelinski, M. Luck, and P. T. Tueller. 2000. Multiscale analysis of landscape heterogeneity: Scale variance and pattern metrics. *Geographic Information Sciences* 6(1): 6–19.

Wulder, M. A., S. M. Ortlepp, J. C. White, and S. Maxwell. 2008. Evaluation of Landsat-7 SLC-off image products for forest change detection. *Canadian Journal of Remote Sensing* 34(2): 93–99.

Xu, L., J. Q. Zhang, and Y. Yan. 2004. A wavelet-based multisensor data fusion algorithm. *IEEE Transactions on Instrumentation and Measurement* 53(6): 1539–1545.

Xu, G., F. Liang, and M. G. Genton 2013. A Bayesian spatio-temporal geostatistical model with an auxiliary lattice for large datasets. *Statistica Sinica* 1–34. http://stsda.kaust.edu.sa/Documents/2014.XLG.SS.pdf (accessed August 25, 2015).

Zeng, Y., J. Zhang, and J. Van Genderen. 2006. *Comparison and analysis of remote sensing data fusion techniques at feature and decision levels.* http://www.isprs.org/proceedings/XXXVI/part7/PDF/014.pdf (accessed August 25, 2015).

Zhang, X., P. Xiao, X. Feng, J. Wang, and Z. Wang. 2014. Hybrid region merging method for segmentation of high-resolution remote sensing images. *ISPRS Journal of Photogrammetry and Remote Sensing* 98: 19–28.

5

Error and Accuracy Assessment for Fused Data: Remote Sensing and GIS

Edwin Chow and Bandana Kar

CONTENTS

Uncertainties and Errors

Common data capture methods for geospatial data include remote sensing, Global Positioning System (GPS), field surveys, and analog-to-digital conversion (e.g., digitizing and scanning) of existing maps. Each and every geospatial data are subject to some degree of error, a quantifiable departure from the "ground truth" or the best-measured reference. During the process of data integration, the errors commonly originate from the sensing device, environmental characteristics, and the postprocessing operations conducted to transform the raw data into useful information for decision-making, such as radiometric correction, rectification, enhancement, conversion, classification, and so on (Jensen 2009). Uncertainty is commonly characterized using statistics based on the generalization of multiple observed errors (e.g., 90% confidence interval). Formally defined, *uncertainty* is the unpredictability of an outcome due to a lack of knowledge about the appropriateness of parameters, models, or factors to be used in representing the complex and chaotic reality (Wang et al. 2005). Uncertainties in any components of the data integration would degrade the accuracy in describing the actual landscape of a geographic phenomenon.

Among the data capture methods, remote sensing is of particular importance because it remains the dominant data capture method that provides primary geospatial data (e.g., elevation, land cover, vegetation, and weather) over a large region. Depending on the instrumental specifications of sensor(s), remote sensing data from a single source have limited spatial, temporal, spectral, and radiometric resolutions. In digital image processing, image fusion techniques combining additional data can provide complementary information to better understand the observation site (Pohl and Genderen 1998; Simone et al. 2002). For example, pan-sharpening, a process that typically merges a finer-scale panchromatic imagery with a coarser-scale multispectral imagery, can enrich the resulting product to enhance feature extraction. While the goal of image fusion is to reduce the uncertainties of single-source data, such as incompleteness, the process of image fusion could introduce other uncertainties (e.g., consistency) and often make lineage documentation harder to trace (Jing et al. 2012).

When remote sensing data are used in geographic information systems (GIS) to model a spatial phenomenon, more errors tend to be introduced and propagated in the process of data representation, transformation, integration, and modeling. During this procedure, a varying amount of errors would be introduced through many pathways (to be further elaborated on in the section "A Framework of Error Propagation for Process Uncertainty"), resulting in a myriad effect in the quality of the final outcome. Therefore, it is pertinent to identify the sources of uncertainty and their corresponding error(s) in order to quantify the confidence level of the modeled product and improve its accuracy.

Thomson et al. (2005) proposed a typology consisting of nine general categories of uncertainties. In the context of integrating remote sensing data in GIS, the nine types of uncertainties can be defined as follows (Thomson et al. 2005):

- *Accuracy/error:* The difference between fused value and true value
- *Precision:* Closeness of fused measurement
- *Completeness:* The degree of coverage in the extent of the phenomenon
- *Consistency:* The degree of agreement in the measurement from fused data to information
- *Lineage:* Documentation of all transformational operations
- *Currency/timing:* The temporal gap between fused data and its use
- *Credibility:* The degree of trust in each data source
- *Subjectivity:* Elements involving personal decision in the process of data integration and its associated parameters
- *Interrelatedness:* The degree of dependence across multiple data sources

The purpose of this chapter is to examine the uncertainties associated with the integration of remote sensing data with GIS and the uncertainties resulting from this integration. The following section first examines different types of uncertainty in GIS and their sources. The discussion then focuses on the propagation of these uncertainties with examples of associated operation(s) throughout the four stages of data integration: field, image, theme, and object. The implications of such uncertainties are further explored from the perspective of the usability and effectiveness of products resulting from data integration. This chapter later reviews the appropriate techniques to mitigate, evaluate, visualize, and report these uncertainties and highlight some challenges for continual research.

Sources of Uncertainty

Broadly speaking, the sources of uncertainty originate from the data (i.e., the intrinsic nature of geospatial objects), user (i.e., the appropriateness of contextual application), and process (i.e., the interactive effect during data integration). This section discusses the first two sources of uncertainty and the next section describes the last one.

Data Uncertainty

Uncertainties are inherently present in geospatial data because of measurement imperfection, poorly defined geographic entities, and model generalization. According to Cheung and Shi (2001), measurement imperfection includes human mistakes, systematic error induced from instrumental limitations, and random errors due to unpredictable imprecision from the true value. In the process of acquiring remotely sensed data, measurement error is dependent on geometric aspect (e.g., terrain relief displacement, illumination variation, and geometric scaling caused by off-nadir tilting), sensor systems (e.g., camera system, multispectral scanning, and passive/active system), sensor platform (e.g., stability and altitude), ground control (e.g., sampling bias and standards), and scene consideration (e.g., atmospheric attenuation and surface bidirectional reflectance) (Lunetta et al. 1991).

In addition to the challenges in measuring observation, many geographic entities, including both tangible features (e.g., soil) and intangible concepts (e.g., near *vs* far), are poorly defined due to their vagueness and ambiguity (Zadeh 1965; Fisher 2005). Geospatial data typically use the classic set theory to define a crisp and exact distinction between well-defined classes but fails to characterize the gradual change of a vague and continuous phenomenon, such as the transition or intermixing of various soil types (Zhu et al. 2001).

Uncertainties associated with poorly defined geographic entities are also ubiquitous in many concepts (e.g., proximity and suitability) that establish the fundamental assumptions in subsequent analyses (Zhu 2005). A challenge related to this uncertainty in remote sensing is the mixed pixel problem, which denotes a confusing spectral signal that averaged from different feature classes within a pixel area (Jensen 2009). Moreover, uncertainty associated with ambiguity arises from discord (e.g., conflicting claims common in territorial disputes and inconsistent schema in class definition) and non-specificity (e.g., overgeneralization of hierarchical attributes, and lexical and semantic difference) especially in multisource data integration (Fisher 2005). Although fuzzy membership and fuzzy logic have been used to address issues related to vagueness and ambiguity, uncertainties associated with poorly defined geographic entities remain an active research area (Zadeh 1965; Fisher 2005; Zhu 2005).

Model generalization includes both model uncertainty and cartographic generalization (Goodchild 1996). The former refers to the uncertainties associated with inadequate representation of GIS data models in preserving their observed properties, whereas the latter describes "the application of both spatial and attribute transformation" for cartographic purposes (McMaster and Shea 1988). Good examples of model uncertainty are the inaccurate representation of three-dimensional (3D) properties (e.g., volume) in existing GIS data models (Cheung and Shi 2001). In contrast, common operations of cartographic generalization, such as selection, simplification, combination, smoothing, and enhancement, would introduce both positional and thematic errors into the transformed product. Depending on the quality of rules for cartographic generalization, the spatial and thematic properties of geographic features, their topology, and spatial relationship might be preserved or distorted in varying degrees.

User Uncertainty

Besides data uncertainty, the uncertainties associated with the user and the quality of his or her decisions determining how the data should be used are also responsible for incorporating uncertainties into the final product (Frank 2008). From a data usage perspective, user uncertainty entails "the degree to which the formalized structure of data is incompatible with the concepts that a data user is trying to analyze" (Gottsegen et al. 1999). Ideally, the user contextualizes the application, its parameters, and data usage, appropriate for problem solving. In reality, however, imperfect decisions, such as the choice of map projection, scale, scope of analysis, and so on, may affect user uncertainty. Exemplary sources of user uncertainty include the modifiable areal unit problem (MAUP), path dependence, and human subjectivity.

The boundaries of many geographic units, such as administrative units and census enumeration, are defined artificially and are often subject to change over time. Moreover, thematic attributes are often acquired at varying spatial resolutions and require resampling to be compatible at the same scale for further analysis. The MAUP describes the ecological fallacy when the data to be analyzed are formed by changing geographic boundaries by either zoning or scaling (Openshaw and Taylor 1981; Fotheringham and Wong 1991). The former effect is a result of different ways to group observations, whereas the latter effect can be attributed to (dis)aggregating observations at another spatial resolution. In essence, different sets of data can be formed by either zoning or scaling, resulting in possible inconsistent findings. In data fusion, this issue requires special attention because multisource data sets are likely to be different in spatial, temporal, radiometric, and spectral resolutions. The adverse impact of MAUP illustrates the uncertainty of ecological fallacy when the user fails to acknowledge the problem and take appropriate measures to account for such effects (Fotheringham and Wong 1991; Hengl 2006).

In addition to ecological fallacy, spatial data are also subject to two other fallacies: locational fallacy and atomic fallacy. Locational fallacy results from spatial characterization of certain objects (Dutton 2014). For instance, human beings or animals tend to be represented as points, which is misleading, when analyzing spatial interaction or disease transmission studies. Likewise, the nighttime imagery used as a proxy for population density distribution represents human distribution as a pixel rather than accounting for the exact location of humans in a study site, thereby introducing locational fallacy (Dutton 2014). Atomic fallacy occurs when the spatial dependency that is part of spatial data is disregarded in the analysis process by considering an object outside of its spatial context.

Path dependence describes the limitations associated with the decisions made in the application design and execution that would affect the flexibility of options available in the remaining procedure of data integration. For example, the number of decimal places allowed in specific fields of an attribute table is critical to the precision and accuracy required in some engineering applications (Frank 2008). While some decisions are amendable in some instances, such as recreating a new field in the previous example, some are "irreversible," such as data specification in a remote sensing data acquisition project. In process modeling, the choice of model parameters would also affect geographic outcomes that are subject to path (in)dependence (Cheung and Shi 2001; Brown et al. 2005; Frank 2008). For example, one challenge in fuzzy modeling for feature extraction is the justification of proper fuzzy memberships (e.g., linear, Gaussian, and sigmoid) and associated class boundaries to be used to describe fuzzy class definitions. Similarly, the use of attribute thresholds for accuracy assessment, such as the buffering distance in epsilon band, could also affect the confidence level of data quality (Frank 2008).

Human subjectivity is the largest potential source of error and can affect every other source of uncertainties. The uncertainty associated with a user is subject to the decisions made based on his or her experience, knowledge, education, training, and judgment. Although objectivity is the ultimate goal of best practice recommendation, protocols, and scientific experiment, human subjectivity is nevertheless inherent in every aspect of data integration. Needless to say, user decision in real-world application is also limited by many practical constraints, such as approved budget, allowable time, available labor, accessible hardware and software, and so on. Despite the uncertainty of human subjectivity throughout the process, Frank (2008) noted positively that high data uncertainty (i.e., one with unknown or noticeable low quality) does not warrant poor user decision. Thus, human subjectivity could either mitigate or compound user uncertainty and its interactive effect in other components related to data integration.

A Framework of Error Propagation for Process Uncertainty

Although uncertainties can be theoretically described and estimated at each source, they are often assessed in the final product at the end of data processing. Thus, it is imperative to understand error propagation within a general framework of uncertainties (Buttenfield and Ganter 1990; MacEachren 1992; Buttenfield and Beard 1994; Gahegan and Ehlers 2000). For illustration purposes, this section describes a general framework to model uncertainties as geospatial data go through the four stages of data transformation: field, image, theme, and object (Gahegan and Ehlers 2000). For a geospatial data set (A), the function $f(A)$ can describe the data set of a specific conceptual model, so that:

$$f(A) = A(D, S, T, \alpha, \beta, \chi, \delta, \varepsilon) \tag{5.1}$$

where the data set consists of a given set of data values (D), spatial extent (S), and temporal extent (T), and there would be uncertainties in the properties of value (α), space (β), time (χ), consistency (δ), and completeness (ε). Geographic data are being measured in the field and captured as a remotely sensed image, such that:

$$A_{\text{field}}(D, S, T, \alpha, \beta, \chi, \delta, \varepsilon) \neq A_{\text{image}}(D', S', T', \alpha', \beta', \chi', \delta', \varepsilon') \tag{5.2}$$

where A_{field} represents the field measurement of a landscape and A_{image} implies the image captured with a slight discrepancy in the data characteristics during data integration. It is noted that the transformation from field to image captured in the above equation is slightly different from that used by Gahegan and Ehlers (2000) to illustrate value difference α, spatial extent (β), temporal gap (χ), inconsistency (δ), and incompleteness (ε).

Following image capture, subsequent operations, such as image registration, would introduce additional spatial uncertainty. Assume that the ground control points (GCPs) are collected at the time of image acquisition (i.e., T) such that:

$$A_{GCP}(D, S, T, \alpha, \beta, \chi, \delta, \varepsilon) \rightarrow \text{registration} \rightarrow A_{image}(D, S', T, \alpha', \beta', \chi, \delta, \varepsilon) \quad (5.3)$$

where S' and β' imply the positional error of GCPs and α' is the value uncertainty associated with image registration that involves translation, rotation, scaling, and skewing (Gahegan and Ehlers 2000). Note that this value uncertainty may include but is not limited to the measurement error as explained earlier. Similarly, spatial interpolation may alter data values (D)

$$A_{image}(D, S, T, \alpha, \beta, \chi, \delta, \varepsilon) \rightarrow \text{interpolation} \rightarrow A'_{image}(D', S', T, \alpha', \beta', \chi, \delta, \varepsilon) \quad (5.4)$$

where A_{image} and A'_{image} are the imageries before and after spatial interpolation, respectively, assuming that it preserves any underlying data distribution consistently and completely.

In the context of data integration, it is important to achieve a high accuracy of spatial co-registration for subsequent operations that utilize multisource data sets. In image fusion, for example, multisource remote sensing imageries are often acquired at different spatial, temporal, spectral, and radiometric resolutions. Depending on the range of wavelengths sensitive to a particular spectral channel, it is often not uncommon to have varying spatial resolutions collected by the same source. In Landsat 8, for example, Band 8 (panchromatic), Bands 1–7 and 9 (optical and near- and middle-infrared), and Bands 10 and 11 (thermal) are collected at 15, 30, and 100 m, respectively.

In the process of image fusion, multisource data are subject to varying degrees of error and introduce additional uncertainties. For example, the differences in the environmental characteristics during independent data acquisition would introduce additional uncertainties in consistency and completeness across the two or more imageries involved in image fusion, such that:

$$A_{image\ 1}(D, S, T, \alpha, \beta, \chi, \delta, \varepsilon) \neq A_{image\ 2}(D', S', T', \alpha', \beta', \chi', \delta', \varepsilon') \quad (5.5)$$

where all parameters and associated uncertainties are different between two images (*Image* 1 and *Image* 2), and:

$$f(A_{image\ 1}) + f(A_{image\ 2}) \rightarrow f(A'_{fused\ image}) \quad (5.6)$$

where the summation sign does not indicate arithmetic addition but rather the fusion procedure (e.g., a practical local subtraction commonly used in change detection). Unlike other image enhancement operations, image fusion involving multisensors with varying spatial and temporal parameters would introduce errors associated with all uncertainties (Pohl and Genderen 1998). For example, contrast stretching would dramatically modify the data values (i.e., D and α) but preserve spatial extent, temporal extent, order of internal consistency, and completeness.

The fused image is typically used to derive thematic product through image classification (i.e., feature extraction and thematic labelling). In pixel-based or object-based classification, the data values (D) are reclassified into a certain class labels (D') without changes to the spatial and temporal extents, such that:

$$A_{image}(D, S, T, \alpha, \beta, \chi, \delta, \varepsilon) \rightarrow \text{classification} \rightarrow A_{theme}(D', S, T, \alpha', \beta, \chi, \delta', \varepsilon') \quad (5.7)$$

Depending on the spatial resolution relative to the size of the geographic object to be classified, uncertainties in α' would include the MAUP as well as the mixed pixel problem. Ideally, a well-trained classifier should preserve the consistency (δ) of any underlying trend. A confusion matrix and kappa statistics would quantify the consistency uncertainty δ'. Note that completeness uncertainty ε' is commonly introduced due to the simplification nature of a predefined class schema that may not fully capture all possibilities. Gahegan and Ehler (2000) further discussed the uncertainties associated with different types of classification that involve single image, multitemporal images, ancillary data sources, and multisensors.

Besides raster data mode, data integration in GIS often involves other geospatial data models, so that the thematic data are converted into vector data model of objects:

$$A_{theme}(D, S, T, \alpha, \beta, \chi, \delta, \varepsilon) \rightarrow \text{conversion} \rightarrow A_{object}(D', S', T, \alpha', \beta', \chi, \delta', \varepsilon') \quad (5.8)$$

where the raster-to-vector conversion (or *vice versa*) may produce α' and β' uncertainties by data shifting and geometry misalignment (Jing et al. 2012). In image classification, consistency uncertainty δ' might be of concern depending on the severity of the commission or omission errors of specific classes. Similarly, the MAUP could introduce completeness uncertainty ε' as it is possible to clump proximate objects into a single object during the process of conversion or object formulation.

Light detection and ranging (lidar) is an excellent example to illustrate the error propagation of fused data between remote sensing and GIS, because lidar typically returns massive three-dimensional (i.e., x, y, and z) values in the vector data model, whereas most remote sensing products are in raster formats. As lidar provides elevation data at very high spatial resolution targeting topographic and surface features, it is complementary to multi- or hyperspectral remote sensing products (Renslow 2013). In order to integrate these multisources, multidata models, and multiscale data, however, the fusion of lidar and other remote sensing products typically involves data co-registration, conversion, interpolation, and classification. These operations would introduce all associated uncertainties as discussed above.

In addition, Hodgson and Bresnahan (2004) presented an error budget to evaluate the error of lidar-derived elevation. The authors reported that the error of lidar-derived elevation is a function of the lidar system itself, horizontal displacement, survey measurement, slope, land cover, and interpolation. A complicating factor is the quality of feature classification (e.g., separating ground and nonground features like trees), which would also be subject to sampling bias, calibration parameters, user experience, terrain complexity, and landscape heterogeneity (Hodgson and Bresnahan 2004). These factors illustrate the uncertainties associated with data, users, and processes at all stages of lidar pre- and postprocessing. It is possible, however, to quantify these errors and predict the associated uncertainties with fused products. For example, Chow and Hodgson (2009) used sensitivity analysis to model the scaling effect of varying lidar postspacing and digital elevation model (DEM) cell size on terrain slope. The researchers found a linear and logarithmic functional relationship between mean slope and lidar post spacing, as well as mean slope and DEM cell size, and discussed their implications on propagating the scaling effect to subsequent hydrologic modeling.

Although it is beyond the scope of this chapter, it is obvious that errors would further propagate in any resulting information, knowledge, or decisions derived from subsequent analyses utilizing these multisource data in GIS. Conclelis (2003) acknowledged the intrinsic nature of uncertainty and further discussed the outcome of data integration to produce geospatial knowledge.

Techniques for Modeling Accuracy

Extensive research has been conducted since the 1980s to assess accuracy in the context of geospatial analysis and modeling (Goodchild and Gopal 1989; Veregin 1996; Kiiveri 1997; Heuvelink 1998; Zhang and Goodchild 2002; Kundzewicz 1995; Mowrer and Congalton 2000; Hunsaker et al. 2001; Odeh and McBratney 2001). Previous research has led to the development of different techniques to assess and communicate uncertainty.

Although *uncertainty* is a widely used term that relays the problems associated with geospatial modeling and data, the term is not very informative and is divided into *ambiguity* and *vagueness*. While *vagueness* results from lack of information or incomplete information, *ambiguity* results from the presence of conflicting information (Manslow and Nixon 2002; Fisher 2005). For instance, in remote sensing images, certain pixels may contain information about multiple land cover features, and depending upon pixel size it might not be possible to classify these pixels into specific classes due to mixing of reflectance values, which will result in vague classification of some pixels. If multiple classes are available for one land cover feature due to similar reflectance values, then it will result in ambiguity and subsequently fizzy classification of the pixels.

As Foody and Atkinson (2002) pointed out, increase in data also adds complexity to meaningful extraction of information, which again increases the uncertainty of geospatial data. Thus, uncertainty is ubiquitous and inherently associated with geospatial data and concepts; it results from the fusion of spatial data sets and spatial relations. The following concepts are used to numerically depict the extent of uncertainty present and quality of the spatial data.

1. *Error* is the amount of deviation or variation from the correct data value or the referenced data value Burrough (1986); Atkinson and Foody (2002). Error relates to individual measures and must be computed at each location; it corresponds to data quality rather than uncertainty. For an observation i measured at time t_i and location x_i, y_i, the error e_i will be the difference between the true value z_i and predicted value z_i', so that:

$$e_i = z_i - z_i' \tag{5.9}$$

   If the observed temperature at a location is 90°C and the estimated temperature at that location is 92°C, the error will be 90 – 92 = –2°C. Note that in this case the error and the predicted value are known perfectly. Thus error exists, but there is no uncertainty because uncertainty is associated with statistical prediction and inference (Atkinson and Foody 2002). This error also represents the variation in data value rather than positional error that is inherent in geospatial data. Because uncertainty is an unknown value, bias, precision, and accuracy are used to determine uncertainty through statistical prediction.

2. *Bias* is defined as the presence of consistent or systematic error in a set of values and is determined by fitting a statistical model through the observed values (Atkinson and Foody 2002). Mean error (ME or $\bar{e}_i$) can be used to represent bias:

$$\bar{e}_i = \frac{1}{n} \sum_{i=1}^{n} \left(z_i' - z_i \right) \tag{5.10}$$

   where n is the number of samples. The bias will be higher if the systematic error is larger. For instance, having an error of +2 units implies overprediction of measurements, but it does not imply bias. In order for bias to occur, the error has to be overpredicted for repeat measurement of the same value or for a set of values. The bias is an expectation that provides information about over- or underprediction of values based on some statistical models, and it cannot be inferred from one measurement (Atkinson and Foody 2002).

3. *Precision* depicts the recorded level of detail present in data and is represented by the spread of error surrounding the mean error for a set of values (Burrough and McDonnell 1998; Atkinson and Foody 2002). Like bias, it is depicted by fitting a statistical model to a set of values, and standard deviation and variance are generally used to express precision. *Standard deviation,* or *prediction error,* can be defined as the deviation of a data value from the mean error for the set of values as follows:

$$S_e = \frac{\sqrt{\sum_{i=1}^{n} (\bar{e} - e_i')^2}}{n-1} \tag{5.11}$$

where S_e is the sample standard deviation, $\bar{e}$ the mean error for the set of data values, and e_i' is the error of the value at location i.

The variance of a sample, which is estimated by squaring the standard deviation, also provides information about error distribution surrounding the mean error for all samples. However, being a square number, it does not provide information about under- and overprediction like standard deviation, which is essential from a modeling and uncertainty perspective.

4. *Accuracy* signifies to what extent the measured value is close to the true value of the data and is a combination of unbiasedness and precision. Like bias and precision, a statistical model is used to predict accuracy, and it provides information about the overall error present in the final data. Another definition of *accuracy* is that it represents the deviation from the *true mean* as opposed to *precision,* which refers to the size of deviation from the mean obtained from repeated observations (Cochran 1953). While error depends on empirical data for assessment, a number of statistical techniques are used to assess accuracy and determine the uncertainty associated with the data itself. A discussion of these techniques that are used to assess attribute and positional accuracy of geospatial data is presented in the following sections.

Using Equation 5.1 to calculate the error, the mean square error (MSE) for this data set can be determined by using Equation 5.12

$$MSE = \frac{\sum_{i=1}^{n} (e_i)^2}{n} \tag{5.12}$$

MSE takes the square of the errors because it is primarily concerned with the deviation, regardless of over- or underestimation. Moreover, it gives more weight to large errors than smaller errors. Therefore, it is useful if the intention is to determine the large errors (e.g., outliers) that might be influencing the accuracy of a data set (Makridakis and Hibon 1995). From a statistical perspective, MSE provides information about the distribution of uncertainty around the accepted predicted value (Makridakis and Hibon 1995). Due to the fact that squared error changes the dimension of original value, caution is necessary that the MSE is not interpreted directly using the same measurement unit (e.g., meters or feet). Therefore, the root mean square error (RMSE) is introduced to overcome this limitation by taking the square root of the MSE as follows:

$$\text{RMSE} = \sqrt{\frac{\sum_{i=1}^{n}(e_i)^2}{n}} \tag{5.13}$$

Both MSE and RMSE measure the degree of uncertainty in prediction and are widely used. Note that RMSE is similar to the calculation of standard deviation (Equation 5.11). Hence, assuming the errors are normally distributed, it can be interpreted as meaning that 68% of the error would fall within the range of the RMSE. However, because these parameters are absolute measures and are influenced by error values, it is difficult to use them in different prediction situations (Armstrong and Collopy 1992; Makridakis and Hibon 1995).

Mean absolute error (MAE) is an absolute measurement of error that does not account for over- or underprediction. It can be computed as follows (Makridakis and Hibon 1995):

$$\text{MAE} = \frac{\sum_{i=1}^{n}|e_i|}{n} \tag{5.14}$$

Unlike MSE and RMSE, MAE is not influenced by the error outliers and it provides information about the average size of the prediction error (Makridakis and Hibon 1995). Mean absolute percentage error (MAPE) depicts error as a percentage of the true data value:

$$\text{MAPE} = \frac{100}{n} * \left(\sum_{i=1}^{n}\left|\frac{e_i}{z_i}\right|\right) \tag{5.15}$$

Because MAPE provides an estimation of percent error with regard to true data value, it is much more informative and easier to use for comparison across predictions. However, when the true data value is significantly different than the predicted value, the MAPE value would increase (Armstrong and Collopy 1992; Makridakis and Hibon 1995). The symmetrical mean absolute percentage error (SMAPE) overcomes the issue of outlier influence on the MAPE by dividing the error with the average of the true and predicted value (Equation 5.15). This approach also provides a symmetric distribution of error with a lower bound of 0% and an upper bound of 200% (Makridakis and Hibon 1995).

$$\text{SMAPE} = \frac{200}{n} * \left(\sum_{i=1}^{n} \left| \frac{e_i}{z_i + \overline{z}_i} \right| \right)$$

(5.16)

The median absolute percentage error (MdAPE) is similar to both the MAPE and SMAPE, but instead of computing the average of the error, the median of the error distribution is computed. Therefore, this approach is not influenced by the outliers, but the information provided is not always informative, as it is difficult to predict on average what the accuracy of the final product is going to be (Armstrong and Collopy 1992; Makridakis and Hibon 1995).

Theil's U statistic (*UStatistic*) is a relative accuracy assessment technique that can be depicted by Equation 5.17:

$$\text{UStatistic} = \sqrt{\frac{\sum_{i=1}^{n} \left(\frac{e_i}{z_i} \right)^2}{\sum_{i=1}^{n} \left(\frac{e_i'}{z_i} \right)^2}}$$

(5.17)

where e_i' depicts the difference between z_i at time (t) and modeled value z_{t+1}' at time ($t + 1$) using a predictive model. The UStatistic value ranges from 0 to infinity, where the value of 1 means the predicted accuracy has not changed with time. A value smaller than 1 means that the accuracy of the predicted data value decreases over time; and a value greater than 1 implies the accuracy of the original data value is lower than that of the predicted value (Makridakis and Hibon 1995). This parameter is greatly influenced by outliers and is difficult to interpret when the statistical values are very close to each other. McLaughlin's batting average (Batting Average) is an improvement over UStatistic because it uses absolute measurement of the numerator and denominator of Equation 5.9 instead of their squares. The Batting Average can be calculated as follows:

$$\text{Batting Average} = \left[4 - \frac{\sum_{i=1}^{n} \left| \frac{e_i}{z_i} \right|}{\sqrt{\sum_{i=1}^{n} \left| \frac{e_i'}{z_i} \right|}} \right] * 100$$

(5.18)

A Batting Average value of 300–400 implies higher accuracy at a different time; a value of 300 means that the accuracy is not different; and less than

300 implies a reduced accuracy. Although this technique has a reduced effect of outliers, when the actual value is close to the predicted value in a different time, it can result in a negative value and its interpretation might not be very useful (Makridakis and Hibon 1995).

The geometric means of square error (GMSE) use a product of square errors rather than their sum as in the MSE (Equation 5.4)

$$GMSE = \left(\Pi_i e_i^2\right)^{\frac{1}{n}} \tag{5.19}$$

The geometric mean root mean square error (GMRMSE) can be written as follows:

$$GMRMSE = \left(\Pi_i e_i^2\right)^{\frac{1}{2n}} \tag{5.20}$$

The geometric mean of relative absolute error (GMRAE) is another accuracy assessment technique that is dependent on geometric means of error and is defined as follows:

$$GMRAE = \left(\Pi_i RAE_i\right)^{\frac{1}{n}} \tag{5.21}$$

where

$$RAE = \left[\frac{\left|\frac{e_i}{z_i}\right|}{\left|\frac{e_i'}{z_i}\right|}\right] \tag{5.22}$$

These geometric mean errors are not influenced by outliers to the same extent as the above-mentioned techniques because they use geometric mean to compare the errors of two data sets (Armstrong and Collopy 1992; Makridakis and Hibon 1995). They are also easier to communicate than UStatistic and Batting Average.

The median relative absolute error is also a representation of absolute error, but it is computed by determining the median of the MAE computed in Equation 5.14 by sorting them in an ascending/descending order (Makridakis and Hibon 1995). Although it is not influenced by outliers like the geometric means, its interpretation is difficult as for MdAPE.

In addition to these techniques that are applied to individual data values, a number of other techniques are used to address the positional accuracy of geospatial features and also the accuracy of matrix data values that are inherent to raster or continuous data sets (e.g., DEM and remote sensing images).

Area error proportion (AEP) is a simple measurement that provides information about the extent of agreement between a set of known proportions in a matrix $m1$ (i.e., an image or a raster layer) and a set of predicted proportions in another matrix $m2$ (second image or raster layer) (Tatem et al. 2002). This statistic informs the bias present in the prediction matrix and is computed for each class of values by using Equation 5.15 as follows:

$$AEP_c = \frac{\sum_{i=1}^{n} (m1_{ic} - m2_{ic})}{\sum_{i=1}^{n} m2_{ic}} \quad (5.23)$$

where c is the class and n is the total number of pixels. The correlation coefficient (r) is another measurement that provides information about the extent of association between an observed value z_i and a predicted value z_i' at time t_i for a class c containing n number of samples (Tatem et al. 2002). The coefficient can be computed as follows:

$$r_c = \frac{\text{cvar}_{ziz'i}}{\text{std}_{ziz'i}} \quad (5.24)$$

and

$$\text{cvar}_{ziz'i} = \frac{\sum_{i=1}^{n} (\overline{z'_{ic}} - z'_{ic}) * (\overline{z}_{ic} - z_{ic})}{n-1} \quad (5.25)$$

Closeness (C) is another measurement of accuracy that computes the Euclidean distance between the observed value z_i in Image 1 and the predicted value z_i' in Image 2 at the same location (x_i, y_i). This measurement provides information about the closeness of the data value based on the relative proportion of each class in a pixel and can be computed as follows:

$$S_i = \frac{\sum_{q=1}^{n} (y_{iq} - a_{iq})^2}{k} \quad (5.26)$$

where y_{iq} is the proportion of class i in a pixel in Image 1, a_{iq} is the measure of strength of membership to class q, and k is the total number of classes (Tatem et al. 2002).

The RMSE discussed above depicts the assessment of accuracy with regard to data values rather than positional accuracy of a spatial feature. The RMSE for a feature can be computed by the following:

$$RMSE_r = \sqrt{RMSE_x^2 + RMSE_y^2} \quad (5.27)$$

where RMSE_x^2 and RMSE_y^2 represent the variance in the x and y directions and can be computed by using the following equations:

$$\text{RMSE}_x = \sqrt{\frac{\sum_{i=1}^{n}(x_i - x_i')^2}{n}} \qquad (5.28)$$

and

$$\text{RMSE}_y = \sqrt{\frac{\sum_{i=1}^{n}(y_i - y_i')^2}{n}} \qquad (5.29)$$

The circular standard error (CSE) and circular map accuracy standard (CMAS) are also used to assess the positional accuracy of a feature and can be computed by using the following equations:

$$\text{CSE} = \frac{\text{RMSE}_r}{\sqrt{2}} \qquad (5.30)$$

and

$$\text{CMAS} = \frac{2.146 * \text{RMSE}_r}{\sqrt{2}} \qquad (5.31)$$

The kappa coefficient (K) is a multivariate statistic used to assess the overall accuracy of remote sensing data sets (Hudson and Ramm 1987; Zhan et al. 2002; Jensen 2009). It is a measure of agreement or accuracy between the reference data and measured data and can be computed by the following:

$$K = \frac{N\sum_{i=1}^{n} x_{ii} - \sum_{i=1}^{k} x_{i+} * x_{+i}}{N^2 - \sum_{i=1}^{n} x_{i+} * x_{+i}} \qquad (5.32)$$

where N is the total number of pixels, n is the total number of classes, x_{ii} is the total number of pixels classified correctly, x_{i+} is the number of pixels classified into class i, and x_{+i} is the number of pixels classified into class i in the reference data set (Zhan et al. 2002; Jensen 2009). Producer's accuracy and user's accuracy are two other major accuracy assessments that can be computed using the following equations:

$$\text{Producer's accuracy} = \frac{N\sum_{i=1}^{n} x_{ii}}{x_{+i}} \qquad (5.33)$$

and

$$\text{user's accuracy} = \frac{N \sum_{i=1}^{n} x_{ii}}{x_{i+}} \qquad (5.34)$$

Producer's accuracy is also known as the *omission error,* which gives information about the total number of correct pixels in a class divided by the total number of pixels for that class as derived from the reference data. It is the probability of a reference pixel being correctly classified. User's accuracy is known as the *commission error,* which gives information about the total number of correct pixels in a class divided by the total number of pixels that were actually classified in that class. It is a probability that a classified pixel represents that category on the ground (Jensen 2009).

The R^2 is another indicator of accuracy that is used in regression analysis and is defined as follows:

$$R^2 = \frac{\sum_{i=1}^{n} (z_i' - \bar{z}_i)}{\sum_{i=1}^{n} (z_i - \bar{z}_i)} \qquad (5.35)$$

where for a data set at time t_i containing n number of samples, z_i is the true value, z_i' is the predicted value, and $\bar{z}_i$ is the average of all true values in the data set (Makridakis and Hibon 1995). The R^2 value varies between 0 and 1, and it provides information about the percentage of the total variation explained by the reference data in relation to the mean. It is informative and easy to understand but suffers from the influence of data that is not normal in distribution (Makridakis and Hibon 1995).

The percentage better (*%Better*) and the average ranking of various methods (RANKS) are two methods used to cross-compare the accuracy estimated by two different methods. As the name suggests, percentage better provides the error percentage of one method compared to another method. Although it is intuitive and not influenced by outliers, it does not account for the size of error associated with a method (Makridakis and Hibon 1995). In contrast, the RANKS method sorts the errors from lowest to highest, and the lowest error is provided the highest rank. This approach is not influenced by outliers but also does not account for the size of the error associated with a method. Therefore, it is not a good indicator of accuracy with regard to a data set (Makridakis and Hibon 1995).

Error propagation: The word *propagation* means "the spreading of an entity into a subsequent procedure." It is a given that geospatial data inherently have some amount of error associated with it. When these data sets are used as input in a spatial model, the error associated with the input data is propagated to the output. Error propagation continues if the output of an operation is used as input in another operation, thereby reducing the accuracy of the final product. If a consistent record of the error associated with each input data is not maintained, it is difficult to assess the accuracy of final product due to error propagation. Uncertainty analysis and sensitivity analysis are two components of error propagation that must be determined in order to assess the accuracy of a model (Crosetto et al. 2000). Uncertainty analysis depicts the uncertainty associated with the model output due to uncertainties associated with the model inputs.

The *analytical integration method* is a method used to determine error propagation in which the probability density function (PDF) for each input variable is combined to determine the output variable's PDF (NRC 2014). This method is useful if a limited number of input variables is used and their PDF is known. However, if the PDF is unknown, it is difficult to implement this technique. The discrete probability distribution (DPD) method determines the DPD of input uncertainties, which are then combined to determine the DPD of the output variable (NRC 2014). The method is simple to implement when the input variables are statistically independent, but for dependent and large number of input variables steps must be taken to manipulate the input variables.

The *Monte Carlo method* is a common method used to assess uncertainty. Let O be the output for a GIS operation g implemented on m input attributes Z_i, such that $O = g (Z_1, Z_2, ..., Z_m)$. The method implements four steps: (1) determine a PDF for each input factor Z_i; (2) select a random sample of size n from the factor distribution n (Z_i; $i = 1 ... n$); (3) compute O for each sample location x_i, y_i, and z_i; and (4) analyze the distribution of the model outputs O_i (Hevulink 2002; Crosetto et al. 2000). The samples can be obtained by using random sampling, stratified sampling, or other sampling techniques (Hevulink 2002; Crosetto et al. 2000). Finally, the mean and variance of the output distribution is computed using Equations 5.2 and 5.3 to assess the accuracy of the model output. The accuracy of the Monte Carlo method is inversely related to the square root of the sample runs (n), which means to increase accuracy more runs must be implemented (Hevulink 2002). A limitation of this method is the requirement of a large number of simulations to increase accuracy, and very often it is unknown how many simulations are needed to provide the desired accuracy, which makes this a time-consuming method (Hevulink 2002). The method also does not provide information about which input variables contribute to higher uncertainty in the output.

The Taylor series method tries to approximate the operation $g(\cdot)$ by truncating the input variables used in the operation (Hevulink 2002). In general, the first derivative of $g(\cdot)$ around a nominal point b is used to determine the input variables and the mean and variance for the first derivative inputs. The variance is the sum of the correlations and standard deviations of the input variables Ai and the first derivative of $g(\cdot)$ at b (Hevulink 2002). The variance indicates the impact of each variable in the total variance. To decrease the approximate error that results from using a set of input variables, higher order derivatives can be used to derive the input variables and compute variance.

Geostatistics has become a popular approach to assess the uncertainty associated with spatial data, because it takes into account the spatial autocorrelation between observations to determine values at unsampled locations (Burrough and McDonnell 1998). The semivariogram generated by this technique depicts the distribution of semivariance between observation pairs with regard to the distance that separates them. This knowledge helps the identification of samples that will have a higher impact to assign the associated weight to estimate the unknown value. This technique also generates a probability and standard error map that helps visualize the spatial distribution of uncertainty, which can be used to increase the accuracy of predicted values. A detailed discussion of this technique and its implementation can be found in chapter 6 of Burrough and McDonnell (1998).

Generalized likelihood uncertainty estimation (GLUE) is a Monte Carlo approach and an extension of the generalized sensitivity analysis (GSA). In GSA, a set of model parameters is extracted from the entire set, and the model output is generated using each subset of parameters. Finally, the model outputs are compared with the original observation derived from using all the model parameters to determine the sensitivity of different parameters on the model output (Stedinger et al. 2008). Like GSA, the parameter subsets in GLUE are derived from any probability distribution using Monte Carlo simulation. The model output for each subset of parameters is derived and compared with the observed output. A goodness-of-fit function is used to determine the likelihood of closeness of the predicted output to the observed output. This step helps to determine the uncertainty of the model inputs (Stedinger et al. 2008).

Unlike the above-mentioned techniques, Bayesian probability assumes that the parameter of interest has a prior probability distribution rather than a fixed value to quantify the uncertainty associated with spatial data sets. By sampling the data distribution, the Bayesian approach calculates the prior probability of the model parameters and the associated probability of resulting outputs. This technique uses Bayes' rule to determine the *a priori* and *a posteriori* distribution for a set of parameters. This distribution helps to determine the probability that the parameter is within an accepted interval (Love 2007). Further discussion of this technique and its implementation with regard to spatial data can be found in a previously published work (Love 2007).

Multiscale analysis is also a popular technique used to assess the uncertainty associated with the model output due to variation in the scale of analysis. Given the presence of the MAUP in spatial data, this technique examines not only the sensitivity of a model to its input parameters but also the scale of analysis used in the model. Multiscale analysis methods can be direct and indirect (Wu et al. 2000). The direct method implements a multiple-scale method applied to each scale (e.g., semivariance analysis, wavelet analysis, spectral analysis, fractal analysis, lacunarity analysis, and blocking quadrat variance analysis). On the other hand, the indirect method employing a single-scale method is used repeatedly at different scales (Wu et al. 2000). The indirect scale method allows establishing a functional relationship between the scale of analysis and the modeled output for the same method so that the relationship can be used to upscale or downscale the model output (Mandelbrot 1967; Chow and Hodgson 2009; Kar and Hodgson 2012).

The fuzzy model is a soft classifier technique used to deal with the fact that many geographic phenomena involve gradual changes among classes (e.g., the ecotone between a grassland and a forest), and the classic set theory lacks a universal justification to classify a specific object belonging to a class based on a set of arbitrarily established class boundaries. Uncertainty due to vagueness in class boundaries is a common challenge in remote sensing images and classifications. The fuzzy classification technique allows the assignment of fuzzy memberships of one object to a number of classes and uses fuzzy logic operations (e.g., union or intersection) to evaluate the likelihood of fuzzy classification (Binaghi et al. 1999; Fisher et al. 2005; Love 2007). Because this approach indicates the presence of some extent of uncertainty, *fuzzy measures, measures of fuzziness,* and *classical measures of uncertainty* are used to assess uncertainty. Other techniques, such as measuring the distance between the reference data and predicted data using *fuzzy distance* and *fuzzy similarity relations,* are also used to determine the accuracy of this soft classification approach (Binaghi et al. 1999; Fisher et al. 2005).

Sensitivity analysis investigates the uncertainty contribution of individual input data sets and decomposes the uncertainty of the output into each input factor. This information provides useful insights into influential factors that need to be measured accurately in order to achieve high accuracy in the final product. This analysis also works as a tool for precalibration analysis, especially when data are limited (Crosetto et al. 2000; Saltelli 2002). Sensitivity analysis techniques can be categorized into local and global methods. The one-at-a-time method is a local method in which only one input value is varied at each time while the other input values are held constant to determine the impact on the output. In case of larger input values, only a sample of the experimental region is used in the analysis. In general, the local method is faster during implementation, but unlike global methods it is unable to assess the impact of large changes in the inputs (Saltelli et al. 2008, 2010). The global methods can be further categorized into correlation-based and variance-based techniques (Saltelli et al. 2008).

The partial correlation coefficients (PCC) technique provides relevant information about the correlations among different input variables. Given that X_1 and X_2 are input variables and Y is the output variable, the PCC measures the correlation between X_1 and Y while accounting for the effects of X_2—and thereby eliminates the indirect correlations that may exist between X_1 and X_2 or X_2 and Y (Hamby and Tarantola 1999). Regression methods can be used to determine any sensitive model inputs based on the magnitude of the regression coefficients in determining the predictor variables (i.e., output). A stepwise regression extends this approach by examining the statistical significance for each input iteratively and only keeps those that are significant at the 0.05 level at the end (Hamby and Tarantola 1999). In the partitioning technique, the input variables are divided into two or more empirical distributions and statistical tests are used to compare the characteristics of the input distributions. Because the input distribution is created by random sampling of all input variables, nonparametric tests, such as the Smirnov test, are used to determine the extent of similarity between the distributions and the degree of sensitivity between the input and output values (Hamby and Tarantola 1999).

The variance method assesses the influence of individual inputs and their interaction with other inputs rather than determining the impact of specific inputs. A technique similar to analysis of variance is implemented such that the total variance V associated with the output is distributed among the input factors X_i, such that $i = 1, 2, \ldots, k$ (Hamby and Tarantola 1999).

$$V = {}_iV_i + {}_{i<j}V_{ij} + {}_{i<j<m}V_{ijm} + \ldots + V_{1, 2, \ldots, k} \tag{5.36}$$

$$V_i = V\left(E(Y|X_i = x_i^*)\right), V_{ij} = V\left(E(Y|X_i = x_i^*, X_j = x_j^*)\right)$$
$$- V\left(E(Y|X_i = x_i^*)\right) - V\left(E(Y|X_j = x_j^*)\right) \tag{5.37}$$

In the above formulas, Y denotes the output variable, X_i denotes the input variables i, and $V\left(E(Y|X_i = x_i^*)\right)$ denotes the conditional variance. The sensitivity index is the ratio of V_i/V, which reveals that if the conditional mean $(E(Y|X_i = x_i^*))$ varies significantly with the value x_i^* for X_i while the effects of other variables are averaged, then X_i is the influential input. When dealing with a large number of input variables, a large number of indices have to be computed to determine the impact of different interactions on the output uncertainty. The Fourier amplitude sensitivity test computes the total sensitivity index (S_i) and the sensitivity (S_i) for one input variable X_i based on its interactions with all other input variables and the output. The variance-based Sensitivity analysis is model-independent, but it captures the influence for each input variable and its interaction with other inputs in the model (Hamby and Tarantola 1999).

Visualizing Error and Accuracy

Despite the available techniques to reduce error and increase overall accuracy, the visualization of spatial and temporal distribution of error is pertinent to better understand the uncertainty of model output(s). For example, a simple display of high–low error distribution can better identify the specific steps to be taken to reduce error at a specific location or point in time. Therefore, visualization is recommended to communicate the extent of error and its uncertainty to stakeholders. In addition to mapping errors, a number of conventional visualization techniques that are part of exploratory data analysis are in use to visually depict the uncertainties associated with spatial data. This section uses a sample of common visualization techniques to illustrate the discrepancy between lidar and the derived DEM. Further discussion of these techniques can be found in works by Tukey (1977), Fotheringham et al. (2000), Rogerson (2006), Nöllenburg (2007), and Brodlie et al. (2012).

- An *error bar* is a simple bar chart that visually depicts the variability of error associated with each observation in a sample data set.
- A *box plot* is a graphic representation of a group of numerical data with the quartiles of corresponding error (Figure 5.1). The vertical lines extending outward from the box indicate the variability beyond the lower and upper quartiles of the specific group, and these lines are known as *whiskers*. The width of the box is determined by placing a line at the central mean of the group and extending outwards in both directions based on the variation or dispersion associated with the group on either side of the mean.

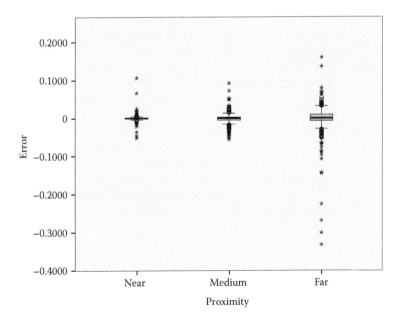

FIGURE 5.1
A box plot visualizing the elevation errors of a lidar-interpolated digital elevation model from the closest lidar points.

- A *violin plot* is similar to a box plot but adds density for each value (i.e., the number of samples used to determine error) along the *x* axis.
- A *stem and leaf plot* depicts the distribution of error. It is easy to extract information and compare the frequency and variation in error among different classes (Figure 5.2). The stem of the plot represents the leading digit(s) of the quantified error, and the leaves represent the next digit of that error and its frequency for each case. For example, each leaf represents two cases (i.e., observation) of elevation error.
- A *histogram* is a common way to represent error. It creates a rectangular bar for each bin (i.e., group) of errors where the area of the graph is proportional to the frequency of that group (Figure 5.3). It is also known as a *frequency graph*.
- A *scatter plot*, also known as a *scatter graph*, depicts the point distribution on a graph such that one axis represents the reference data and the other axis depicts the true data value. A 3D scatter plot can be created to add in a third dimension, where the third dimension can depict another variable or the temporal error for the same observations (Figure 5.4).
- A *parallel coordinate plot* is used to visualize multidimensional geometry and multivariate data (Figure 5.5). For instance, in the case of

```
Frequency    Stem & Leaf

   81.00 Extremes    (=<-.016)
    7.00       -1 .   455
   18.00       -1 .   22233333
   22.00       -1 .   0001111111
   24.00       -0 .   88888999999
   38.00       -0 .   666666666666777777
   76.00       -0 .   44444444444444455555555555555555555555
  108.00       -0 .   222222222222222222222222222222333333333333333333333333
  158.00       -0 .   000000000000000000000000000000000000000000000000000001
                      11111111111111111111111111
  202.00        0 .   000000000000000000000000000000000000000000000000000000
                      000000001111111111111111111111111111111111111111111111
   76.00        0 .   222222222222222222223333333333333333333
   55.00        0 .   44444444444455555555555555555
   39.00        0 .   66666666667777777777
   21.00        0 .   8888889999
   21.00        1 .   0000011111
   12.00        1 .   222233
    5.00        1 .   44&
  104.00 Extremes    (>=.016)

Stem width:     .0100
Each leaf:      2 case(s)
```

FIGURE 5.2
A stem-and-leaf plot cataloging the frequency of elevation errors.

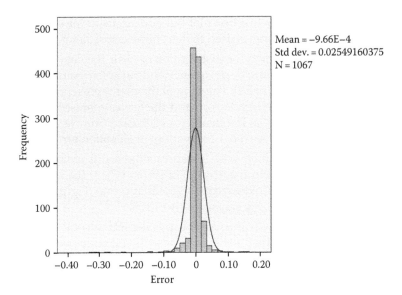

FIGURE 5.3
A classic histogram using frequency to describe the error distribution relative to a hypothetical normal distribution curve.

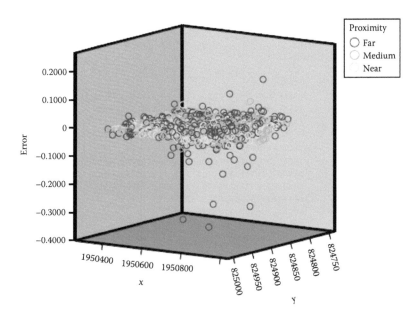

FIGURE 5.4
A three-dimensional scatter plot illustrating the error landscape and color-coded by the three proximity classes.

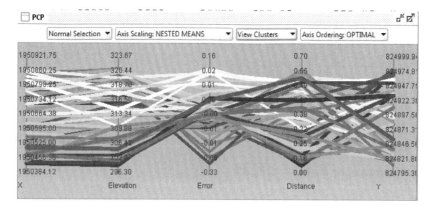

FIGURE 5.5
A parallel coordinate plot of elevation along with multiple attributes.

two-dimensional spatial data, two parallel lines equally spaced are drawn and the errors associated with each observation on each dimension are plotted on separate lines. In the case of multivariate data, the error associated with each variable is plotted on a line corresponding to each variable.

- A *glyph* is a mapping technique used to depict uncertainties in multidimensional geometry and multivariate data. A star plot is a type of glyph where the length of each ray of the star depicts the error associated with a specific attribute or dimension. Glyphs can be combined with maps where each star can be placed at a specific coordinate on the map to visualize the quantity of error.

- A *space–time cube* is a 3D depiction of error where the first two dimensions represent space and the third dimension depicts time. It is useful for smaller data sets, but with larger data sets containing too many observations or variables it becomes illegible points.

- *Maps*: In addition to these techniques, maps are used to depict the spatial distribution of errors. A simple choropleth map can represent the error at each point or a polygon over space. Figure 5.6 illustrates a pair of maps featuring lidar elevation and the interpolated DEM. Symbols can be used to represent the error and each symbol's fill clarity, size, and resolution can be changed to depict the extent of the uncertainty (MacEachren 1992). Map pairs (two maps side by side, one depicting the data and the other representing uncertainty at a specific location), bivariate maps (combining the data and uncertainty together), probability maps (depicting the probability that the variable of interest is above or below a specified threshold), and standard error maps (depicting the spatial distribution of standard deviation or variance) are also used to visually depict errors two-dimensionally (MacEachren 1992).

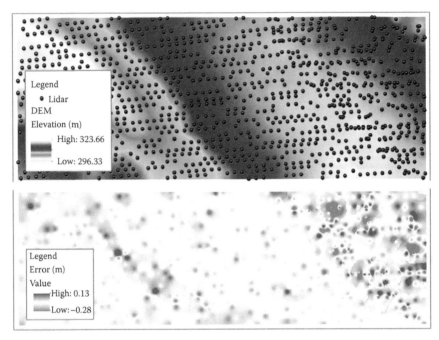

FIGURE 5.6
An error map of referencing a lidar-derived digital elevation model in the same area.

Conclusion

This chapter introduced nine types of uncertainties (Thomson et al. 2005) and examined their sources in terms of data, users, and processes during the course of RS and GIS integration. A framework of error propagation was highlighted to demonstrate various examples of associated operation(s) throughout the four stages of data integration: field, image, theme, and object. This chapter also reviewed some appropriate techniques to mitigate, evaluate, visualize, and report these uncertainties and highlight some challenges for continual research.

It is apparent that spatial data sets are rife with errors. These errors can result from different sources, which require knowledge of these sources to reduce them from the beginning. Spatial data sets can be subject to spatial, temporal, and attribute errors. Spatial/positional error can result if the coordinates (x, y) representing a location are recorded or measured incorrectly. Likewise, attribute error can result due to measurement or recording error. Finally, due to the dynamic nature of physical and social systems, the same object can have a different property over time, thereby adding temporal inaccuracy.

In summary, the main causes of these errors are as follows: measurement (the attribute measured is erroneous, which could be due to instrument or human error), assignment (the attribute entered is erroneous or the object is assigned to the wrong class due to measurement error or human error), class generalization (the object is grouped either wrongfully or intentionally into a group with different properties), spatial generalization (the spatial characteristics of an object are generalized due to simplification, displacement, etc.), entry (miscoding during data entry), temporal (change in data characteristics over time), and processing (error is introduced during data transformation) (Fisher 2005). In addition to these errors, spatial data sets also have inherent properties (spatial representation and spatial autocorrelation) that influence the accuracy and uncertainty of the final outputs. Likewise, certain data analysis techniques have to be implemented on spatial data (spatial interpolation and spatial interaction) before any interpretation of the results can be made. These techniques also introduce uncertainty and error.

1. *Spatial characterization or representation:* Spatial data can be represented in object view as a point, line, or a polygon or in field view as a raster layer. Although statistical techniques are prevalently used to analyze points, it is difficult to implement the same analysis on linear or areal features and on rasters. This simplified depiction of the real world introduces uncertainty that cannot be eliminated.

2. *Spatial autocorrelation:* The first law of geography states that "everything is related to everything else, but near things are more related than distant things" (Tobler 1979). Despite spatial dependence providing useful information that is essential for spatial analysis, it violates the requirement to have independent observations to conduct conventional statistical techniques (Griffith 2003). Therefore, specific techniques—Moran's *I*, Geary's coefficient, Getis's coefficient, geographically weighted regression, Ripley's *K*, and join count analysis (Rogerson 2006; Fotheringham et al. 2000)—are used to analyze the degree of dependency among observations distributed across geographic space. These techniques develop a spatial weighted matrix to decide the influence each location has on an observation, thereby reducing the impact of dependency on the final outcome.

3. *Spatial interpolation:* Spatial interpolation techniques are used to compute values in unobserved locations in geographic space based on observed values in other locations. The basic interpolation techniques include inverse distance weighting, spline, Kriging, and so on (Burrough and McDonnell 1998). These techniques depend on the implementation of a statistical model fitting a set of samples to determine the interpolated value. Because there is no guideline about the sample size to be used, the directionality of sample distribution,

or the statistical technique to be used, the outcomes of these techniques have inherent uncertainties.

4. Spatial interaction techniques, such as the gravity model, are used to understand the interaction among spaces (i.e., origin and destination) to explore the flow of information in geographic space. These techniques require knowledge of origin, destination, topological relationships, travel time, and distance to understand the interaction. Again, the representation of these input parameters impacts the uncertainty of the output.

This chapter briefly summarized the sources of common error and uncertainties associated with each step in the process of geospatial data fusion and integration. Geospatial data producers have the responsibility to utilize the appropriate techniques to reduce, quantify, and communicate the associated uncertainties resulting from any data processing. In contrast, users should understand the implications of these errors, design appropriate measures for subsequent analysis, and visualize the confidence level of the resulting outputs. Although errors are inevitable and ubiquitous, more research and tools are needed to better understand how to reduce, quantify, and visualize these uncertainties and their implications for the decision-making process.

References

Armstrong, J. S., and F. Collopy 1992. Error measures for generalizing about forecasting methods: Empirical comparisons. *International Journal of Forecasting*, 8(1): 69–80.

Atkinson, P. M., and G. M. Foody. 2002. Uncertainty in remote sensing and GIS: Fundamentals, in *Uncertainty in Remote Sensing and GIS*, G. M. Foody and P. M. Atkinson (Editors), John Wiley & Sons, London. p. 1–18.

Binaghi, E., P. A. Brivio, P. Ghezzi and A. Rampini. 1999. A fuzzy set-based accuracy assessment of soft classification. *Pattern Recognition Letters*, 20: 935–948.

Brodlie, K., R. A. Osorio and A. Lopes. 2012. A review of uncertainty in data visualization, in *Expanding the Frontiers of Visual Analytics and Visualization*, J. Dill, R. Earnshaw, D. Kasik, J. Vince, P. C. Wong (Editors), Springer, London, p. 81–109.

Brown, D. G., S. Page, R. Riolo, M. Zellner and W. Rand. 2005. Path dependence and the validation of agent-based spatial models of land use. *International Journal of Geographic Information Science*, 19(2): 153–174.

Burrough, P. A. 1986. *Principles of Geographical Information Systems for Land Resources Assessment*, Clarendon Press, Oxford.

Burrough, P. A., and R. A. McDonnell. 1986. *Principles of Geographical Information Systems*, Oxford University Press, Oxford.

Buttenfield, B. P., and M. K. Beard. 1994. Graphical and geographical components of data quality, in *Visualization in Geographic Information Systems*, D. Unwin and H. Hearnshaw (Editors), Wiley, London. p. 150–157.

Buttenfield, B. P., and J.H. Ganter, 1990. Visualization and GIS: What should we see? What might we miss? In *Proceedings of the 4th International Symposium on Spatial Data Handling.*

Cheung, C. K., and W. Shi, 2001. Measuring uncertainty of spatial features in a three-dimensional Geographic Information System based on numerical analysis. *Annals of GIS*, 7(2): 124–130.

Chow, T. E., and M. E. Hodgson. 2009. Effects of Lidar post-spacing and DEM resolution to mean slope estimation. *International Journal of Geographical Information Science*, 23(10):1277–1295.

Cochran, W. G. 1953. *Sampling Techniques*, Wiley, New York.

Conclelis, H. 2003. The certainty of uncertainty: GIS and the limits of geographic knowledge. *Transects in GIS*, 7(2): 165–175.

Crosetto, M., S. Tarantoal and A. Saltelli. 2000. Sensitivity and uncertainty analysis in spatial modelling based on GIS, *Agriculture, Ecosystems and Environment*, 81: 71–79.

Dutton, J. 2014. *Understanding spatial fallacies*, Penn State University, WWW document: https://www.e-education.psu.edu/sgam/node/214. Accessed on 26 October 2014.

Fisher, P., J. Wood and T. Cheng. 2005. Fuzziness and ambiguity in multi-scale analysis of landscape morphometry, in *Fuzzy Modeling with Spatial Information for Geographic Problems*, F. E. Petry, V. E. Robinson and M. A. Cobb (Editors), Springer. pp. 209–232

Fisher, P. F. 2005. Models of uncertainty in spatial data, in *New development in Geographic Information Science*, P. A. Longley, M. F. Goodchild, D. J. Maguire and D. W. Rhind (Editors), West Sussex: Wiley, London. p. 191–205.

Foody, G. M., and Atkinson, P. M., 2002. *Uncertainty in Remote Sensing & GIS*. Wiley & Sons, Ltd., pp. 307.

Fotheringham, A. S., C. Brunsdon, and M. Charlton. 2000. *Quantitative Geography: Perspectives on Spatial Data Analysis*, Sage, London.

Fotheringham, A. S., and D. W. S. Wong. 1991. The modifiable area unit problem in multivariate statistical analysis. *Environment and Planning A*, 23: 1025–1044.

Frank, A. U. 2008. Analysis of dependence of decision quality on data quality, *Journal of Geographical Systems*, 10(1): 71–88.

Gahegan, M., and M. Ehler. 2000. A framework for the modelling of uncertainty between remote sensing and geographic information systems. *ISPRS Journal of Photogrammetry and Remote Sensing*, 55: 176–188.

Goodchild, M. and S. Gopal. (Editors), 1989. *Accuracy of spatial databases*, Taylor and Francis, London.

Goodchild, M. F. 1996. Generalization, uncertainty, and error modeling, In *Proceedings of GIS/LIS 1996*, Denver, CO, pp. 765–774.

Gottsegen, J., D. Montello, and M. Goodchild. 1999. A comprehensive model of uncertainty in spatial data. In *Spatial accuracy assessment: Land information uncertainty in natural resources*, Lowell, K., and Jaton, A. (Editors), Chelsea, Michigan: Ann Arbor Press.

Griffith, D. 2003. *Spatial Autocorrelation and Spatial Filtering: Gaining Understanding Through Theory and Scientific Visualization*, Springer Science & Business Media, New York.

Hamby, D. M., and S. Tarantola. 1999. Exploring sensitivity analysis techniques for the assessment of an environmental transport model. In *Proceedings of ESREL 99*, Balkema, Rotterdam, 1193–1198.

Hengl, T. 2006. Finding the right pixel size. *Computers & Geosciences*, 32(9): 1283–1298.

Heuvelink, G. B. M. 1998. *Error Propagation in Environmental Modelling with GIS*, Taylor and Francis, London.

Heuvelink, G. B. M. 2002. Analyzing uncertainty propagation in GIS: Why is it not that simple? In *Uncertainy in Remote Sensing and GIS,* Foody, G. M. and Atkinson, P. M. (Editors), West Sussex: Wiley & Sons, Ltd., pp. 155–165.

Hodgson, M. E., and P. Bresnahan. 2004. Accuracy of airborne lidar-derived elevation: Empirical assessment and error budget. *Photogrammetric Engineering and Remote Sensing*, 70(3), 331–339.

Hudson, W. D. and C. W. Ramm. 1987. Correct formulation of the Kappa coefficient of agreement. *Photogrammetric Engineering and Remote Sensing*, 53: 421–422.

Hunsaker, C. T., M. F. Goodchild, M. A. Friedl and T. Case. (Editors), 2001. *Spatial Uncertainty in Ecology*, Springer-Verlag: Berlin.

Jensen, J. R. 2009. *Introductory Digital Image Processing: A Remote Sensing* Perspective, Prentice Hall, Upper Saddle River, NJ.

Jing, L., Q. Cheng, H. Guo and Q. Lin. 2012. Image misalignment caused by decimation in image fusion evaluation. *International Journal of Remote Sensing*, 33(16): 4967–4981.

Kar, B., and M. E. Hodgson. 2012. Relationship between observational scale and modeled potential residential loss from a storm Surge. *GISRS*, 49(2): 202–227.

Kiiveri, H. T. 1997. Assessing, representing and transmitting positional uncertainty in maps. *International Journal of Geographical Information Science*, 11, 33–52.

Kundzewicz, Z. W. 1995. Hydrological uncertainty in perspective, in *New Uncertainty Concepts in Hydrology and Water Resources*, Z. W. Kundzewicz (Editor), Cambridge University Press, Cambridge, pp. 3–10.

Love, K. R. 2007. *Modeling Error in GIS*, doctoral dissertation, Department of Statistics, Virginia Polytechnic Institute and State University.

Lunetta, R. S., R. G. Congalton, L. K. Fenstermaker, J. R. Jensen, K. C. McGwire and L. R. Tinney, 1991. Remote sensing and geographic information system data integration: Error sources and research issues, *Photogrammetric Engineering & Remote Sensing*, 57(6): 677–687.

MacEachren, A. M. 1992. Visualizing uncertain information. *Cartographic Perspectives*, 13: 10–19.

Makridakis, S., and M. Hibon. 1995. *Evaluating Accuracy (or error) Measures*, WWW document: http://www.insead.edu/facultyresearch/research/details_papers. cfm?id=2265. Accessed 25 October 2014.

Mandelbrot, B. 1967. How long is the coast of Great Britain? Statistical self-similarity and fractional dimensions. *Science,* 156: 636–638.

Manslow, J. F., and M. S. Nixon. 2002. On the Ambiguity induced by a remote Sensor's, in *Uncertainty in Remote Sensing and GIS*, G. M. Foody and P. M. Atkinson (Editors), Wiley, London. p. 37–58.

McMaster, R. B., and K. S. Shea. 1988. Cartographic generalization in a digital environment: A framework for implementation in a geographic information system. In *Proceedings of GIS/LIS 1988*, San Antonio, TX, 1: 240–249.

Mowrer, H. T., and R. G. Congalton. 2000. *Quantifying Spatial Uncertainty in Natural Resources: Theory and Applications for GIS and Remote Sensing*, Ann Arbor Press, Chelsea, MI.

Nöllenburg, M. 2007. Geographic visualization, in *Human-Centered Visualization Environments*, A. Kerren, A. Ebert and J. Meyer (Editors), Springer-Verlag Berlin Heidelberg, 4417: 257–294.

Nuclear Regulatory Commission (NRC), 2014. *Uncertainity and Sensitivity Analysis*, NRC. WWW document: http://www.nrc.gov/reading-rm/doc-collections/nuregs/contract/cr2300/vol2/cr2300v2-c.pdf. Accessed on 29 October 2014.

Odeh, I. O. A., and A. B. McBratney. 2001. Estimating uncertainty in soil models (Pedometrics'99), *Geoderma*, 103: 1.

Openshaw, S., and P. J. Taylor. 1981. The modifiable areal unit problem, in *Quantitative Geography: A British view*, N. Wrigley and R. Bennett (Editors), Routledge and Kegan Paul, London, pp. 60–69.

Pohl, C., and J. L. V. Genderen. 1998. Multisensor image fusion in remote sensing: Concepts, methods and applications. *International Journal of Remote Sensing*, 19(5): 823–854.

Renslow, M. 2013. *Manual of Airborne Topographic Lidar*, American Society of Photogrammetry and Remote Sensing (ASPRS): Kindle edition, pp. 884.

Rogerson, P. A. 2006. *Statistical Methods for Geography*, Sage, London.

Saltelli, A. 2002. Sensitivity analysis for importance assessment. *Risk Analysis*, 22(3): 579–590.

Saltelli, A., M. Ratto, T. Andres, F. Campolongo, J. Cariboni, M. Saisanaand S. Tarantola. 2008. *Global Sensitivity Analysis: The Primer*, West Sussex: Wiley Publishing.

Saltelli, A., P. Annoni, I. Azzini, F. Campolongo, M. Ratto and S. Tarantola. 2010. Variance based sensitivity analysis of model output: Design and estimator for the total sensitivity index. *Computer Physics Communications*, 181: 259–270.

Simone, G., A. Farina, F. C. Morabito, S. B. Serpico and L. Bruzzone. 2002. Image fusion techniques for remote sensing applications. *Information Fusion*, 3: 3–15.

Stedinger, J. R., R. M. Vogel, S. U. Lee and R. Batchelder. 2008. Appraisal of the generalized likelihood uncertainty estimation (GLUE) method. *Water Resources Research*, 44. doi:10.1029/2008WR006822.

Tatem, A. J., H. G. Lewis, P. M. Atkinson, and M. S. Nixon. 2002. Super-resolution land cover mapping from remotely sensed imagery using a Hopfield Neural Network, in *Uncertainty in Remote Sensing and GIS*, G. M. Foody and P. M. Atkinson (Editors), John Wiley & Sons: London. p. 77–98.

Thomson, J., E. Hetzler, A. MacEachren, M. Gahegan and M. Pavel. 2005. A typology for visualizing uncertainty. *Conference proceeding in Visualization and Data Analysis 2005*, 146–157.

Tobler, W. R. 1979. Smooth pycnophylactic interpolation for geographical regions. *Journal of the American Statistical Association*, 74: 519–536.

Tukey, J. W. 1977. *Exploratory Data Analysis*, Addison-Wesley, Reading.

Veregin, H. 1996. Error propagation through the buffer operation for probability surfaces. *Photogrammetric Engineering and Remote Sensing*, 62: 419–428.

Wang, G., G. Z. Gertner, S. Fang and A. B. Anderson. 2005. A methodology for spatial uncertainty analysis of remote sensing and GIS products. *Photogrammetric Engineering & Remote Sensing*, 71(12): 1423–1432.

Wu, J., D. E. Jelinski, M. Luck and P. T. Tueller. 2000. Multiscale analysis of landscape heterogeneity: Scale variance and pattern metrics. *Geographic Information Sciences*, 6(1): 6–19.

Zadeh, L. A. 1965. Fuzzy sets. *Information and Control*, 8: 338–353.

Zhan, Q., M. Molenaar and A. Lucieer. 2002. Pixel unmixing at the sub-pixel scale based on land cover class probabilities: Application to urban areas, in *Uncertainty in Remote Sensing and GIS*, G. M. Foody and P. M. Atkinson (Editors), Wiley, London. p. 59–76.

Zhang, J., and M. F. Goodchild. 2002. *Uncertainty in Geographical Information,* Taylor and Francis, London.

Zhu, A. X., 2005. Research issues on uncertainty in geographic data and GIS-based analysis, in *A Research Agenda for Geographic Information Science*, R. McMaster and L. Usery. (Editors), CRC Press, New York, pp. 197–223.

Zhu, A. X., B. Hudson, J. Burt, K. Lubich and D. Simonson. 2001. Soil mapping using GIS, expert knowledge and fuzzy logic, *Soil Science Society of America Journal*, 65: 1463–1472.

6

Remote Sensing Techniques for Forest Fire Disaster Management: The FireHub Operational Platform

Charalampos Kontoes, Ioannis Papoutsis, Themistocles Herekakis, Emmanuela Ieronymidi, and Iphigenia Keramitsoglou

CONTENTS

Introduction

Wildfires have always been present in Mediterranean ecosystems and thus constitute a major ecological and socioeconomic concern. During the last decades, both the number and average size of large fires have experienced an increasing trend, causing extensive economic and ecological losses and often human casualties (Dimitrakopoulos and Mitsopoulos 2005). Increased wildland fire activity over the last 30 years has had profound effects on the budgets and operational priorities of the forest services, civil protection agencies, fire brigades, and local entities with wildland fire management responsibilities (Giannakopoulos et al. 2009; Dimitrakopoulos et al. 2011; Koutsias et al. 2013). Significant alterations in the fire regime have occurred in recent decades, primarily as a result of socioeconomic changes, increasing dramatically the catastrophic impact of wildfires. Despite the recent advances in firefighting tactics and means and the increased amount of resources allocated for fire suppression, the efficiency of the adopted strategy has been decreasing over the last four decades, with both number of fires and burnt area increasing (Bassi et al. 2008).

In this context, the development of appropriate fire suppression strategies for wildfires is challenging. A careful reconsideration of the current wildfire management strategy is necessary in order to reduce the devastating impacts of wildfires on an ecosystem's ecological integrity, society, and economic activity in the future. Fire managers are required to consider and balance threats to multiple socioeconomic and environmental resources and need to identify, in real time, the probability that a wildfire will affect valuable resources and disrupt activities, as well as to estimate the level of damage in ecosystems. The development of more effective wildfire management strategies is a real necessity and requires the availability of accurate and spatially explicit data in order to support evidence-based decision-making.

Earth observation (EO) technology can provide such evidence, through the systematic and standardized processing of satellite imagery. In this context, a large number of EO images of different spectral and spatial resolutions are exploited by the National Observatory of Athens (NOA) through BEYOND (Building a Centre of Excellence for EO-Based Monitoring of Natural Disasters; www.beyond-eocenter.eu), in order to derive thematic products that cover a wide spectrum of wildfire management applications. These products address the requirements of crises occurring before, during, and after fires and follow the Copernicus (GMES) Emergency Response and Emergency Support standards (http://emergency.copernicus.eu/). The NOA has developed a portfolio of similar products, including early fire detection, fire monitoring, and rapid fire mapping, as well as weekly, seasonal, and diachronic burn scar mapping (BSM) and land use/land cover damage assessments over the affected areas.

The concept is to rely on the effective integration of satellite imagery with auxiliary geospatial information and meteorological data, based on statistical and rule-based methods. Input satellite data are comprised of multispatial, multitemporal, and multispectral remote sensing data from EUMETSAT, NASA, NOAA, and European Space Agency missions, and the incorporated processing chains are scalable *via* the exploitation of array database and semantic Web technologies (Koubarakis et al. 2012).

The FireHub real-time fire monitoring service is operated on a routine basis by the BEYOND Center of Excellence, which provides continuous information on active fires detected from EO satellites. The system ingests raw satellite images of coarse spatial resolution from the SEVIRI instrument on board the Meteosat Second Generation (MSG) series of satellites, providing data every 5 minutes. In addition, medium resolution images captured by the moderate-resolution imaging spectroradiometer (MODIS) onboard the Earth Observing System (EOS) Aqua and Terra satellites, the Visible Infrared Imaging Radiometer Suite (VIIRS) onboard the Suomi National Polar-Orbiting Partnership (NPP) satellite, the advanced very high resolution radiometer (AVHRR) onboard the EUMETSAT MetOp, and NOAA Polar Operational series of satellites, with a revisiting capacity of a few hours a day, are automatically ingested into the system by the time of acquisition. Finally, the FireHub system design foresees that in the immediate future, high-resolution Sentinel-2 data will become available in real time through the Hellenic National Sentinel Data Mirror Site (http://sentinels. space.noa.gr), which is part of the ESA's Collaborative Ground Segment in Southeastern Europe. The workflow integrates a number of geospatial layers and *in situ* data representative of the area's fuel model, the topography, and the dynamic meteorological forecasts relevant to wind speed and wind direction. The system provides on a 5-minute basis, and with a time interval of less than 6 seconds after the satellite image acquisition, a fine-grained classification of fire occurrence in subpixels of 500×500 m wide, thus improving the initial MSG/SEVIRI raw observation by about 50 times.

In addition to early fire detection and monitoring, the identification and recording of the burnt areas is routinely achieved through the implementation of a remote sensing method explicitly developed at the NOA for BSM (the BSM-NOA method). The applied BSM-NOA method (Kontoes et al. 2009) was developed and deployed in the framework of the Copernicus (GMES) European flagship program. It aims to contribute to a standardized and homogeneous mapping of burnt areas and related vegetation damage in the European Union member states. The system ensures timely production of burnt area maps, from 1 day (for specific fires that need rush-mode mapping) to a few days (for emergency support), or up to 2 months after the end of the fire season to cover the national scale demands with high thematic and spatial accuracy. This activity supports the reporting and planning needs of the operational users nationwide.

Today, after several development phases, it is delivered through the BEYOND Center of Excellence to the wide institutional user community—ministries of environment, forestry services, and civil protection authorities—and it has been approved as a robust and accurate method. The method has a high spatial precision (0.5–1 ha), at desirable mapping scales ranging from 1:10,000 to 1:50,000. Specifically for Greece, the service is provided *via* a Web GIS application. It serves a yearly updated geodatabase that contains the results of the diachronic burnt area mapping over the country since 1984. Its production was based on analysis of the full USGS archive of Landsat Thematic Mapper (TM) images, since the first satellite image was ever recorded over Greece.

This chapter describes the theoretical background, architecture, and performance characteristics of these two fully automated Web GIS–based systems (fire monitoring and fire mapping) that are designed to assist land managers in wildfire suppression planning and in postfire damage assessment. They consist of the two basic modules of the so-called FireHub

Platform (http://ocean.space.noa.gr/FireHub), which was awarded first prize for Best Challenge Service in the Copernicus Masters Awards Competition 2014.

Theoretical Background

Real-Time Fire Monitoring

Real-time fire activity has shown great potential to be detected from polar orbiters (Giglio et al. 2003) and geostationary satellites (Calle et al. 2006). Polar orbiters are capable of providing data at moderate to high resolution, whereas data from geostationary satellites have proven to be useful for the detection of fire activity at continental and global scales and offer broad direct broadcast capabilities. Polar orbiters provide only four observations per day of approximately 1 km spatial resolution at nadir. High variance of the detectable hotspots and temporal sampling issues related to the diurnal fire cycle have been reported. In contrast, geostationary satellites offer great advantages in filling in the gaps in spatial coverage worldwide at high temporal rates (5–15 minutes), although with a much coarser spatial resolution (approximately 4–5 km) (Prins and Menzel 1996).

In the literature we found EO-based fire-detection studies that were mainly based on the use of radiometers, such as the AVHRR—a space-borne sensor onboard the NOAA family of polar-orbiting platforms that measures the reflectance of the Earth in five relatively wide spectral bands (Chuvieco and Martin 1994). Another well-documented and tested sensor, widely used in active fire detection, is MODIS, which is equipped on the EOS and operates on both the Terra and Aqua spacecrafts (Kaufman et al. 1998). Several operational systems have been developed worldwide using the two abovementioned sensors for active fire-detection purposes. The Global Fire Information Management System delivers MODIS hotspot/fire location information to natural resource managers and other stakeholders around the world (Justice et al. 2002). In Europe, the European Forest Fire Information System (EFFIS) maps active hotspots using MODIS and provides a synoptic view of current fires in Europe as a means to assist the subsequent mapping of burnt area perimeters. Information on active fires is nominally updated on a daily basis and, when needed, made available in EFFIS within 2–3 hours of the MODIS image acquisition (San-Miguel-Ayanz et al. 2005).

Regional operational active fire-detection systems also exist. The German Remote Sensing Data Center of the German Aerospace Center offers an operational service on fire detection from space. Based on data obtained from the experimental satellite BIRD and from MODIS, wildfires are detected and mapped (Brieb et al. 1996). In Canada, the Canadian Fire Monitoring, Mapping, and Modelling System uses infrared imagery from NOAA/AVHRR for the

daily monitoring of active fires and smoke across the country. This information is further used to derive estimations of fire impact and fuel consumption
at a national scale (Li et al. 2000). In Australia, the FireWatch Map Service provides emergency services personnel with an online mapping application to
help in fire management over the continent. The data sets include fire hotspots
from MODIS and NOAA imagery (Steber et al. 2012). The Remote Sensing
Laboratory of the University of Valladolid in Spain provides public operational
information on fires detected from geostationary MSG/SEVIRI in some countries of Western Europe and North Africa, with 15-minute information updates
and disseminates the results over the Internet (Pennypacker et al. 2013).

Despite its coarse spatial resolution, several studies have demonstrated
the capabilities of the SEVIRI instrument for the detection of fires with a
size much smaller than the resolution cell. Two of SEVIRI's spectral bands
are operative in the shortwave infrared (SWIR) (3.9 μm) and thermal infrared
(10.8 μm) wavelengths, and they are sensitive to fire and to Earth's surface
radiative temperature. Laneve et al. (2006) reported that MSG/SEVIRI can be
used to detect fires up to a relatively small size (0.1 ha) with a synoptic view
of their distribution on a large scale, thus allowing for a more efficient and
operational fire-suppression component. In the same context, Van den Bergh
and Frost (2005) employed multitemporal approaches to detect fires based
on the high update rate of MSG/SEVIRI, while Umamaheshwaran et al.
(2007) investigated the potential application of an image mining method for
monitoring and analyzing fire behavior in high-resolution scale in order to
improve the information extracted from MSG/SEVIRI.

The potential of MSG/SEVIRI was in fact promptly explored, namely
within the scope of characterizing the spatiotemporal distribution of wildfire activity on the African continent (Amraoui et al. 2010), as well as estimating the amounts of released fire intensity and fuel consumption (Roberts
and Wooster 2008). In Europe, MSG/SEVIRI images were incorporated in
processing workflow in order to develop a real-time detection system for
Greek territory (Sifakis et al. 2011). MSG imagery has shown good results
when used for generating fire risk maps based on fire weather indexes for
the Mediterranean basin (Amraoui et al. 2013).

Burn Scar Mapping

Several studies have shown that remotely sensed imagery acquired in various spatial, spectral, and temporal resolutions is an effective means to delin-
eate the burnt areas and to determine the species affected and the degree of
damage caused (Sifakis et al. 2004; Quintano et al. 2006). Burn scars can be
clearly identified on a variety of satellite image acquisitions like those from
NOAA/AVHRR, Landsat TM and Enhanced TM+ (ETM+), MODIS, the
medium-resolution imaging spectrometer Satellite Pour l'Observation de la
Terre, and Indian Remote Sensing satellites (e.g., Fung and Jim 1998; Koutsias
2000; Koutsias and Karteris 2000; Rogan and Yool 2001; Chuvieco et al. 2002;

Fraser and Li 2002; Pu and Gong 2004; Gong et al. 2006). In practice, satellite-based BSM takes advantage of the distinctive spectral response of burnt vegetation. While healthy, living vegetation reflects near-infrared (NIR) radiation and absorbs red light in the visible (VIS) part of the spectrum, burnt areas reflect comparatively more radiation in the VIS and SWIR parts of the spectrum and absorb radiation in the NIR. This is attributed to the destruction of the plant and leaf structure (Rogan and Yool 2001). Subsequently, elimination of healthy green vegetation and the inevitable presence of charcoal or bare soil in the fire zone result in a change of radiation recorded by satellite sensors in the relevant spectral bands. These spectral discrepancies between pre- and postfire image acquisitions allow for a clear identification of the burnt area boundaries.

For automatic fire mapping, different methods are employed. The choice is largely dependent on the types of satellite data (spectral and spatial resolutions), the area landscape characteristics (mixed land cover classes, fragmented landscape, and mixed forests with agriculture), and the size of the study area (region, country, and continent). These methods may include fixed thresholding algorithms, adaptive thresholding contextual algorithms (Li et al. 2001), or an integration of the two (Gong et al. 2006) applied to image spectral bands and/or computed indices derived from uni- or multitemporal image acquisitions. Apart from data thresholding techniques, there exist diverse methods, employing logistic regression, exploiting image-derived indices (e.g., vegetation indices) coupled with geographic data (Koutsias 2000), approaches using linear and/or nonlinear spectral mixture analysis techniques (Sa et al. 2003; Ustin 2004), rule-based tree classification (Simard et al. 2000), and neural network (Pu and Gong 2004) methods.

Extraction of burnt land information from remotely sensed data can be performed by using either uni- or multitemporal image acquisitions. Three different approaches have been reported including the following: (1) application of multiple tests on spectral values and indices derived from unitemporal data; (2) multitemporal change analysis of spectral and biophysical indices; and (3) image segmentation and classification techniques using uni- or multitemporal data (Arino et al. 1999). In the first approach, the identification of burnt areas is performed by analyzing the spectral differences of image bands and image-derived indices (e.g., Normalized Burn Ratio Index; Key and Benson 2003) using a single postfire image (Pereira 1999). In certain projects, this approach is preferred to a multitemporal one, as it makes the analysis straightforward. In the second approach, the temporal changes of spectral and/or biophysical parameters due to fires are detected using two images, pre- and postfire (Martin and Chuvieco 1995; Miller and Yool 2002; Fisher et al. 2003). Analyzing the postfire decrease of vegetation vigor (e.g., multitemporal change analysis of vegetation indices), the changes depicted in multitemporal principal component analysis (PCA) vectors (Fisher et al. 2003), or even the changes of brightness, greenness, and wetness components introduced by the so-called tasseled cap Kauth–Thomas transform (Collins and Woodcock 1996), the burnt areas can be identified and mapped more effectively than using a single image. In addition, this approach minimizes the spectral confusion of burnt areas with other land cover types such as permanent crops, open agricultural fields, shadows, and urban and water surfaces. The third method involves conventional image classification and postclassification of uni- or multitemporal satellite data and image-derived indices.

NOA's FireHub Real-Time Forest Fire Detection and Monitoring Service

The System Architecture

The real-time fire monitoring platform delivers an integrated fully automatic processing chain. This module is part of the FireHub service of the BEYOND Center of Excellence. It is divided into dedicated subsystems that offer stakeholders online access to robust, accurate, and fully operational Web-accessible products to assist in fire management and decision-making. The system is enhanced *via* the integration of innovative information technologies for the effective storage and management of the large amount of EO and GIS data, the postprocessing refinement of fire products using semantics (Kyzirakos et al. 2014), and the timely creation of fire extent and damage assessment thematic maps (Figures 6.1 and 6.3).

The architecture of this fully automated forest fire–monitoring application consists of the following parts:

1. Satellite Ground Segment facilities (Block 1 of Figure 6.1) comprise the following:
   a. The high-throughput *MSG/SEVIRI ground-based receiving antenna* (DVB-S2), which collects all spectral bands from any available Meteosat satellite every 5 or 15 minutes, depending on the satellite platform.
   b. The *X-/L-band receiving antenna*, which provides real-time acquisitions from NASA, NOAA, and third-party satellite missions such as the Earth Observing System (EOS), NPP, JPSS, NOAA/AVHRR, MetOp, and FengYun systems.
   c. The *ESA's Sentinel Collaborative Ground Segment (mirror site) infrastructure*: The so-called mirror site of the NOA provides real-time acquisitions of the ESA Sentinel 1, 2, and future 3 and 5P missions, covering the geographic area of Southeastern Europe, the Balkans, North Africa, and the Middle East. The mirror site has been designed to connect with the backbone of the GEANT network (http://www.geant.net/) for fast access to the image data from ESA's core ground segment.

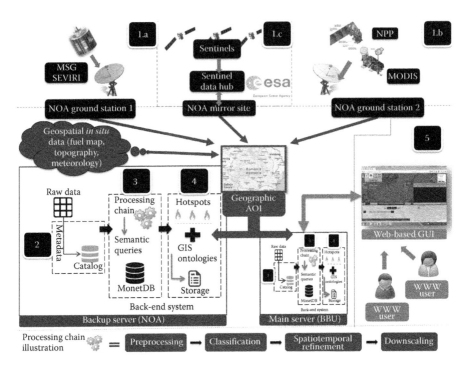

FIGURE 6.1
Architecture of the real-time forest fire detection and monitoring module of the FireHub tool.

2. The raw image data sets that are decoded and temporarily stored in the data vault. This system is responsible for the ingestion policy and enables the efficient access to large archives of image data and metadata in a fully transparent way, regardless of their format, size, and location (Block 2 of Figure 6.1).

3. The back end of the system (Figure 6.3). The back end relies on array image processing solutions such as MonetDB (https://www.monetdb. org/) for two tasks: (1) the implementation of the fire hotspot detection processing chain (using the SciQL scientific query language, https:// en.wikipedia.org/wiki/MonetDB#SciQL) and (2) the evaluation of semantic queries for improving the accuracy of the products and rapidly generating thematic maps (using the semantic spatiotemporal Resource Description Framework [http://www.w3.org/RDF] store Strabon [http://www.strabon.di.uoa.gr/]) (Block 3 of Figure 6.1).

4. A geospatial ontology that links the generated hotspot products (probable active fire pixels) with stationary GIS data (Corine Land Cover, Coastline, Administrative Geography) and open geospatial data available on the Web (e.g., LinkedGeoData—http://linkedgeodata.org/,

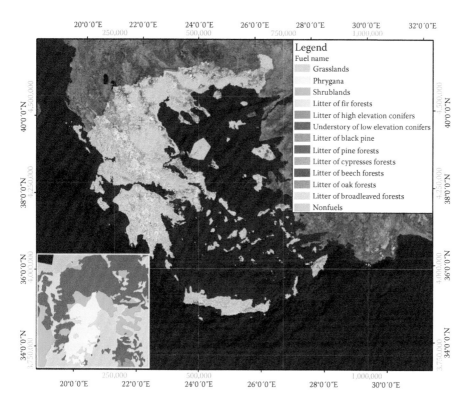

FIGURE 6.2
Fuel map of Greece, generated from the update and fusion of detailed governmental land use databases with the Corine Land Cover data.

GeoNames—http://www.geonames.org/). This ontology is expressed in Web Ontology Language (Block 4 of Figure 6.1 and Block 2 of Figure 6.3).

5. The spatial resolution refinement process, which employs a complex model for the improvement of the spatial accuracy of the satellite-based observations by approximately 50 times, thus downscaling the hotspot spatial resolution from cells of 3.5 × 3.5 km to cells of 500 × 500 m. The algorithms behind this process are currently being evaluated for awarding a patent (Block 3 of Figure 6.3).

6. The submodule for the ingestion of meteorological model forecasts. It consists of a 52-hour wind speed and wind direction prediction, with a 4-km spatial resolution at a fixed grid and a temporal resolution of 1 hour (Block 4 of Figure 6.3).

7. The sun module that feeds large-scale and high-specificity fuel information, as depicted in Figure 6.2. This map was derived through the

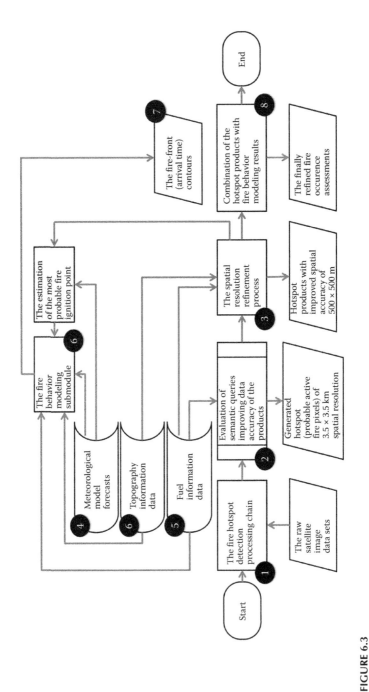

FIGURE 6.3
The main algorithmic blocks of FireHub's fire monitoring process.

fusion of the major vegetation classes represented in existing land use databases with the Corine Land Cover 2000 database. The fuel-type generation was based on a reclassification procedure and conversion of the vegetation types into custom fuel models representative for Greek ecosystems (Block 5 of Figure 6.3).

8. The module that processes the 30 × 30 m ASTER Global Digital Elevation Map tiles for deriving the slope and aspect information parameters in each 500 × 500 m cell (Block 5 of Figure 6.3).

9. The fire behavior modelling submodule, which automatically invokes in specific time frames the FlamMap (Finney 2006) fire model (Block 6 of Figure 6.3); the output of the model (Block 7 of Figure 6.3) is combined with the real-time satellite observations for deriving refined fire occurrence assessments (Block 8 of Figure 6.3).

10. The front-end interface, for controlling the back-end functionality with user-friendly tools, controlling the appearance of the information layers in the monitor, and disseminating the products to the end-user community through the Web (Block 5 of Figure 6.1).

The graphical user interface provides several functionalities for serving the hotspot and the smoke dispersion forecasts *via* the Web GIS interface. These are as follows: (1) the systematic provision every 5 minutes of the fire extent; (2) the retrieval and display of past fire events; and (3) the systematic provision on an hourly basis of smoke plume dispersion in 2D and 3D. Active fires are displayed in (1) refined mode (500 × 500 m wide cells) and (2) raw mode (~3.5 × 3.5 km wide cells). Three background map layers can be selected as background maps: (1) the LSO/VLSO Orthophotos of Ktimatologio S.A. (http://www.ktimatologio.gr), which is a detailed raster basemap with a spatial resolution of 1 m; (2) Google Earth tiles; and (3) the CORINE Land Cover (2000).

Methodology and System Operations

The following operations are invoked on a routine basis every 5 minutes, as soon as a new MSG/SEVIRI satellite image is ingested into the system from the receiving station. It should be acknowledged that the different processing steps, the description of which follows, have been developed and validated in the frameworks of the TELEIOS ICT (http://www.earthobservatory.eu/) and BEYOND EC projects (http://beyond-eocenter.eu/). For more detailed information, a rich compilation of related publications stemming from research work in the framework of these projects is available to the reader through the projects' websites.

Figure 6.3 shows the general methodology used for the incorporation of the minimum travel time (MTT) algorithm in NOA's FireHub real-time fire detection and monitoring system.

First, each pixel of a new satellite acquisition is classified either as *fire*, *potential fire*, or *nonfire*. However, the inherent coarse resolution of the MSG/SEVIRI instrument results in false alarms and omission errors, which reflect on the product's accuracy. The accuracy of the algorithm is enhanced by combining the first classification outcome with external information from linked geospatial data.

For example, a typical shortcoming of the original classification is false alarms at locations with inconsistent land use, such as urban or agricultural areas. This problem is overcome by using a data set that describes the Greek ecosystems in terms of land use/land cover classes and removing those early detected hotspots in nonvegetated areas. The hotspot product, generated every 5 minutes, is subsequently passed through a series of refinement steps to increase its accuracy and robustness by respecting several spatiotemporal fire behavior rules. Indicatively, the temporal persistence of a fire pixel over a period of, say, half an hour increases the confidence level (CL) that it is correctly classified as fire. Therefore, these operations primarily focus on updating the CL of each hotspot pixel and thus moving from the three-flag approach (fire, potential fire, and nonfire) to a real CL value. In addition, the refined hotspot is annotated with the region name it belongs to as attribute information.

Finally, the requirement to generate added-value thematic maps is addressed at this processing level. The Linked Open Data Cloud supplies an abundance of data sets, ranging from fine-grained geometric objects like fire stations to coarser ones like countries. Therefore, instead of manually combining heterogeneous data, the user can design a semantic query to integrate and overlay information layers, generate maps, and export data in well-established formats (Kyzirakos et al. 2012). Although this service was designed for Greece, it can be applied to any geographic area due to the open technologies adopted.

The next step is particularly important because it improves the spatial resolution of fire detection. At this processing phase, each MSG/SEVIRI pixel corresponding to a fire or potential fire event is divided into a 7×7 grid, that is, to subpixels of 500×500 m. For each of these 500-m wide cells, a new CL for fire occurrence is calculated with the use of specific fire hazard weight factors. Such factors take into account the probability of fire occurrence and ease of fire propagation, for example, topography and vegetation characteristics. The new CL is the product of the raw CL (CL_{raw}), with the weight factors derived from the fuel type (W_{FT}), elevation (W_E), slope (W_S), aspect (W_A), and fuel cover (W_{FD}) weights. The normalization of the weight factors and therefore their contribution to the calculation of follows the suggestions of Kontoes et al. (2013a).

$$CL = CL_{raw} \times W_E \times W_S \times W_A \times W_{FT} \times W_{FD} \qquad (6.1)$$

Moreover, the MTT algorithm (Finey 2002) is used for modeling the fire propagation in each event, as embedded in the FlamMap fire behavior

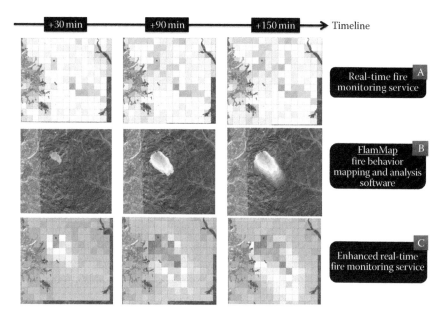

FIGURE 6.4
A typical fire evolution example at the different processing levels of FireHub.

software. FlamMap calculates the fire size and shape from an ignition point. The output of the model is compared in terms of its spatial distribution and temporal evolution to the fire event's pixel observations with the MSG/SEVIRI image. From this point on, a complex modeling scheme is implemented that fuses the information from the fire pixel classification CL and the fire propagation model output to derive the refined fire occurrence evidences in each 500×500 m subpixel.

A typical example of a fire evolution event in 2013, on the island of Rhodes, Greece, is presented in Figure 6.4. The products generated at the various processing levels of the FireHub system are shown. The first row in Figure 6.4 shows the downscaled hotspots after the first satellite imagery classification (CL), the second row corresponds to the integration of the FlamMap dispersion model results, and the last row is the final result delivered to the end-user community, which is the outcome of the combination of the satellite observation after integrating the simulated fire dispersion forecast product.

NOA's FireHub BSM and Damage Assessment Service

This is a fully automatic single or multidate processing chain that takes as input multispectral satellite images of any spatial resolution and produces precise burnt area polygons and wildland area damage assessments over

Greek territory (Kontoes et al. 2013a). The service follows the Copernicus (GMES) accuracy and validation standards and it has been successfully evaluated over different territories in Southeastern Europe. As such, it has been qualified and is transferable to any place over Europe at the regional, national, and continental levels. The burn scar mapping (BSM-NOA) service was initially developed in the framework of the ESA GMES Service Element program called *Risk-EOS*, the so-called BSM-NOA service (Kontoes et al. 2009), and has been fine tuned to become a fully operational processing chain.

The BSM-NOA service is activated on a user-demand basis, and the burnt area products are delivered to end users either in rush mode for emergency response purposes or in nonrush mode within a few days after the suppression of the fire event for emergency support purposes, and also immediately after the end of the fire season to meet recovery needs for the entire region/country. Depending on the input satellite data, the service provides BSMs at high spatial resolution (20–30 m pixel size, minimum detected fire size of 1 ha) and very high spatial resolution (2–8 m pixel size, spatial accuracy of 4–10 m, detected fire size of 0.5 ha), as well as damage assessments at the landscape level.

Based on the BSM-NOA core processing algorithm, a multitemporal analysis is feasible to estimate the annual burnt areas spanning several years. Such an analysis provides a diachronic mapping product that can be exploited for further statistical analyses, fire behavior cyclic patterns, climate change studies, and so on. For the production of the diachronic BSM of Greece, the entire USGS Landsat TM imagery archive over Greece since 1984 was used—that is, the first year when Greece was captured by the Landsat TM sensor. Figure 6.5 depicts the main steps of the BSM-NOA production chain.

The processing chain is divided into three stages, each one containing a series of modules:

1. The preprocessing stage:
   a. Identifying appropriate satellite data (spatial/spectral resolution, coverage, and acquisition dates), downloading, and archiving (Block 1a of Figure 6.5).
   b. *Radiometric normalization, registration, and georeferencing:* A fully automatic procedure wherein the input raw satellite images are calibrated, pixel values are converted from digital counts to radiometric values, and automatic image orthorectification is performed (Gao et al. 2009) (Block 1b of Figure 6.5).
   c. *Cloud/water masking:* The generation of a mask to exclude from subsequent processing pixels "contaminated" by clouds, as well as pixels representing water areas. This is done using NASA's LEDAPS algorithm (http://ledaps.nascom.nasa.gov) (Block 1c of Figure 6.5).

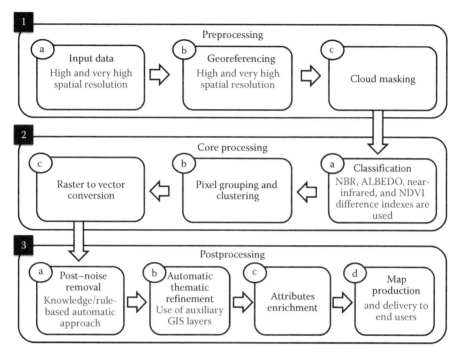

FIGURE 6.5
The FireHub BSM-NOA service chain.

2. The core processing stage:

The focus on the core processing phase is the burnt area classification algorithm (Block 2a of Figure 6.5). The algorithm aims at identifying burnt and nonburnt sets of pixels within the georeferenced satellite image. Each image pixel is basically a vector of intensities that correspond to emissions from different frequency channels. Using the Landsat 5 TM as an example, a raw image consists of seven spectral bands. Classification to burnt and nonburnt areas relies on the fact that the emissions from different frequency bands have a physical interpretation. Simple band algebra can lead to the derivation of physical indexes. The main criteria used within the BSM-NOA process to correctly classify pixels are as follows: (1) the Normalised Burn Ratio (NBR) Index, (2) the Albedo Index, (3) the Normalized Difference Vegetation Index (NDVI), and (4) the $NDVI_{MULTI}$, which is the difference of the two NDVIs calculated before and after a fire event over the same area. Then, a decision tree is formed where the adopted indexes are compared to site-specific thresholds. These image indexes are as follows:

a. *The NBR*: It is one of the most widely used image enhancements for mapping wildfires worldwide. Key and Benson (2003) introduced this index as a variation of the NDVI. They replaced the red reflectance value in the NDVI with the mid-infrared reflectance value:

$$\text{NBR} = \frac{(R_{\text{NIR}} - R_{\text{MIR}})}{R_{\text{NIR}} + R_{\text{MIR}}} \tag{6.2}$$

with R_{NIR} and R_{MIR} denoting the reflectance values recorded in the NIR and mid-infrared channels of the satellite image (Bands 4 and 7 of Landsat TM), respectively.

Several researchers have proposed this index for burnt area mapping (Cocke et al. 2005; Roy et al. 2006), as reflectance values in the red and mid-infrared ranges exhibit the greatest reflectance change in response to a fire. Although NBR has been effective in many burnt areas mapping studies, it has not been widely tested for Greek ecosystems. In the south Mediterranean zones and especially within the Greek pine and shrubland ecosystems, land abandonment has resulted in intense fuel accumulation. Because of this, a significant reduction in green vegetation is reported inside the burnt areas after a fire occurrence. In these areas, the NBR index can differentiate accurately between burnt and unburnt areas. Forest ecosystems, however, are much diversified in Greece and forest canopy density decreases from north to south. Therefore, the forest stands become less uniform and are interrupted by the presence of agricultural fields, dispersed settlements, roads, open fields, abandoned farms, or permanent crop cover. This high mixture of classes makes automatic image segmentation with the sole use of NBR problematic. Indeed, because the burnt vegetation is characterized by an increase in reflectance in the VIS, a decrease in the NIR, and a slight increase in the mid-infrared, the spectral response of burnt forests tends to be "flatter" than that of healthy vegetation, which may cause confusion with nonvegetation classes like open agricultural fields, bare soils, water surfaces, urban areas, or permanent crops. This type of confusion between charcoal and other soil colors in highly fragmented ecosystems was also reported by Rogan and Yool (2001), who suggested the use of the Kauth–Thomas tasselled cap transformation (Kauth and Thomas 1976) to resolve the reported confusion. In order to cope with this problem, the BSM-NOA approach integrates two additional spectral indices complementary to NBR.

b. *The uni- or multitemporal NDVI (NDVI and NDVI$_{MULTI}$)*: The NDVI is a common spectral vegetation index derived by dividing the difference between reflectance in the NIR and the VIS red channels by the sum of the two (Rouse et al. 1974):

$$NDVI = \frac{(R_{NIR} - R_{RED})}{R_{NIR} + R_{RED}} \tag{6.3}$$

with R_{NIR} and R_{RED} denoting the reflectance values recorded in the NIR and red channels of the multispectral satellite image, respectively.

NDVI has long been used in the Mediterranean for assessing the vegetative health and moisture content of an area and resolving ambiguities in the discrimination between healthy and dead or removed vegetation (Marsh et al. 1992; Tappan et al. 1992; Lyon et al. 1998). Moreover, NDVI has been used to demonstrate the extent of vegetation removal associated with a fire event, as it exhibits a sharp postfire drop (Li et al. 2000; Diaz-Delgado and Pons 2001; Vafeides and Drake 2005). Depending on the number of acquisitions, the NDVI analysis can be unitemporal (calculated only at the postfire level) or multitemporal. The multitemporal difference of NDVI adopted in BSM-NOA is denoted as $NDVI_{MULTI}$ and is calculated using the following equation:

$$NDVI_{MULTI} = NDVI_{PREFIRE} - NDVI_{POSTFIRE} \tag{6.4}$$

with $NDVI_{PREFIRE}$ and $NDVI_{POSTFIRE}$ denoting the NDVI values calculated before and after a fire occurrence over the affected area, respectively.

The multitemporal NDVI approach is preferred to a unitemporal one, as it better resolves the confusion between classes. Several studies have differenced prefire and postfire NDVI images to discern fire scars fast and efficiently (Cahoon et al. 1992; Kasischke et al. 1993; Kasischke and French 1995; Li et al. 1997; Leblon et al. 2001).

c. *The albedo index*: In highly diversified Mediterranean ecosystems, the NDVI might put limitations on the detection and delineation of burnt from unburnt surfaces. Pereira (1999) and Elmore et al. (2000) concluded that the NDVI is affected by soil color and is therefore not always comparable across a heterogeneous area. Due to this issue, BSM-NOA integrates the empirical approximation of the surface albedo (Saunders 1990; Lasaponara 2006), which is an indicator of the surface brightness. The albedo index is calculated using the following equation:

$$ALBEDO = \frac{R_{NIR} + R_{RED}}{2} \tag{6.5}$$

with R_{NIR} and R_{RED} denoting the reflectance values recorded in the NIR and red channels of the multispectral satellite image, respectively.

From the above, it is shown that none of the proposed image indexes by themselves can be considered sufficient to efficiently resolve the problem of burnt area mapping in south Mediterranean ecosystems. Hence, the BSM-NOA approach suggests the appropriate thresholding and combined use of the three image indexes, with appropriate classification refinement (noise removal) processes, which is performed at the postprocessing level (Kontoes et al. 2009).

Upon deciding on the burnt and nonburnt pixels of the image, the neighboring pixels are grouped together (Block 2b of Figure 6.5) since they constitute the same fire event, and then the raster is converted to vector (ESRI polygons) (Block 2c of Figure 6.5) to proceed to the postprocessing phase (Figure 6.5).

3. The postprocessing stage:

   a. Noise removal, the process necessary to eliminate isolated pixels that have been wrongfully classified as burnt. The minimum mapping unit depends on the spatial resolution of the input satellite data and ranges from 0.5 to 1 ha. Hence, a rectangle group of pixels with an edge of three or fewer pixels (for the case of Landsat TM with 30-m spatial resolution) should not be classified as burnt. This filtering is performed with the appropriate spatial functions using the Geospatial Data Abstraction Library *via* the Python programming language API (Application Programming Interface) (Block 3a of Figure 6.5).

   b. In addition, a set of logical classification rules is applied, using evidence from a series of auxiliary GIS layers, to ensure product thematic accuracy and consistency with the underlying land use/land cover conditions and landscape morphology. The basic operations performed are (1) refinement of the polygons to comply with certain restrictions, similar to those applicable for the fire monitoring scenario (burnt areas in the sea, or inconsistent underlying land cover types) and (2) normal GIS processes such as classification polygon aggregation and polygon boundary smoothing. The final refinement stage relies on the employment of visual checks to resolve any remaining classification inconsistencies and uncertainties. The aforementioned approach was developed to minimize any manual (visual interpretations) operations that are laborious and time consuming (Block 3b of Figure 6.5).

   c. Attribute enrichment of the BSM product by overlaying the polygons with geoinformation layers (e.g., Greek Administrative Geography, CLC, open data, etc.) (Block 3c of Figure 6.5).

d. Generation of thematic maps that include damage assessments, that is, the land cover types and quantities of burnt areas per prefecture, at a national level (Block 3d of Figure 6.5).

Figure 6.6 provides a more detailed view of the main algorithmic step of the BSM-NOA approach, based on multidate (prefire and postfire) image acquisition to generate burnt scar maps in vector and raster format.

The BSM product is ideal for use in further environmental time series analyses, production of statistical indexes (frequency of fire occurrence, geographical distribution, and number of fires over the studied territory) and applications, including change detection and climate change models, urban planning, and correlation with manmade activities. The BSM-NOA service is freely provided through the FireHub platform, allowing end users to search, view and retrieve (1) the annual BSM records at a fully detailed scale, (2) a single map layer depicting the areas affected for the last 30 years, (3) the number of times a certain area has been affected by fires, and (4) information and statistics on the impact of forest fires on the natural and built environment at the prefecture, regional, and country levels.

Figure 6.7 shows cases of BSMs and damage assessments derived for selected wildfires in Greece during recent years.

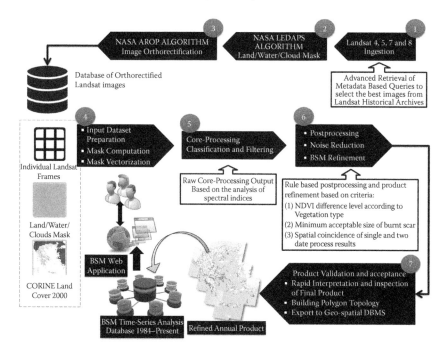

FIGURE 6.6
BSM-NOA algorithmic workflow.

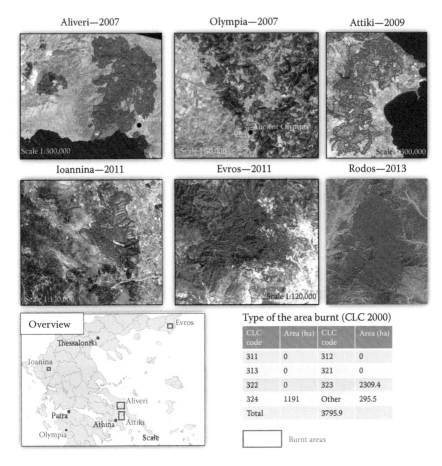

FIGURE 6.7
Selected cases of BSM-NOA mapping and damage assessments over Greece.

Evaluation of Fire Products and Services

Real-Time Fire Monitoring Service

During the summer of 2013, active fire occurrence data were collected in order to validate the fire detection process. The data attributes taken under consideration were as follows: (1) fire locations (at the commune level), (2) ignition time (first alarm), (3) time of first intervention by the fire brigade, (4) burnt area type (forest and nonforest), and (5) burnt area in hectares per area type. These data were found in reports provided by the Hellenic Fire Brigade on a daily basis. The information regarding the area burnt and the burnt area type were *ad hoc* estimations provided by firefighters, submitted during or after suppression of the active fire event. For cross-validation purposes, additional information from the local media was taken under consideration. The evaluation process entailed the matching of active fire events recorded in the reports with the data returned by the hotspot detection algorithm delivered every 5 minutes.

The approach focuses on evaluating the fire detection accuracy of the algorithm as a percentage of the events obtained from the fire brigade that were also successfully detected by the algorithm. Additionally, the capability of the algorithm to provide early warnings about fire events was investigated, by comparing the ignition time provided by the algorithm with the ignition time provided by the fire brigade. To estimate the algorithm's performance concerning the fire's location in real time, the active hotspots returned by the algorithm within the 500 × 500 m wide cells were compared with the BSM polygons, the latter referring to the entire fire season over Greek territory. The BSM polygons used as reference were generated from the full Landsat TM data set acquired over Greece immediately after the end of the fire season 2013. Due to the much higher spatial resolution of the Landsat TM data compared to the MSG/SEVIRI-based fire detections, the derived BSM polygons were considered an ideal validation data set to assess the fire detection algorithm's robustness. Therefore, the accuracy of the MSG/SEVIRI active fire detection was assessed through the estimation of the commission error, that is, returned as hotspots by the MSG/SEVIRI processing chain but not included in the BSM polygons, and the omission error, that is, the number of hotspot locations from the reports included in the BSM polygon but not returned by the MSG/SEVIRI-based algorithm outputs.

Table 6.1 presents a comparison between the total BSM area and the area that was successfully detected by the real-time MSG/SEVIRI-based wildfire detection algorithm. Ninety-three percent of the burnt areas' surface mapped over the entire country matches with the returned MSG/SEVIRI-based fire polygons detected, and only a percentage of 7% of the burnt areas was missed.

An additional criterion for assessing the algorithm's accuracy is shown in Table 6.2. This table represents the number of fire events reported by the fire brigade log files (Column 2 of Table 6.2) that were successfully matched with the returned active fire detections (Column 3 of Table 6.2) in relation to the size of the affected area.

TABLE 6.1

Comparison of the BSM Total Area, with the BSM Area Detected by the MSG/SEVIRI Wildfire Detection Algorithm at National Scale

	Area Size (ha)
Nationwide *BSM* area as mapped from the Landsat TM imagery (*BSM-NOA*)	20,100
Nationwide *BSM* area returned by the MSG/SEVIRI-based real-time detection algorithm	18,727 (93%)
Nationwide omitted *BSM* area not returned by the MSG/SEVIRI-based real-time detection algorithm	1,373 (7%)

TABLE 6.2

Comparison of the Reported Fire Events in the Fire
Brigades Log Files, with the Active Fire Events Returned
by the MSG/SEVIRI Forest Fire Detection Algorithm

Affected Area (ha)	Total Fire Events (Fire Service)	Reported by Fire Brigades and Matched by MSG/SEVIRI	Level of Matching (%)
0–50	57	18	32
50–100	12	9	75
100–150	3	3	100
150–200	4	4	100
≥200	11	11	100
Summary	87	45	52

Smaller size fires, with active burning areas of less than 50 ha, could not be detected by the system with adequate accuracy. This is because the fire radiation emitted is not at the required level to saturate the corresponding low spatial resolution MSG/SEVIRI pixel. Another reason relates to the fact that the fire detection system has an internal control mechanism, which returns a first fire occurrence only after it has been detected in two out of the three consecutive observations. Therefore, for small fires that are rapidly controlled by the firefighting mechanism, that is, within the first 15–30 minutes, there is not enough time for the system to confirm that its first detections match with the subsequent two to three observations. This control mechanism prevents the system from sending fire alarms to the fire brigades that are not certain.

Thus, for the above reasons the detection efficiency of the system for small fires is limited to the order of 32%. However, as shown in Table 6.2, larger fire events with sizes greater than 50 ha were adequately detected by the system with a level of accuracy ranging between 75% and 100%. There are two main reasons for the omitted detections: the first one relates to the presence of sparse clouds in the field of view of the sensor, while the second is because, for a few cases, the algorithm thresholds were not appropriate to detect the wildfire. However, lowering the thresholds would lead to increasing the false alarm rate; as expected, there is always a trade-off between false positives and false negatives.

To evaluate the capability of the real-time detection process to timely detect a fire event, the first fire alarm returned by the system was compared with the ignition time provided in the fire brigade log files. Table 6.3 summarizes the outcome of this validation. Out of the total of 45 fire events used for validation, 7 were first detected earlier than their announcement from the fire brigade control room. For the remaining events, 11 were detected with a delay of 0–15 minutes, 6 with a delay of 15–30 minutes, and 18 with a delay of 30–45 minutes.

TABLE 6.3

Comparison of the Time an Event Was First Detected by the Forest Fire Detection Algorithm with the Time Given by the Fire Brigades Service Report

Time Difference (minutes)	Number of Events
−15, 0	7
0, 15	11
15, 30	6
30, 45	18

BSM-NOA Service

The entire BSM-NOA service chain was extensively evaluated by subjecting it to a standardization procedure using several criteria (thematic accuracy, user support, sustainability of the means used, transferability, timeliness, etc.). The validation was done in the framework of the RISK-EOS/ESA/GSE and SAFER EC/GMES projects, which aimed to establish qualified and validated emergency response and emergency support services based on EO technology to meet the operational needs of the end-user communities. The validation experiments were internal in NOA and external from the Joint Research Centre, using high accuracy reference data over various European test sites (Greece, Portugal, Spain, and Corse). Scientific and technical validation of the product was carried out both in terms of vector data and map layout. The validation experiments compared the service BSM-NOA products against ground-truth data, the latter generated through dedicated *in situ* field campaigns. The surface accuracy figures are expressed in terms of detected area efficiency, skipped area rate (omission error), and false area rate (commission error). These accuracy figures were calculated on the basis of the following formulae:

$$\text{Detected area efficiency} = \frac{\text{DBA}}{\text{DBA} + \text{SBA}} \tag{6.6}$$

$$\text{Commission error (flase area rate)} = \frac{\text{FBA}}{\text{DBA} + \text{FBA}} \tag{6.7}$$

$$\text{Ommission error (skipped area rate)} = \frac{\text{SBA}}{\text{DBA} + \text{SBA}} \tag{6.8}$$

where DBA is the detected burnt area (common area between the generated burn scar polygon and the reference *in situ* polygon), FBA is the false burnt area (area included in the generated burn scar polygon but not in the reference *in situ* polygon), and SBA is the skipped burnt area (area included in the reference *in situ* polygon but not in the generated burn scar polygon).

The results of the validation experiments showed that the BSM-NOA service, and its subsequent evolution as a module of the NOA FireHub platform, is capable of processing images with different spectral and spatial resolutions and can effectively exploit data from different acquisition modes (uni- and multitemporal). In a multitemporal approach using a pair of Landsat 5 TM images, the method performed better than using a single-date image in identifying the postfire decrease of vegetation vigor and minimizing the spectral confusion of burnt areas with classes such as permanent crops, bare soil, shadows, urban fabric, and water. The minimum burnt area size detected is approximately 0.9–1.0 ha nonetheless, and the method performs well in delineating small fires of ~0.8–2.5 ha located in the alpine zones of the Mediterranean mountains.

The overall burnt area detection accuracy returned in the different evaluation experiments conducted reached levels of 85%–91%, with omission errors at the level of 9%–15% and commission errors as low as 6%–4% (Kontoes et al. 2013a). In fact, the service was qualified in the framework of the SAFER EC/GMES project—top of its class—as an end-to-end service for fire-related emergency support activities for integration into operational scenarios all over Europe. Figure 6.8 illustrates the BSM polygons for the Penteli Mt (2007)

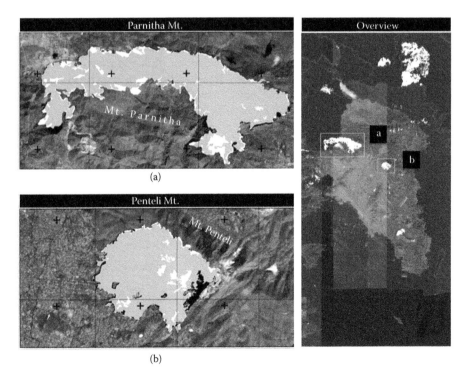

FIGURE 6.8
Burn scar mapping in the Mt. Parnitha and Penteli Mt. fires using the BSM-NOA method.

and Mount Parnitha (2007) fires in Greece, as well the omission (white areas) and the commission errors (black areas) reported with respect to the reference validation data used (Kontoes et al. 2009).

Discussion and Conclusions

The application of the MSG/SEVIRI active fire detection methodology over Greece has provided objective and accurate detections of wildfire spots with satisfactory accuracy on a 5-minute basis. The reported results provide insights into the method's flexibility, timeliness, and efficiency, especially when applied to very large areas that extend beyond the national scale. Moreover, the active fire maps generated when used in combination with highly accurate fuel maps can provide a useful overall fire situation awareness picture for the effective deployment of fire suppression resources and promoting evidence-based decision-making.

Based on the specific end-user demands, the fire detection algorithm was expanded further during the development of the FireHub platform operated by NOA's BEYOND Center of Excellence for EO-based monitoring and management of natural disaster. This patent-pending approach results in a subpixel approximation (500 × 500 m wide) of wildfire presence within the initial MSG/SEVIRI pixel. Several elements and information layers were taken under consideration to achieve such a level of precision. A sophisticated data fusion approximation is used, combining satellite fire detections with updated fuel data and fire spread models using real-time weather information. The hotspot detection methodology was assessed against reference information on real cases of fire events, with the conclusion of accurate fire estimation, with certain restrictions when it comes to small fire events.

According to the feedback received from the fire brigade and civil protection authorities, the FireHub approach with its enhanced spatial resolution is exceeding the EO-based real-time active hotspot detection standards established by the fire and forestry bodies for supporting actions relating to wildfire suppression management. Following the Copernicus (GMES) standards, the method is characterized by high flexibility and transferability; that is, it is applicable to other geographic areas in Europe, featuring an interactive approach for the definition and fine tuning of the spectral thresholds for active fire spot detection. Moreover, the dynamic integration of medium resolution satellite images that are acquired every 2–3 hours at the NOA reception stations, such as NOAA/AVHRR, MODIS, Suomi NPP, and MetOp, can update the hotspot products derived from the MSG/SEVIRI low spatial but high temporal resolution sensor, constituting a suitable and robust solution for operational active fire monitoring at the European, national, and regional levels.

This chapter also provided concrete evidence that the NOA's FireHub platform offers advanced burnt area mapping capabilities to meet rush and nonrush fire mapping needs for emergency response, emergency support, and recovery operations at the regional, national, or continental levels. It requires limited effort from an operator and returns higher mapping accuracies, compared to conventional mapping approaches (e.g., field surveys, aerial photo interpretation, GPS campaigns, etc.), as shown by Kontoes et al. (2009). The mapping accuracy of the developed remote sensing method was assessed in a very challenging environment, namely the accentuated relief and highly diversified ecosystems of the mountainous terrain of Greece. The approach proved highly sensitive in detecting burnt areas and avoiding spectral confusion with other classes such as bare soil, urban fabric, water, and permanent crops. Finally, the methodologies presented here, as well as the overall experience gained through several Copernicus (GMES) projects, suggest that the satellite-based mapping methods can certainly replace previous mapping methods, providing accuracies that exceed the end-user's operational requirements.

The NOA Web services (fire detection and monitoring, as well as BSM-NOA) and the generated products are delivered to institutional end users (e.g., Hellenic Fire Brigade, General Secretariat of Civil Protection, Copernicus EMS, Forestry Services, etc.) and are now part of the everyday decision-making processes of these institutions. As such, the FireHub platform was awarded first prize for Best Challenge Service in the Copernicus Masters Competition of 2014.

Acknowledgments

The research mentioned in this chapter and the relevant results were supported by the following European Community and ESA projects:

1. BEYOND—Building a Centre of Excellence for Earth Observation–Based Monitoring of Natural Disasters in Southeastern Europe, EC Seventh Framework Programme (FP7-REGPOT-2012-2013-1), Grant Agreement 316210.
2. SAFER—Services and Applications for Emergency Support. Program for Global Monitoring for Environment and Security (GMES)/EC/ FP7-2007-SPACE-1/GMES Collaborative Project. Grant Agreement 218802.
3. TELEIOS—Virtual Observatory Infrastructure for Earth Observation Data. Program FP7-ICT-2009-5 Collaborative Project. Grant Agreement 257662.

4. RISK-EOS Extension to Greece—Promotion of the GSE RISK-EOS fire services portfolio in Greece. Program for Global Monitoring for Environment and Security. Service Consolidation Actions of EarthWatch GMES Services Elements, ESA/GSE Program. Contract Notice: Ref. ITF-Aj-0043-08.

We are grateful to Dr Ioannis Mitsopoulos for his support and effort in reviewing the manuscript and providing valuable comments raising the scientific value of the chapter.

References

Amraoui, M., DaCamara, C., and J. Pereira. 2010. Detection and monitoring of African vegetation fires using MSG-SEVIRI imagery. *Remote Sensing of Environment* 114: 1038–1052.

Amraoui, M., Liberato, M., Calado, T., DaCamara, C., Coelho, L., Trigo, R., and C. Gouveia. 2013. Fire activity over Mediterranean Europe based on information from Meteosat-8. *Forest Ecology and Management* 294: 62–75.

Arino, O., Piccolini, I., Siegert, F., et al. 1999. Burn scars mapping methods, forest fire monitoring and mapping: A component of global observation of forest cover. In *Report of a Workshop*, F. Abern, J-M. Gregoire and C. Justice (eds.), Joint Research Centre, Ispra, Italy, pp. 198–223.

Bassi, S., Kettunen, M., Kampa, E., and S. Cavalieri. 2008. Forest fires: Causes and contributing factors in Europe; IP/A/ENVI/ST/2007-15 PE 401.003; Policy Department, Economic and Scientific Policy, European Parliament, Brussels, Belgium.

Brieb, K., Jahn, H., Ginati, A., and K. Sommer. 1996. A small satellite solution for the fire detection and monitoring from space. *Acta Astronautica* 39: 447–457.

Cahoon, D., Stocks, B., Levine, J., Cofer, W., and C. Chung. 1992. Evaluation of a technique for satellite-derived area estimation of forest fires. *Journal of Geophysical Research* 97: 3805–3814.

Calle, A., Casanova, J., and A. Romo. 2006. Fire detection and monitoring using MSG spinning enhanced visible and infrared imager (SEVIRI) data. *Journal of Geophysical Research* 111: G04S06.

Chuvieco, E., and M. Martin. 1994. A simple method for fire growth mapping using AVHRR channel 3 data. *International Journal of Remote Sensing* 15: 3141–3146.

Chuvieco, E., Martin, M. P., and A. Palacios. 2002. Assessment of different spectral indices in the red near-infrared spectral domain for burned land discrimination. *International Journal of Remote Sensing* 23(16): 5103–5110.

Cocke, A. E., Fule, P. Z., and J. E. Crouse. 2005. Comparison of burn severity assessments using differenced normalized burn ratio and ground data. *International Journal of Wildland Fire* 14(2): 189–198.

Collins, J. B., and C. E. Woodcock. 1996. An assessment of several linear change detection techniques for mapping forest mortality using multitemporal LandsatTM data. *Remote Sensing of Environment* 56: 66–77.

Diaz-Delgado, R., and X. Pons. 2001. Spatial patterns of forest fires in Catalonia (NE of Spain) along the period 1975–1995. Analysis of vegetation recovery after fire. *Forest Ecology and Management* 147: 67–74.

Dimitrakopoulos, A., and I. Mitsopoulos. 2005. Thematic report on forest fires in the Mediterranean Region. Forest Fire Management Working Paper 8, FAO. In *Global Forest Resources Assessment, 2005*, A.P. Vuorinen (ed.), 43 p.

Dimitrakopoulos, A., Vlahou, M., Anagnostopoulou, Ch., and I. Mitsopoulos. 2011. Impact of drought on wildland fires in Greece; Implications of climatic change? *Climate Change* 109: 331–347.

Elmore, A. J., Mustard, J. F., Manning, S. J., and D. B. Lobell. 2000. Quantifying vegetation change in semiarid environments: Precision and accuracy of spectral mixture analysis and the normalized difference vegetation index. *Remote Sensing of Environment* 73: 87–102.

Finney, M. 2002. Fire growth using minimum travel time methods. *Canadian Journal of Forest Research* 32: 1420–1424.

Finney, M. 2006. An overview of FlamMap modeling capabilities. In *Fuels Management - How to Measure Success: Conference Proceedings*, USDA, Forest Service, Rocky Mountain Research Station, General Technical Report, RMRS-P-41, P. Andrews, and B. Butler (eds.), pp. 213–219.

Fisher, R., Vigilante, T., Yates, C., and J. Russel-Smith. 2003. Patterns of landscape fire and predicted vegetation response in the North Kimberley region of Western Australia. *International Journal of Wildland Fire* 12(3–4): 369–379.

Fraser, R. H., and Z. Li. 2002. Estimating fire related parameters in boreal forests using SPOT VEGETATION. *Remote Sensing of Environment* 82: 95–110.

Fung, T., and C. Y. Jim. 1998. Assessing and modelling hill fire impact in country parks with SPOT HRV images and GIS. *Geocarto International* 13(1): 47–58.

Gao, F., Masek, J., and R. E. Wolfe. 2009. Automated registration and orthorectification package for Landsat and Landsat-like data processing. *Journal of Applied Remote Sensing* 3(1): 033515.

Giannakopoulos, C., Le Sager, P., Bindi, M., Moriondo, M., Kostopoulou, E., and C. Goodess. 2009. Climatic changes and associated impacts in the Mediterranean resulting from a 2° C global warming. *Global Planetary Change* 68: 209–224.

Giglio, L., Descloitres, J., Justice, C., and Y. Kaufman, 2003. An enhanced contextual fire detection algorithm for MODIS. *Remote Sensing of Environment* 87: 273–282.

Gong, P., Pu, R., Li, Z., Scarborough, J., Clinton, N., and L. M. Levien. 2006. An integrated approach to wildland fire mapping of California, USA using NOAA/AVHRR data. *Photogrammetric Engineering and Remote Sensing* 72(2): 139–150.

Justice, C., Giglio, L., Korontzi, S., et al. 2002. The MODIS fire product. *Remote Sensing of Environment* 83: 244–262.

Kasischke, E., and H. French. 1995. Locating and estimating the areal extent of wildfires in Alaskan Boreal Forest, using multiple season AVHRR NDVI composite data. *Remote Sensing of Environment* 51: 263–275.

Kasischke, E., French, H., Harrell, P., Christensen, N., Ustin, S., and D. Barry. 1993. Monitoring of wildfires in Boreal forests using large area AVHRR NDVI composite image data. *Remote Sensing of Environment* 45: 61–71.

Kaufman, J., Justice, C., Flynn, L., et al. 1998. Potential global fire monitoring from EOS-MODIS. *Journal of Geophysical Research* 103: 32215–32238.

Kauth, R., and G. Thomas. 1976. The Tasseled Cap—A graphic description of the spectral–temporal development of agricultural crops as seen by Landsat.

Proceedings of the Symposium on Machine Processing of Remotely Sensed Data, Purdue University, West Lafayette, IN, pp. 41–51.

Key, C., and N. C. Benson. 2003. *The Normalised Burn Ratio (NBR): A Landsat TM radiometric measure for burn severity.* http://www.nrmsc.usgs.gov/research/ndbr.html

Kontoes, C., Keramitsoglou, I., Papoutsis, I., Sifakis, N., and P. Xofis. 2013a. National scale operational mapping of burnt areas as a tool for the effective development of wildfire management strategy. *Sensors* 18: 11146–11166.

Kontoes, C. C., Poilvé, H., Florsch, G., Keramitsoglou, I., and S. Paralikidis. 2009. A comparative analysis of a fixed thresholding vs. A classification tree approach for operational burn scar detection and mapping. *International Journal of Applied Earth Observation and Geoinformation* 11(5): 299–316. ISSN 0303 2434, DOI: 10.1016/j.jag.2009.04.00.

Koubarakis, M., Kyzirakos, K., Karpathiotakis, M., et al. 2012. TELEIOS: A database-powered virtual earth observatory. In *Proceedings of International Conference on Very Large Data Bases 2012, 38th International Conference on Very Large Databases (VLDB 2012).*

Koutsias, N. 2000. Burned area mapping using logistic regression modelling of a single post-fire Landsat-5 Thematic Mapper image. *International Journal of Remote Sensing* 21(4): 673–687.

Koutsias, N., and M. Karteris. 2000. Burned area mapping using logistic regression modeling of a single post-fire Landsat-5 Thematic Mapper image. *International Journal of Remote Sensing* 21(4): 673–687.

Koutsias, N., Xanthopoulos, G., Founda, D., et al. 2013. On the relationships between forest fires and weather conditions in Greece from long-term national observations (1894–2010). *International Journal of Wildland Fire* 22: 493–507.

Kyzirakos, K., Karpathiotakis, M., and G. Garbis. 2014. Wildfire monitoring using satellite images, ontologies and linked geospatial data. *Journal of Web Semantics* 24: 18–26, DOI: 10.1016/j.websem.2013.12.002.

Kyzirakos, K., Karpathiotakis, M., Garbis G., et al. 2012. Real Time fire monitoring using semantic web and linked data technologies. *Presented in the 11th International Semantic Web Conference.* Boston, MA. November 11–15, 2012.

Laneve, G., Castronuovo, M., and G. Cadau. 2006. Continuous monitoring of forest fires in Mediterranean Area using MSG. *IEEE Transactions on Geoscience and Remote Sensing* 44: 2761–2768.

Lasaponara, R. 2006. Estimating spectral separability of satellite derived parameters for burned areas mapping in the Calabria region by using SPOT-Vegetation data. *Ecological Modelling* 196(1–2): 265–270.

Leblon, B., Alexander, M., Chen, J., and S. White. 2001. Monitoring fire danger of Northern Boreal Forests with NOAA-AVHRR NDVI images. *International Journal of Remote Sensing* 22(14): 2839–2846.

Li, Z., Cilar, J., Moreau, L., Huang, F., and B. Lee. 1997. Monitoring fire activities in the Boreal ecosystem. *Journal of Geophysical Research* 102: 29611–29624.

Li, Z., Kaufman, Y. J., Ichku, C., Fraser, R., Trishchekno, A., Giglio, L., Jin, J., and X. Yu. 2001. A review of AVHRR-based active fire detection algorithms: Principles, limitations, and recommendations. In *Global and Regional Vegetation Fire Monitoring from Space: Planning and Coordinated International Effort.* A. Abern., J. G. Goldammer and C. Justice (eds.), SPB Academic Pub., The Hague, The Nederlands, pp. 199–225.

Li, Z., Nadon, S., and J. Cihlar. 2000. Satellite detection of Canadian boreal forest fires: Development and application of the algorithm. *International Journal of Remote Sensing* 21: 3057–3069.

Lyon, G. J., Ding, Y., Lunetta, R. S., and C. D. Elvidge. 1998. A change detection experiment using vegetation indices. *Photogrammetric Engineering and Remote Sensing* 64(2): 143–150.

Marsh, S., Walsh, J., Lee, C., Beck, L., and C. Hutchinson. 1992. Comparison of multi-temporal NOAA-AVHRR and SPOT XS satellite data for mapping land cover dynamics in the West African Sahel. *International Journal of Remote Sensing* 13: 2997–3016.

Martin, M. P., and E. Chuvieco. 1995. Mapping and evaluation of burnt land from multi-temporal analysis of AVHRR NDVI images. *EARSEL Advance Remote Sensing* 4: 7–13.

Miller, J. D., and S. R. Yool. 2002. Mapping forest post-fire canopy consumption in several overstory types using multi-temporal Landsat TM and ETM data. *Remote Sensing of Environment* 82: 481–496.

Pennypacker, C., Jakubowski, M., Kelly, M., et al. 2013. FUEGO—Fire Urgency Estimator in Geosynchronous Orbit—A proposed early-warning fire detection system. *Remote Sensing* 5: 5173–5192.

Pereira, J. M. C. 1999. A comparative evaluation of NOAA/AVHRR vegetation indexes for burned surface detection and mapping. *IEEE Transactions on Geoscience and Remote Sensing* 37: 217–226.

Prins, E. M., and W. P. Menzel. 1996. *Investigation of biomass burning and aerosol loading and transport utilizing geostationary satellite data.* Biomass and Global Change, MIT Press, Cambridge, MA, pp. 67–72.

Prins, E. M., Schmidt, C. C., Feltz, J. M., Reid, J. S., Wesphal, D.L., and Richardson, K. 2003. A two year analysis of fire activity in the Western Hemisphere as observed with the GOES Wildfire Automated Biomass Burning Algorithm. *Preprints, 12th Conf. on Satellite Meteorology and Oceanography, Long Beach, CA, Amer. Meteor. Soc.*, CD-ROM, p. 2.28

Pu, R., and P. Gong. 2004. Determination of burnt scars using logistic regression and neural network techniques from single post-fire Landsat 7 ETM+ image. *Photogrammetric Engineering and Remote Sensing* 70(7): 841–850.

Quintano, C., Fernandez-Manso, A., Fernandez-Manso, O., and Y.E. Shimabukuro. 2006. Mapping burned areas in Mediterranean countries using spectral mixture analysis from a uni-temporal perspective. *International Journal of Remote Sensing* 27(4): 645–662.

Roberts, G., and M. Wooster. 2008. Fire detection and fire characterization over Africa using Meteosat SEVIRI. *IEEE Transactions on Geoscience and Remote Sensing* 46: 1200–1218.

Rogan, J., and S. Yool. 2001. Mapping fire induced vegetation depletion in the Peloncillo Mountains, Arizona and New Mexico. *International Journal of Remote Sensing* 22(16): 3101–3121.

Rouse, J.W., Haas, R.H., Schell, J.A., and Deering, D.W. 1974. Monitoring vegetation systems in the Great Plains with ERTS. In: S.C. Freden, E.P. Mercanti, and M. Becker (eds) *Third Earth Resources Technology Satellite–1 Syposium. Volume I: Technical Presentations*, NASA SP-351, NASA, Washington, DC, pp. 309–317.

Roy, D. R., Boscheti, L., and S. N. Trigg. 2006. Remote sensing of fire severity: Assessing the performance of the normalized burn ratio. *IEEE Geoscience and Remote Sensing Letters* 3(1): 112–116.

Sa, A. C. L., Pereira, J. M. C., Vasconcellos, M. J. P., Silva, J. M. N., Ribeiro, N., and A. Awasse. 2003. Assessing the feasibility of sub-pixel burned area mapping in miombo woodlands of northern Mozambique using MODIS imager. *International Journal of Remote Sensing* 24: 1783–1796.

San-Miguel-Ayanz, J., Ravail, N., Kelha, V., and A. Oller. 2005. Active fire detection for fire emergency management: Potential and limitations for the operational use of remote sensing. *Natural Hazards* 35: 361–376.

Saunders, R. W. 1990. The determination of broad band surface albedo from AVHRR visible and near-infrared radiances. *International Journal of Remote Sensing* 11(1): 49–67.

Sifakis, N., Iossifidis, C., Kontoes, C., and I. Keramitsoglou. 2011. Wildfire detection and tracking over Greece using MSG-SEVIRI satellite data. *Remote Sensing* 3: 524–538.

Sifakis N., Paronis D., and I. Keramitsoglou. 2004. An application of AVHRR imagery used in combination with CORINE Land Cover data for forest-fire observations and consequences assessment. *International Journal of Applied Earth Observation and Geoinformation* 5(4): 263–274.

Simard, M., Saatchi, S. S., and G. De Grandi. 2000. The use of decision tree and multiscale texture for classification of JERS-1 SAR data over tropical forest. *IEEE Transactions on Geoscience and Remote Sensing* 38 (5): 2310–2321.

Steber, M., Allen, A., James, B., and K. Koss. 2012. Enhancing the capabilities of VLandgate's FireWatch with fire spread simulation. In *Proceedings of Bushfire CRC & AFAC 2012 Conference Research Forum Perth Australia*, R. Thornton, and L. Wright (eds.), Bushfire Cooperative Research Center, Australia, pp. 115–123.

Tappan, G., Tyler, D., Wehde, M., and D. Moore. 1992. Monitoring rangeland dynamics in Senegal with advanced very high resolution radiometer data. *Geocarto Internationala* 1: 87–98.

Umamaheshwaran, R., Bijker, W., and A. Stein. 2007. Image mining for modeling of forest fires from Meteosat images. *IEEE Transactions on Geoscience and Remote Sensing* 45: 246–253.

Ustin, S. L. 2004. *Manual of Remote Sensing*, vol. 4. Remote sensing for natural resource management and environmental monitoring, Wiley, NJ.

Vafeidis, A. T., and N. A. Drake. 2005. A two-step method for estimating the extent of burnt areas with the use of coarse-resolution data. *International Journal of Remote Sensing* 26(11): 2441–2459.

Van den Bergh, F., and P. Frost. 2005. A multi temporal approach to fire detection using MSG data. In *Proceedings of International Workshop on the Analysis of Multi-Temporal Remote Sensing Images*, Biloxi, MS, pp. 156–160.

7

Geomorphometry and Mountain Geodynamics: Issues of Scale and Complexity

Michael P. Bishop and Iliyana D. Dobreva

CONTENTS

Introduction

Mountain geodynamics refers to multiple aspects of mountain building that include relief production and polygenetic landscape evolution caused by complex interactions of atmospheric, surface, and tectonic processes (Montgomery 1994; Koons 1995; Burbank et al. 1996; Bishop and Shroder 2000; Zeitler et al. 2001; Bishop et al. 2003, 2012). Numerous pathways and feedback mechanisms exist, as the topography partially governs climate, surface processes, and rock strength and uplift (Koons et al. 2002; Bishop et al. 2010). Researchers have long recognized that topography represents the structural manifestation of mountain geodynamics and have therefore focused on the analysis of digital elevation models (DEMs) to extract information about landforms, erosion and deposition, lithology, uplift and relief production, tectonic zones, and the nature of polygenetic landscape evolution. Spatial and morphometric information is vital for understanding the role of specific surface processes in mountain geodynamics (e.g., mass movement, fluvial erosion, and glaciation) and for investigating the controversies associated with climatic *versus* tectonic forcing (Raymo and Ruddiman 1992; Montgomery 1994; Snyder et al. 2000; Montgomery et al. 2001; Reiners et al. 2003; Jamieson et al. 2004). Furthermore, such information is critical for understanding the alpine critical zone and its ability to support and sustain ecosystems and resources in these rapidly changing environments (Baudo et al. 2007; Bishop et al. 2015).

The difficult topic of scale represents a central theme in understanding geomorphological systems (Bishop et al. 2012). Earth scientists have long recognized scale issues associated with (1) land-cover mapping, land-use change modeling, and topographic complexity (Bishop and Shroder 2000; Tate and Wood 2001; Veldkamp et al. 2001; Weng 2014); (2) the operational spatial scale dependencies of mass movements, fluvial erosion, glaciation, and uplift (Bishop et al. 1998, 2003, 2012; Napieralski et al. 2007); and (3) spatiotemporal scale dependencies associated with rates of processes and system couplings (Jantz and Goetz 2005). Scale linkages associated with parameters, processes, feedback mechanisms, and system couplings are thought to govern the complexity of mountain geodynamics and may help to explain the concepts of instability, chaos, self-similarity, steady states, and equilibrium (Willett et al. 2001; Phillips 2004). Consequently, research has focused on examining scale issues and the topics of scale-dependent parameter variation, topographic complexity, and process–form relationships in landscape evolution modeling (e.g., Beven 1995; Roy et al. 2014).

The topic of scale has received considerable attention in numerous disciplines including landscape ecology, geography, geology, remote sensing, geographic information science, geomorphometry, and hydrology. Numerous books and articles have addressed the topic from multiple perspectives that include philosophical, conceptual, representation, statistical analysis, and numerical modeling treatments (e.g., Gallant and Hutchinson 1997; Quattrochi and Goodchild 1997, 2011; Chou et al. 1999; Peuquet 2002; Bishop et al. 2003, 2012; Sheppard and McMaster 2004). Nevertheless, there do not appear to be comprehensive definitions or theories of *scale*, and quantitative treatments of scale concepts are generally lacking (Goodchild 2011). Furthermore, there are a multitude of scale concepts that appear to be highly interrelated, and some are not interchangeable (Dungan et al. 2002).

It is very plausible that mathematical formalization of spatial concepts may be required as criteria for the formalization of others. For example, distance and direction are needed to define concepts of homogeneity and heterogeneity (i.e., texture), as they ultimately govern the localized nature of spatial variability. Many spatial concepts, however, have not been formalized (e.g., indeterminate boundaries). In general, researchers have identified different aspects of scale that include cartographic, geographical, measurement, operational, and computational perspectives (Quattrochi and Goodchild 1997; Bishop et al. 2012). Nevertheless, there seems to be an overemphasis on (1) spatial concepts *versus* temporal concepts; (2) raster data representation and patterns *versus* spatial organization and morphography (i.e., mapped description of landform features); (3) scale-dependent analysis and modeling *versus* formalization of scale concepts and the integration of multiscale landscape information; and (4) pattern recognition and spatial mapping *versus* operational scale dependencies and the understanding of scale linkages with respect to process and structure that governs the spatiotemporal dynamics of phenomena.

While scale issues are a central theme in understanding mountain geodynamics and system complexity, it is essential that we characterize spatiotemporal variation in relation to the magnitude of parameters, spatial structure and organization of the landscape, rates of processes, and the collective influence of scale-dependent parameters, processes, and systems that govern its response to forcings (Grimm et al. 2008). We should not expect that an empirical treatment of geospatial data can effectively address scale linkages, dependencies, and disconnects related to complex system dynamics, given the range of spatial and temporal frequencies of variation associated with climate, landscape, and geological/tectonic systems. Therefore, we need to critically examine a multitude of issues and perspectives related to scale, so that we can begin to understand and formalize the concepts of scale in a meaningful and quantitative way.

Consequently, our objective is to identify, characterize, and discuss various concepts of scale in order to address the notoriously difficult issues associated with understanding surface processes, topographic complexity, and mountain geodynamics. Specifically, we examine scale and complexity from numerous perspectives including space–time representation, spatial extent and parameter magnitude, anisotropy, structure and spatial organization, semantic modeling, indeterminate boundaries, and spatiotemporal dynamics. Our conceptual and quantitative treatments of these concepts of scale demonstrate the importance of morphometric information (i.e., quantitative description of the topography), its linkage to process and polygenetic (i.e., multiple processes) topographic evolution, and the significance of a systems perspective that highlights the role of topography in characterizing the spatiotemporal scale dependencies of various aspects of mountain geodynamics.

Background

Geomorphometry

Terrain information is critical for studying mountain geodynamics, as topography represents the integration of atmospheric, surface, and geological processes (Bishop et al. 2010; Wilson and Bishop 2013). Remote sensing has long been used to generate topographic information *via* the collection of stereoscopic photography and imagery, radar imagery, and lidar data. Stereophotogrammetric methods have been routinely used to generate DEMs from Satellite Pour l'Observation de la Terre (SPOT) and Advanced Spaceborne Thermal Emission and Reflection Radiometer (ASTER) data. Radar imagery such as the Shuttle Radar Mapping Mission (SRTM) and the TerraSAR-X can be used to assess global topographic conditions and assess deformation characteristics of the landscape, given earthquakes and human-induced subsidence (Hensley and Farr 2013). More recently, high-resolution lidar systems and terrestrial laser scanners now generate millions of 3D point measurements. The "point clouds" must be processed to produce versions of the Earth's surface that need to be semantically defined and mapped. In general, DEMs enable the quantification of the Earth's surface (geomorphometry) to support investigations of mountain geodynamics.

Geomorphometry plays a central role in a variety of academic disciplines and is essential for studying surface processes and mapping and characterizing mountain environments. Geomorphometry specifically addresses the issues of (1) sampling attributes of the land surface; (2) geodesy and digital terrain modeling (DTM); (3) DEM preprocessing and error assessment; (4) generation of land-surface parameters, indices, and objects; and (5) terrain information production and problem-solving using parameters and objects. As indicated by Hengl and Reuter (2009) and Bishop et al. (2012), each aspect of geomorphometry represents a separate subdiscipline that contributes to the development of software tools and geospatial technology. This point is especially important, as most geographic information systems (GISs) do not have adequate software for robust analysis of DEMs (e.g., limitations with respect to land-surface metrics, addressing anisotropy, and production of land-surface objects). Ultimately, geomorphometry is a rapidly evolving field that can contribute towards understanding of surface and subsurface processes, process–form relationships, and the functionality of terrain units that govern matter and energy fluxes and resource availability (Hengl and Reuter 2009; Wilson and Bishop 2013).

Remote sensing technologies serve as the foundation for collecting topographic information at a multitude of spatial and temporal scales. Change detection studies require temporal data that adequately sample the surface given variations in rate of processes. Appropriate scale-dependent coverage is usually lacking given that temporal operational-scale dependencies can exist at a higher frequency than repeat spatial coverage allows. Engineering-related studies involving robotics, dynamic spatiotemporal coverage of the terrain, and unmanned aerial systems are rapidly changing our ability to better collect data from rapidly changing landscapes (Ramanathan 2007; Firpo et al. 2011; Niethammer et al. 2012; Immerzeel et al. 2014).

DTM represents a fundamental part of geomorphometry. Currently, most DEMs constitute a 2.5D representation of the terrain. True 3D analysis of the terrain will require accounting for variations in the geoid. Nevertheless, DTM can be very complicated, and users must determine which concept of land surface they wish to represent. Error and uncertainty assessment is another important part of DTM, and great care must be exercised when comparing multitemporal DEMs generated from different data sources (Paul et al. 2004; Nuth and Kääb 2011). Most DEMs are known to contain a variety of nonsystematic and systematic errors due to positional accuracy, sensor noise, and preprocessing and algorithm errors (Eckert et al. 2005; Höhle and Höhle 2009; Nuth and Kääb 2011).

The development and evaluation of geomorphometric parameters and objects is an active research area (Hengl and Reuter 2009). First and second derivatives characterize the fundamental properties of the topography including slope gradient, slope orientation, and surface curvature. These properties partially regulate many surface processes and control the direction and velocities of mass fluxes over the landscape (i.e., water and sediment transport). These and many other scale-dependent parameters also govern surface energy fluxes, surface temperature, evapotranspiration, ablation, and meltwater production. Such parameters are used to characterize and map the functional and scale-dependent organization of the landscape that includes topography, geological structure, and spatial constraints for surface processes, serving as a basis for mapping soils, landforms, and other landscape units.

Geomorphometric parameters and objects are also essential inputs into GIS-based simulations of the surface energy budget, soil erosion, plant dispersion, and landscape evolution (Pike 2000; Haboudane et al. 2002; Dorsaz et al. 2013). Numerical simulations that link process and form can improve our understanding of complex systems related to geochronology, process mechanics, landscape mapping, resource assessment and inventory, and many other scientific and applied problems. Nevertheless, we must go beyond GIS-based empiricism and address the complex issues of scale (Bishop et al. 2012).

Scale Concepts

The topic of scale has been recognized as fundamental in the mapping and Earth sciences (Bishop and Shroder 2004; Goodchild 2011; Hensley and Farr 2013). The treatment of space and time is inherently associated with concepts of scale, although disciplinary treatments of scale-dependent phenomena limit our comprehensive treatment of the topic. Consequently, there are issues associated with semantics, concept usage, representation, and formalization. Much has been written on absolute *versus* relative space, and most scientists do not typically incorporate quantitative spatiotemporal topological information, although qualitative descriptions of space and time concepts are widely utilized in the Earth sciences (Bishop and Shroder 2004; Bishop et al. 2012). Time appears to be more effectively addressed in finite-element numerical models (process perspective), although recently researchers have begun to explore space–time trajectories and incorporate spatial context and temporal concepts into GIS-based analysis (Wu 1999; Shaw et al. 2008).

The dominant perspectives of scale that have been thoroughly discussed in the literature include the following: (1) *cartographic scale*, which defines map spatial coverage and information detail; (2) *geographic scale*, which represents the spatial extent of a study area or the geographic size and extent of features; (3) *measurement scale*, which can refer to the scale of data collection and representation over space and time or to the nature of quantitative representation of information; (4) *operational scale*, which refers to the scale at which processes operate over space and time; and (5) *computational scale*, which represents the spatial and temporal constraints by which data are analyzed to produce spatial or temporal information. These generalized perspectives have served to formalize scale issues; however, they do not adequately address various concepts of scale.

A multitude of concepts have been brought forward in an attempt to understand and formalize scale issues in the mapping sciences. The most widely cited concepts include distance, extent, direction, homogeneity, heterogeneity, self-similarity, scale dependence, texture, and autocorrelation, although there are many others (Table 7.1). There has been a significant effort to attempt to quantify and model these scale concepts using metrics or geospatial statistical analysis techniques including texture features, fractal dimension, semivariogram analysis, wavelet analysis, and pattern metrics (Bishop et al. 1998, 2003; Wu et al. 2000). Nevertheless, a rigorous examination of concepts like *extent* and *direction* clearly reveals interdependencies and semantic dependencies, such that *spatial extent* depends upon a meaningful semantic definition that may need to account for homogeneity, heterogeneity,

TABLE 7.1

Spatial Scale Concepts

Adjacency	Gradient
Affinity	Heterogeneity
Anisotropy	Homogeneity
Autocorrelation	Neighborhood
Complexity	Networking
Concentration	Patterns
Connectivity	Region
Convergence	Scale dependence
Context	Scale independence
Density	Self-similarity
Diffusion	Shape
Direction	Spatial organization
Distance	Spatial variation
Distribution	Structure
Entities	Texture
Entropy	Topology
Extent	Zones

autocorrelation, or other concepts of scale. In addition, thought experiments also reveal that spatial extent depends upon the theme or phenomena under investigation, such that scale dependencies vary for different parameters given the same semantic definition. In addition, scale-dependent processes are thought to define scale-dependent spatial structure, which may or may not be linked to the scale of variation associated with topographic parameters, given different processes or different sets of processes. Such is the case with mass movements, fluvial erosion, glaciation, and polygenetic landscape evolution in mountain environments. Furthermore, the *extent* of scale linkages may define the spatial complexity and organization of mountain systems, although the boundaries of process regimes and coupled systems are most likely indeterminate and change rapidly given high-magnitude rates of processes and nonlinear variation over time. Consequently, there is a need for a theoretical framework that incorporates scale concepts to permit formalization of mountain spatiotemporal dynamics. Any formalization must account for scale issues related to representation, spatiotemporal variation and parameters, spatial organization, process regimes, feedback mechanisms, systems coupling, and spatiotemporal variation of systems. This can be examined from an individual concept perspective, but ultimately scale concepts must be integrated and formalized to evaluate and understand scale issues, principles, and/or hypotheses discussed in the literature.

If one assumes that there may be principles of scale governing the framework of space–time and phenomena, new hypotheses and formalizations are required. For example, does an increase in area always increase the magnitude of a landscape variable? Are there different scale ranges that confine the magnitude, and is this constant across space and time or can it vary everywhere in space–time? We will demonstrate that the answer is sometimes yes and sometimes no, depending on the variable and the complexity of the landscape. It is also clear that it is possible for no significant magnitude variation to occur spatially (i.e., slope angle) depending upon the operational scale dependencies of coupled systems. Such is the case with threshold slopes due to active river incision and uplift, where there is a spatial constraint or limit in terms of computing the slope angle in a meaningful way (e.g., hillslopes). Other principles may relate to spatial extent and the number of variables that are required for modeling, multiscale interaction and emergent system dynamics, multiscale distance decay and boundaries, spatial organization and matter and energy flow, structural and process linkages, scale-change and system dynamics, and determination of fundamental levels of investigation. It may be that scale is a unifying concept for addressing process–form relationships and studying the polygenetic nature of mountain geodynamics. Formalizing scale concepts may also be the only way to examine and represent indeterminate boundaries and zones associated with active deformation, uplift, erosion, and relief production.

From another point of view, scale and complexity appear to be somewhat interrelated. As with scale, there may be dimensions of complexity related to the structural and functional aspects of the topography and mountain systems. It should be noted, however, that complexity has not been rigorously defined, and most researchers have used simplistic metrics to characterize the concept. These include the fractal dimension, the surface area–planimetric area ratio, shape indices, and pattern and diversity metrics and parameters (Bishop et al. 1998, 2003; Wu et al. 2000; Lausch and Herzog 2002; Uuemaa et al. 2005). Complexity is a concept that has been used extensively in the mapping and Earth science literature, although its use and description is mostly experience-based (Ode et al. 2010). There are also practical applications that must account for terrain complexity. For example, Band (1989) addressed the topic of spatial aggregation of complex terrain to select scales for distributed hydrological modeling.

Although most mountain systems have been described as being complex, complexity can be described spatially in terms of diversity in elemental forms and landscape elements, spatial organization of patterns, and variations in the shape of elements and spatial entities. This includes concepts of spatial density, shape, and spatial topological relationships. Nevertheless, which concept of complexity or scale does any particular metric represent? Most metrics are not specifically designed or computed to formalize concepts of complexity and scale. Furthermore, it is important to differentiate the spatial complexity of a geomorphometric parameter *versus* spatial variation in rates of processes or coupled systems, *versus* the complexity of system interactions *via* variables and their scale dependencies and linkages. Similarly, what is the degree of coherence associated with variables, spatial structure, and emergent dynamics? Do scale and complexity represent the notion of the number of scale-dependent variables that are required to adequately simulate space–time dynamics, or is spatial topological variation a form of spatial complexity? Is there a difference between the geometric and topological complexity of the topography? Does the multiscale coupling of climate, surface, and geological processes create a spatial structural fabric in the topography that can provide a framework for formalizing scale and geomorphological systems? These and other scientific questions require us to quantitatively formalize concepts of scale.

Given our lack of theory about scale and complexity, it is difficult to establish a framework that formalizes the long list of concepts and hypotheses about scale, scale changes, and scale linkages. Fractal analysis and geostatistics have been used extensively in order to better understand the nature of topographic complexity and the resulting land cover, erosion, and relief production patterns in mountain environments (Bunyavejchewin et al. 1998; Kellogg et al. 2008; Ibanez et al. 2014). Nevertheless, most research represents methodological attempts to characterize concepts of spatial variation, self-similarity, scale dependencies, and process–form and process–pattern relationships tied to specific variables and features including vegetation, altitude, slope, relief, surface roughness, and elemental forms. Such approaches do not address a multitude of concepts and issues related to representation, parameters, anisotropy, morphodynamics, morphographics, rates of processes, space–time variation, and system coherence. We investigate and provide examples of these in the context of mountain geodynamics, to account for the various dimensions of spatiotemporal dynamics (Meentemeyer and Box 1987; Goodchild 2011).

Addressing Scale and Complexity

A multitude of scale and complexity issues have been presented in the literature; however, semantic modeling and quantitative treatments are just beginning to be discussed and evaluated from a scientific perspective (Dehn et al. 2001; Wu 2004; Drăguţ and Eisank 2011; Bishop et al. 2012). Although empirical and statistical approaches can be used to address spatial variation (e.g., Gallant and Hutchinson 1997; Chou et al. 1999), numerous issues need to be formalized in a meaningful way. Therefore, we discuss various issues and provide concrete examples.

Spatial Representation

A fundamental topic in geographic information science involves representation. It is now widely recognized that spatial phenomena can be represented using a discrete entity or object conceptualization, whereby spatial entities with discrete boundaries can be represented as points, lines, or polygons (vector data structure). This works well for some phenomena depending upon accuracy requirements, although complex phenomena in mountain environments usually require a field conceptualization, whereby entities and/or properties of the landscape are represented as tessellation fields (raster data structure). The field data model can also be used to account for indeterminate boundaries and complex spatial variation of phenomena using fuzzy membership (Burrough and Frank 1996; Fisher et al. 2005; Deng and Wilson 2007). Both approaches, however, have their advantages and disadvantages, and researchers are beginning to acknowledge that hybrid representations are required to address a multitude of issues (Cova and Goodchild 2002; Bishop et al. 2012).

According to Goodchild (2011), these conceptualizations appear to be independent of scale, although digital representations incorporate scale to some degree. In the raster case, scale is related to the size and shape of the tessellation and, in the vector case, scale is related to size, shape, and generalization associated with representing features and boundaries. Furthermore, most commercially implemented data structures do not address the issue of the third spatial dimension, as GIS-based representation, spatial analysis, and numerical modeling is essentially 2.5D. A 3D representation is required to formalize concepts of scale including distance, direction, extent, and curvature.

Given our dependence on projection systems for depicting spatial variation, projection distortion will determine the accuracy of distance, area, and shape computations. Earth scientists use DEMs (2.5D representation) to examine topographic variation over a wide range of spatial scales that frequently exceed distances of over 60 km (i.e., the need to study isostatic and tectonic uplift). We demonstrate how 3D spatial representation of mountain topography using geocentric coordinates over two mountain landscapes in Alaska and the Himalaya fundamentally govern concepts of scale. We also demonstrate that geography, or place, influences scale independent of using a 3D coordinate system, as undulations in the Earth's gravitational field vary with location, due to rock density and latitudinal mass variations caused by the shape of the Earth (Figures 7.1 and 7.2).

We compute the 3D geocentric coordinates (X, Y, Z) as follows (Iliffee and Lott 2008):

$$X = (\upsilon + h) \cos(\varphi) \cos(\lambda) \tag{7.1}$$

$$Y = (\upsilon + h) \cos(\varphi) \sin(\lambda) \tag{7.2}$$

$$Z = \{(1 - e^2)\upsilon + h\} \sin(\varphi) \tag{7.3}$$

where φ and λ are the geodetic coordinates, latitude, and longitude, respectively. The ellipsoidal height (h) is the height above the ellipsoid that accounts for the height above the geoid and the separation, and e is the eccentricity of the reference ellipsoid:

$$e^2 = \frac{a^2 + b^2}{a^2} \tag{7.4}$$

where a and b represent the semimajor and semiminor axes of the Earth, respectively, and υ is the radius of the curvature in the prime vertical, such that

$$\upsilon = \frac{a}{\left(1 - e^2 \sin^2(\varphi)\right)^{1/2}} \tag{7.5}$$

We then computed the difference in the cumulative distance along transects depicted in Figures 7.1 and 7.2, comparing the differences between 2.5D *versus* a 3D representation of the topography (Figure 7.3). We also computed the difference in the surface area as a function of spatial extent, given a 2.5D *versus* a 3D representation of the topography at a point centered within both regions (Figure 7.4). The same datum (WGS-84) and geoid (EGM96) were used for both 2.5D and 3D coordinates.

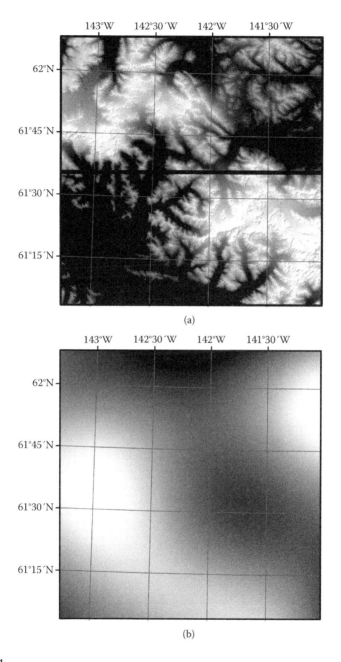

(a)

(b)

FIGURE 7.1
(a) Hill-shade-relief map of the Wrangell–St. Elias National Park and Preserve (WSENPP) in Alaska, derived from an ASTER GDEM version 2, digital elevation model. The horizontal transect was used to compute cumulative distance for 2.5-D and 3-D representations. (b) Geoid undulations are depicted using the Earth Gravitational Model (EGM) 1996, where separation values depicted as white (45.82 m) to black (45.41 m).

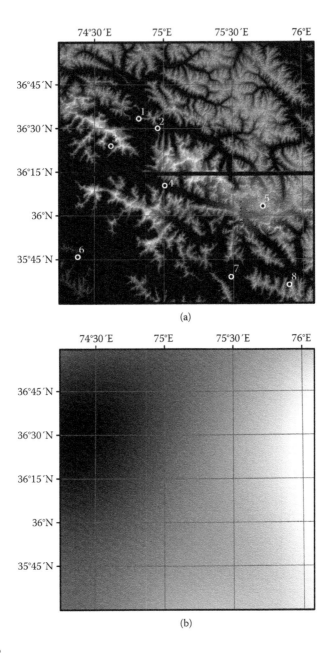

(a)

(b)

FIGURE 7.2
(a) Hill-shade-relief map of a portion of the Karakoram in Pakistan, derived from an ASTER GDEM version 2, digital elevation model. The horizontal transect was used to compute cumulative distance for 2.5-D and 3-D representations. Points depict the locations that were used for computing slope angle, slope azimuth angle, sky-view coefficient and surface roughness parameters. (b) Geoid undulations are depicted using Earth Gravitational Model (EGM) 1996, where separation values are depicted as black (−30.6 m) to white (28.6 m).

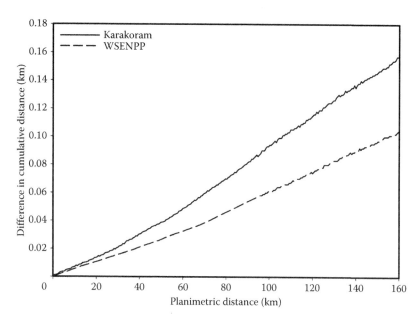

FIGURE 7.3
Difference in cumulative distance between 2.5D and 3D coordinate systems for transects across the Wrangell–St. Elias National Park and Preserve (WSENPP) (Figure 7.1) and the Karakoram region (Figure 7.2).

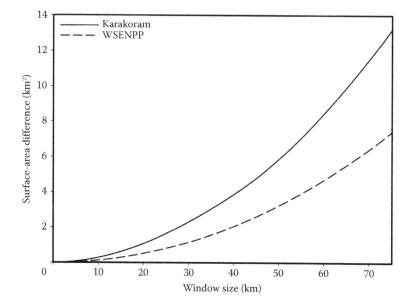

FIGURE 7.4
Difference in surface area using 2.5D and 3D coordinate systems for the Wrangell–St. Elias National Park and Preserve (WSENPP) and the Karakoram region (Figures 7.1 and 7.2).

As expected, we see an increase in the difference of cumulative distance based upon representation, given an increase in planimetric distance. The magnitude of the difference is greatest for the Karakoram Himalaya region, due to differences in relief, coupled with a different pattern and magnitude of the degree of separation between the geoid and the ellipsoid. These results clearly demonstrate that the curvature and the shape of the Earth influence distance measurements and analyses at larger distances. For example, tectonic geomorphological investigations attempt to characterize surface deformation, relief production, and forcings including isostatic and tectonic uplift. In general, the operational scale dependencies of uplift are thought to occur over distances beyond 60–100 km. Therefore, distance measurements and topographic profiling over this scale need to be as accurate as possible. This is required in order to provide empirical evidence in support of modeled rates of uplift.

The increase in the surface-area difference with spatial extent (Figure 7.4) also accounts for the curvature and shape of the Earth. Accurate estimates of distance and area over large geographic areas are required in order to investigate complex erosion/uplift dynamics. For example, surface area is highly correlated with relief and erosion, and we need to estimate rates of erosion and sediment fluxes and relate mass input (uplift) to mass output (erosion) to assess steady-state conditions. The use of a 2.5D coordinate system does not accurately account for the undulations in the geoid (i.e., x and y directions) to provide accurate distance and area measurements for producing erosion and relief estimates. Collectively, these results demonstrate that spatial representation issues must be accounted for when addressing concepts of scale, as the computation of parameters, mapping of features, and extraction of quantitative information based upon distance, direction, and area can introduce error and uncertainty into multiscale analysis.

Failure to account for these differences can result in significant errors related to estimating rates of erosion and uplift.

Parameter Magnitude

The magnitude of many geomorphometric parameters is governed by concepts of distance, extent, and direction. Researchers have exploited this to systematically examine mountain landscape hypsometry, dissection, relief production, and the role of surface processes in terms of the dominant processes responsible for erosion (Gilchrist et al. 1994; Burbank et al. 1996; Montgomery et al. 2001; Bishop et al. 2003). Numerous metrics and terrain parameters are routinely computed, including basic summary statistics, relief, hypsometric integral, slope angle, slope azimuth, curvature metrics, surface roughness, and fractal dimension. Other inherently scale-dependent metrics such as topoclimate-related parameters and geophysical parameters can also be used for mapping and modeling applications (Bishop 2013; Wilson and Bishop 2013).

A critical issue in geomorphometry and geocomputation is that of computational scale (Bishop et al. 2012; Wilson and Bishop 2013). This definition of *scale* refers to the selection of computation parameters that govern the sampling of the data for the estimation of parameter magnitude. It is important to note that different parameters may have similar or different scale dependencies compared to other parameters and that the scale dependencies associated with computing a parameter can change with location. Therefore, it is necessary to select the appropriate computational scale for accurately estimating a parameter. Unfortunately, information is not always available to adequately address this issue, and therefore it must be resolved using automated geocomputational solutions. Furthermore, the spatial structure of the topography can represent a limit to the extent to which data are included or excluded in the estimation of a parameter, as the magnitude of some parameters have little value or meaning when data outside of a particular feature or boundary are used to estimate the parameter.

We demonstrate these issues by examining three parameters that characterize important aspects of mountain geodynamics. We start by examining the scale-dependent computation of slope angle and slope azimuth based upon a linear regression analysis at point locations in the Karakoram Himalaya (Figure 7.2). We plotted the magnitude of the parameter *versus* the computation scale utilized (Figures 7.5 and 7.6). These parameters regulate water runoff, sediment transport, and flow velocity and direction and ultimately govern local erosion and deposition dynamics.

Slope-angle variations increase, decrease, or fluctuate depending upon what computational scale is used to estimate the parameter. Similarly, the

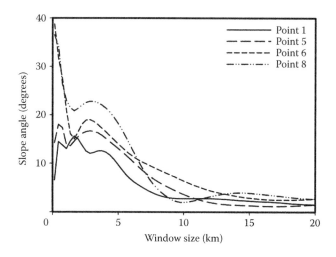

FIGURE 7.5

Slope-angle variations for different computational scales, given four randomly selected points distributed over the Karakoram region (see point locations in Figure 7.2).

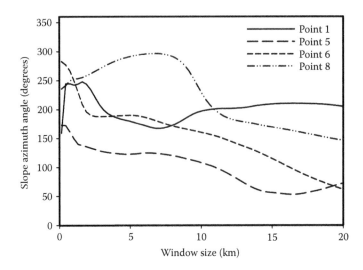

FIGURE 7.6
Slope azimuth variations for different computational scales given four randomly selected points distributed over the Karakoram region (see point locations in Figure 7.2).

slope azimuth angle significantly varies and depicts the orientation structure of the topography. Different sampling locations clearly have different patterns of parameter change depending upon the spatial complexity of the topography. Ultimately, the computation of a parameter at a particular scale must have meaning related to characterizing the properties of the topography or be representative of landform features or aspects of mountain geodynamics. For example, the dynamic coupling between rapid river incision, uplift, and mass movements produces threshold slopes that exhibit relatively steep slopes oriented perpendicular to a rapidly incising stream. Depending upon the resolution of the DEM, we would expect the slope angles and azimuth angles to be constant in magnitude across the spatial extent of the valley walls on both sides of the stream. Scale-dependent analysis of the DEM could be used to characterize and map this type of system coupling and identify uplift zones, as long as the spatial structural organization of basins, valley walls, and ridges was first identified and valley walls spatially segmented to define the spatial constraints needed to compute the slope angle and slope azimuth angle appropriately. It is not meaningful to just choose a scale and compute parameters but is also necessary to evaluate the parameters across computational scales and limits to ensure the presence of threshold slopes.

Another important topographic parameter is the sky-view coefficient (Dozier and Frew 1990), which partially regulates surface irradiance and effectively represents topographic shielding of incoming radiation. This parameter is required to estimate the diffuse-skylight irradiance, which contributes to the total surface irradiance and surface-energy budget.

The surface irradiance is required to estimate glacier ablation rates and meltwater production, which influences the magnitude of glacier ice-flow velocity, ice thickness, and therefore the magnitude of glacier erosion. Furthermore, basin erosion rates can be estimated by examining the concentration of cosmogenic radioactive isotopes such as ^{10}Be and ^{26}Al present in alluvial sediments, which is governed by altitude, latitude, and the topographic shielding of the cosmic irradiant flux (Small et al. 1999; Safran et al. 2005). Similarly, geochronological dating of strath terraces, glacial erratics, and moraine deposits can provide information about the rates of river incision and glacial chronologies, as the isotopic concentrations are related to the time of exposure and knowledge of the cosmogenic isotope production rates (Owen et al. 2002; Schildgen et al. 2002; Zahno et al. 2009).

We demonstrate the scale dependence associated with the magnitude of the sky-view coefficient for sample locations in the Karakoram Himalaya (Figure 7.2). We compute the parameter as follows (Wilson and Bishop 2013):

$$V_f = \sum_{\phi_t}^{360} \cos^2 \left[\theta_{max} \left(\phi_t, d \right) \right] \frac{\Delta \phi_t}{360} \qquad (7.6)$$

where V_f is the skyview factor, ϕ_t is the azimuth angle direction in which the maximum horizon angle is searched for, θ_{max} is the maximum horizon angle given the azimuth angle for a specified distance (d) and search direction (ϕ_t), and $\Delta\phi_t$ is the sampling azimuth angle interval.

Variations in the magnitude of the sky-view coefficient are dependent upon the distance parameter and the directional dependence of the maximum horizon angle. Therefore, the computational scale at which the parameter does not vary is dependent upon the spatial complexity and structure of the topography (Figure 7.7). It is evident that the spatial extent at which this occurs is potentially different for every location on the landscape, as the spatial structure of the basin and ridge orientations govern the distance at which the maximum angle is computed. Lower sky-view coefficients indicate higher relief angles surrounding a particular location on the landscape, which signifies more topographic shielding of radiation. High-magnitude erosion associated with uplift and relief production will generate topography that exhibits relatively low coefficient values. These zones are associated with rapid river incision and uplift, precipitation anomalies that generate high-magnitude slope failures, and high magnitude glacier erosion near glacier equilibrium line altitudes (Figure 7.8).

Anisotropy

Numerous properties of the topography are highly dependent upon the directional sampling of data for computing geomorphometric parameters. Anisotropic variation of morphometric properties is extreme given polygenetic landscape evolution and directionally dependent deformation and uplift.

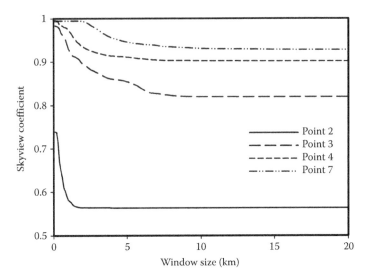

FIGURE 7.7
Sky-view coefficients displayed as a function of spatial extent in computing the parameter for point locations randomly selected over the Karakoram region (see point locations in Figure 7.2).

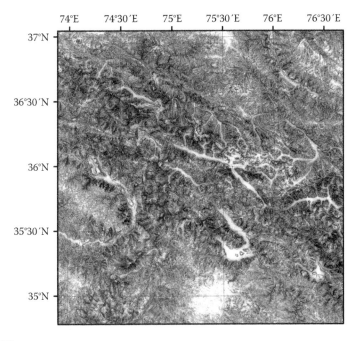

FIGURE 7.8
Sky-view coefficient map over the Karakoram Range in Pakistan. White grey tones represent relatively high sky-view coefficient values, while dark grey tones represent lower coefficient values. The coefficient can range between 0.0–1.0.

For example, average local slope angles can be significantly different than the local slope angle computed in a particular direction. In addition, sampling distance represents another controlling factor. Nevertheless, topographic variation and complexity is highly anisotropic in nature. There is usually a distinctive structural fabric to the topography that is controlled by lithology, surface processes, and tectonics. Furthermore, complex geodynamics control relief production and the orientation of the topography, such that it is highly anisotropic in nature and depicts complex spatial patterns and organization. Such complexity makes it difficult to determine what directions, azimuth intervals, and distances should be accounted for to accurately characterize the properties of the topography. Given such spatial complexity, each location on the landscape may exhibit unique anisotropic variation.

We demonstrate this issue with the simple parameter of surface roughness, computed in all directions around a point. We use the standard deviation as the metric to account for altitude variability, which is a proxy for surface roughness (Glenn et al. 2006; Arrell and Carver 2009; Grohmann et al. 2011). Figure 7.9 clearly reveals periodic variation in altitude as a function of direction, as erosion, deposition, tectonics, and relief production generate highly variable altitude and surface roughness conditions.

Another important example of anisotropic variation related to the spatial organization of mountain topography is presented in Figure 7.10. Here we depict the planimetric distance, as a function of azimuth that is associated with finding the maximum relief angles that were used in the computation of the sky-view coefficient at different locations. Our analysis clearly depicts unique anisotropic variation at each location that is governed by the spatial organization of the topography (mesoscale relief) that defines and characterizes topographic shielding.

Scale Dependencies

A popular theme in research involves the investigation of scale dependencies. This type of research occurs in multiple disciplines including geomorphology, landscape ecology, remote sensing, and GIScience. There has been an emphasis on multiresolution or multiscale analysis (e.g., Burnett and Blaschke 2003; Benz et al. 2004) and the use of geostatistics to characterize the scale dependencies of geomorphometric parameters and landscape biophysical properties (e.g., Sumfleth and Duttmann 2008; Drăguţ et al. 2011).

Numerous studies have evaluated pixel and grid-cell resolution variations for accurately representing spatial variation and depicting patterns that are related to processes (Lam et al. 1998; Montgomery and Brandon 2002; Montgomery 2003). Research has focused on the advantages and disadvantages of using multiresolution representations for a variety of different applications. The spatial theory related to process–form and process–pattern studies dictates the selection of the most appropriate scale at which to represent the landscape or phenomena, such that patterns emerge at that scale,

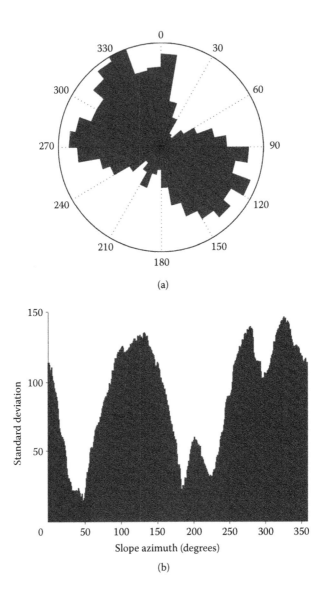

(a)

(b)

FIGURE 7.9
Standard deviation of altitude is used as a proxy of surface roughness and calculated for all directions around Point 1 depicted in Figure 7.2.

caused by processes that operate at that scale (i.e., an operational scale dependency). In geomorphology, it is important to note that multiple processes can produce similar forms and patterns over the landscape and that the landscape is inherently the result of polygenetic landscape evolution. Furthermore, the topography contains only a limited amount of information regarding process–pattern and process–form relationships, as the

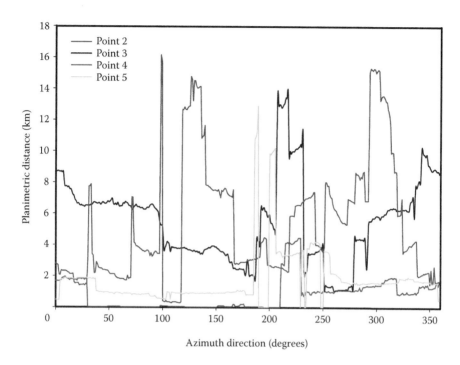

FIGURE 7.10
Planimetric distance associated with the maximum horizon angle in an azimuth direction for four randomly selected points in the Karakoram region (see point locations in Figure 7.2).

magnitude of denudation restricts the amount of time that patterns or information can be inherently encoded on the landscape or in the topography. For example, process–form relationships in the Karakoram Himalaya are thought to be "preserved" in the topography accounting for conditions up to 60,000–75,000 years BP, although no one knows for sure (Bishop and Shroder 2000; Shroder and Bishop 2000).

Finally, the spatial complexity of the landscape needs to be considered, as it is widely known that large grid-cell sizes cannot adequately represent high spatial frequency variation. In general, the spatial resolution dimensions need to be less than the geographic scale of landform elements and features and the operational scale of processes and feedback mechanisms. Given that these characteristics vary in their spatial complexity, it is not reasonable to utilize one spatial resolution to study the multiple dimensions of complex mountain geomorphological systems.

Another aspect of scale-dependent research in mountains involves utilizing geostatistics to characterize geomorphometric parameters. Bishop et al. (2003) demonstrated this across process regimes, although issues involving anisotropy and spatial structural limits were not accounted for. Similarly, other researchers have used geostatistics and variogram and fractal analysis to characterize and understanding spatial variation and scale dependencies (e.g., Bian and Walsh 1993; Atkinson and Tate 2000). Nevertheless, such empirical analysis is still primarily based upon univariate statistics and may be very technique dependent (Bishop et al. 2012). In addition, multiple parameters may exhibit unique scale dependencies depending upon the number of variables that are required to estimate the parameter. Furthermore, how do we integrate, represent, and interpret scale-dependent information? There is a growing body of knowledge that indicates that the spatial structure and organization of mountain topography must be accounted for, as parameters and processes are spatially and temporally constrained in a variety of ways. GIS-based empirical analysis may not be able to address hierarchical complexities of scale-dependent process–form relationships and system dynamics (Burrough 1996; Bishop et al. 1998, 2012).

Spatial Mapping and Organization

Much has been written on geomorphological mapping, and there are numerous approaches for attempting to deal with the issues and difficulties associated with mapping spatial entities, assessing process domains, characterizing polygenetic overprinting, and mapping high-magnitude erosion and tectonic zones (Bishop et al. 2010, 2012). Approaches include characterizing and mapping fundamental form elements and units (MacMillan et al. 2000; Bolongaro-Crevenna et al. 2005; Minár and Evans 2008; Romstad and Etzelmüller 2012), accounting for fuzzy boundaries (Bragato 2004; Fisher et al. 2004; Schmidt and Hewitt 2004), accounting for scale dependencies (Hájek 2008; Bishop et al. 2012), addressing semantic definitions of landforms (Dehn et al. 2001), mathematical characterization of geomorphometric parameters (Pike 2000; Rasemann et al. 2004), and accounting for different requirements related to mapping specific landforms *versus* characterizing the overall landscape (Miliaresis 2001; Gerçek et al. 2011). Bishop et al. (2012) proposed that geomorphological mapping could be integrated into process-based landscape evolution modeling to account for a multitude of scale issues and dependencies involving space and time. Nevertheless, there is a paucity of research related to effectively representing and mapping the spatial organization of complex mountain topography that incorporates spatial topology concepts and structural constraints on processes that continually modify the spatial organization of the topography. In other words, there has been an emphasis on morphometry rather than on morphography (Bishop et al. 2012).

Empirical spatial analysis provides for numerous possibilities, but there is a need to carefully consider the scientific validity and utility of spatial mapping results. GIS-based mapping is prominently dependent upon data manipulations such as thresholding, spatial weighting, prototypical characterization, and subjective criteria (Bishop et al. 2012). Pattern recognition approaches also exist, and indeterminate boundaries represent an important issue in geomorphological mapping.

The excellent work of Deng and Wilson (2008) illustrates issues in mapping that must be addressed in order to facilitate the study of mountain geodynamics. Their research specifically focused on multiscale and multicriteria mapping of mountain peaks as fuzzy entities. The mapping and assessment of mountain peaks is related to relief production and zones of tectonic or isostatic uplift. It is thought that high-magnitude erosion is spatially coincident with such uplift zones and peaks (Finlayson et al. 2002).

Deng and Wilson (2008) first identified summit points accounting for variation in computational scale. They then attempted to characterize peaks using semantic meanings of a prototype peak that was based upon four geomorphometric properties. The properties were then summarized into the peak prototypicality (0–1) across a variety of scales. They indicated that peaks appear differently when the observation scale varies, and they made use of weights across scales, as they indicated that larger scales are more important than smaller scales. The approach is extremely flexible and permits multiscale and multicriteria mapping capabilities while addressing the problems associated with indeterminate boundaries.

Their mapped results are highly variable, and it is difficult to imagine how such an approach can be of value in understanding mountain geodynamics. Rock mass distribution and strength are important concepts in mountain geodynamics, as relief production and system equilibrium concepts are thought to control topographic evolution and mass influx (rock uplift) and mass output (erosion rates and sediment flux) (Selby 1982; Montgomery 2001; Sklar and Dietrich 2001; Montgomery and Brandon 2002). Therefore, peak mapping can be insightful and must include ridges and not just mountain summits, as total rock mass must be differentiated and delineated *versus* those locations on the landscape where erosion and sediment transport dominate. Consequently, mapping must incorporate issues of scale that are based on topographic organization, scale dependencies of rock strength, and scientific principles, rather than attempting to provide empirical flexibility in addressing scale based simply upon the concept of extent and being able to generate a multitude of spatial patterns that may not have any scientific meaning. We acknowledge, however, that this approach may be useful for other applications.

Spatiotemporal Dynamics

Scale issues must also be examined from a temporal perspective that inherently couples concepts of space and time. Bishop and Shroder (2004) identified and discussed a multitude of temporal concepts that are routinely used in the Earth sciences but have not been formalized from a geospatial technology perspective. Concepts of space and time are fundamental for characterizing and understanding mountain geodynamics, where it is realized that fundamental systems (e.g., climatic, hydrologic, and tectonic processes) operate over a wide range of temporal scales and that rates of processes and the status of system dynamics change over time.

Understanding the spatiotemporal dynamics of relief production requires the formalization and modeling of scale dependencies related to parameters, processes, process–form relationships, feedback mechanisms, and system couplings. We demonstrate this point by deterministic modeling of the surface irradiant flux and Alpine Valley glacier erosion. We account for the magnitude of surface irradiance that is dependent upon orbital dynamics, atmospheric conditions, spatial organization of the topography, and landscape evolution (glacier erosion), while the magnitude of glacier erosion is dependent upon surface irradiance, topographic conditions, and landscape evolution.

Surface irradiance fluxes in mountain environments are extremely complex and controlled by a multitude of scale-dependent parameters. We have developed a new GIS-based model that accounts for Earth–Sun orbital dynamics, as these models should be able to be utilized for postdiction and prediction associated with paleoclimate investigations and future climate scenarios, respectively. Therefore, we account for the primary orbital parameters such as eccentricity, the longitude of perihelion, Earth–Sun distance, and obliquity that control the magnitude of the exoatmospheric irradiance (Berger 1976, 1977). We also account for the wavelength dependencies associated with the attenuation of radiation due to atmospheric constituents. These include the processes of Rayleigh and aerosol scattering and absorption due to ozone, homogeneously mixed gases, and water vapor (Gueymard 1995). We also account for global-scale 3D altitudinal and latitudinal variations in gravity and atmospheric pressure to collectively improve the spatial and temporal prediction of atmospheric transmittance that governs the direct and diffuse-skylight irradiance components. Finally, we account for local and mesoscale topographic effects that significantly alter the magnitude of surface irradiance. This includes local slope angles and slope azimuth angles and the regional relief structure that causes cast shadows. We also account for topographic shielding of irradiance by the topography through the use of the sky-view coefficient. Our model utilizes a DEM, and we run simulations over the Nanga Parbat massif in northern Pakistan to demonstrate the spatiotemporal dynamics associated with predicting a critical climate parameter than controlling a multitude of surface processes, including evapotranspiration, ablation, and other surface energy-budget parameters.

Simulation results for total surface irradiance (direct and diffuse-skylight irradiance) at different times during the day on July 15, 2013, are presented in Figure 7.11. Examination of the spatial irradiance patterns clearly reveals the tremendous influence of the topography on the magnitude and distribution of irradiance, such that the relief structure is clearly visible over the Nanga Parbat massif.

Examination of surface irradiance simulations in India (Figures 7.12 and 7.13) reveals the large difference in magnitude between direct and diffuse-skylight

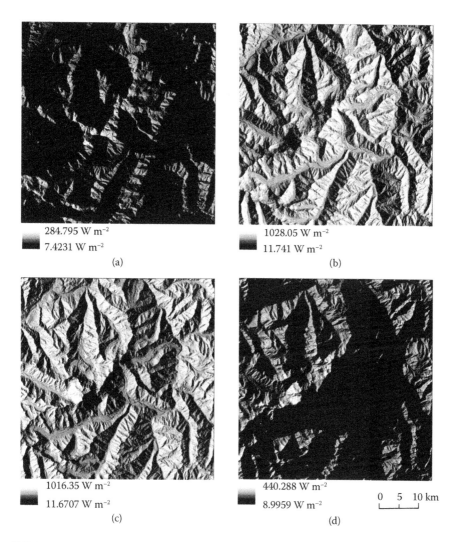

284.795 W m⁻²
7.4231 W m⁻²
(a)

1028.05 W m⁻²
11.741 W m⁻²
(b)

1016.35 W m⁻²
11.6707 W m⁻²
(c)

440.288 W m⁻²
8.9959 W m⁻²
(d)

0   5   10 km

FIGURE 7.11
Simulations of total surface irradiance (direct and diffuse-skylight irradiance) on July 15, 2013.
(a) 5:30 AM. (c) 8:30 AM. (c) 4:00 PM. (d) 6:30 PM.

irradiance and highlights scale-dependent influences (regional relief and cast shadows *versus* topographic shielding) on the temporal distribution of irradiance components. Different locations have different diurnal patterns depending upon local and regional topographic conditions (i.e., dependent upon variations in the spatial organization structure of the topography).

Seasonal variations in total surface irradiance are portrayed for the same two locations in India in Figure 7.14. As we would expect, the overall magnitude in surface irradiance decreases from June to September, and the nature

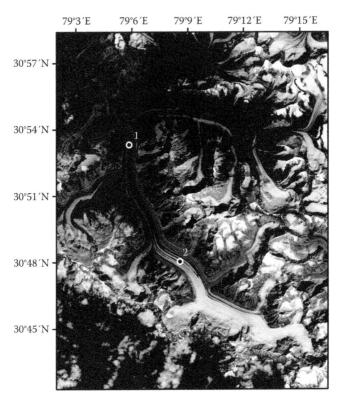

FIGURE 7.12
Point locations on Gangotri Glacier in India that were used to examine diurnal irradiance partitioning of simulation results.

of the scale-dependent topographic effects also change with time due to varying terrain–solar geometry relationships. For example, the influence of cast shadows in any particular location can be seen to vary in terms of magnitude and duration, as represented by the sharp increase and decrease in irradiance near the beginning and end of the day. Such spatiotemporal variations in irradiance are not widely accounted for in understanding the influence of climate forcing on surface processes and mountain geodynamics, as global and regional climate simulations do not utilize high-resolution topographic data, accounting for high spatial frequency variability. This variability effectively governs weathering and regolith production, surface moisture content and sediment redistribution, and ablation and meltwater production, which governs ice depth, ice-flow velocities, and therefore glacier incision rates. Consequently, we still do not understand the role of topography in influencing glacier sensitivity to climate change and thus do not have good estimates of modern-day glacier-erosion rates.

The aforementioned spatiotemporal scale dependencies can significantly influence spatial and temporal variations in the rates of glacier erosion.

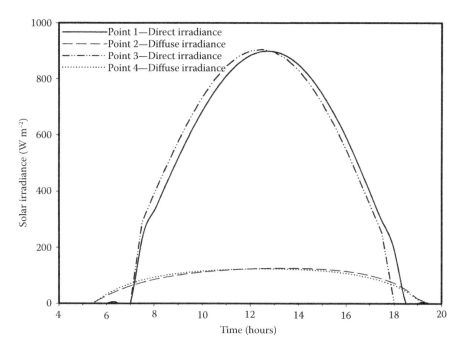

FIGURE 7.13
Simulations of diurnal variation in direct and diffuse-skylight irradiance for two locations on the Gangotri Glacier depicted in Figure 7.12.

We utilize the parameterization schemes of Harbor and Warburton (1992) and MacGregor et al. (2000) to account for glacier valley incision over a time period of 200,000 years. Specifically, we account for basal shear stress, which is dependent on the glacier shape factor, ice depth, and ice-surface slope angle (shallow-ice approximation). The ice deformation velocity, effective stress, and basal sliding are accounted for to determine the magnitude of glacier incision. Computations were based on a 1-year temporal interval, and glacier incision was computed perpendicular to the slope orientation of the topography.

A hypothetical simulation scenario was used, and the initial topography was preset to a shallow V-shaped valley, which attempts to represent a typical mountain landscape at a lower altitude before the onset of glaciation. We then placed an ice mass on the valley floor and set the uplift rate to be 5 mm yr^{-1}. We ran the simulation from 200,000 BP to the present. Figure 7.15 depicts cross-section graphics of the simulation depicting the original surface (yellow line), the lithosphere (brown), the alpine glacier (cyan), and the atmosphere (blue) at the intervals 150,000 BP; 100,000 BP; 50,000 BP; and present day. It can be clearly seen that glacier erosion can start to produce U-shaped valleys after about 100,000 years of erosion (Harbor and Warburton 1992). It is also important to note that the amount of relief production increases over time, as glacier erosion coupled with regional isostatic uplift generates

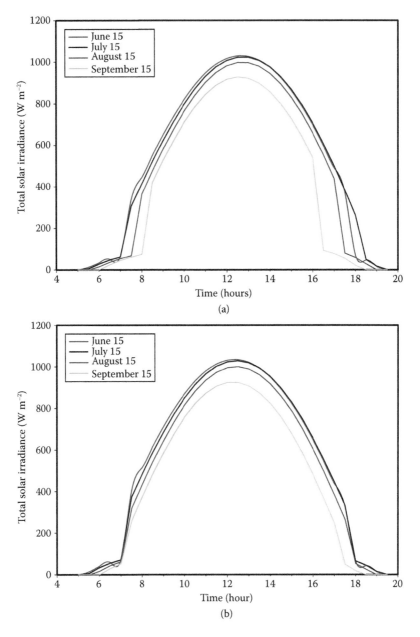

FIGURE 7.14
Simulations of seasonal variation in the total surface irradiance for two locations on the Gangotri Glacier depicted in Figure 7.12. (a) Sample point one depicted in Figure 7.12. (b) Sample point 2 depicted in Figure 7.12.

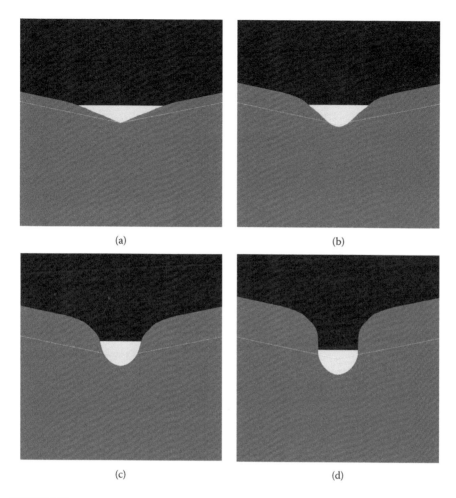

FIGURE 7.15
Glacier-erosion simulation cross-sections depicting the evolution of a U-shaped valley. The uplift rate was set to be 5 mm yr^{-1}. Simulation results ranging from 200 k yrs BP to the present include: 150 k years (a), 100 k years (b), 50 k years (c), and present (d).

high-relief U-shaped valleys that are typically found in all the major glaciated mountain regions around the world.

The glacial incision rate at a particular location underneath the ice is dependent upon the thickness of the ice and the basal-sliding velocity, as defined in governing equations. Nevertheless, researchers need to account for the overall magnitude of glacier erosion, which represents the spatial averaged incision rate. We computed this spatially constrained rate for each time interval, as erosion modifies topographic parameters, which controls the vertical incision rate (Figure 7.16a). It is evident that the vertical glacier incision rate rapidly increases after we place an ice mass on the topography and that the

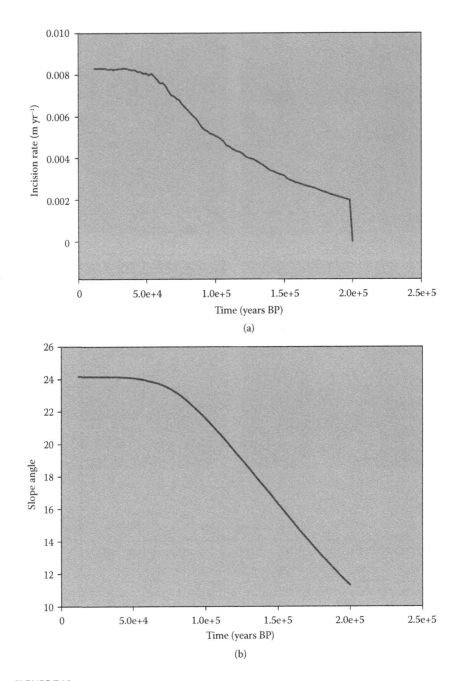

FIGURE 7.16
Simulated glacier incision rate and basal slope angles: (a) spatially averaged vertical incision
rate through time and (b) spatially averaged basal slope angle through time.

incision rate varies nonlinearly with time. In general, it increases and reaches a maximum at approximately 50,000 years. Figure 7.16b depicts the spatially averaged basal slope angle that also changes nonlinearly with time in response to erosion. This simulation example clearly demonstrates a strong coupling between a process (erosion) and a form (slope angle) and shows why extrapolation of rates of processes through time based upon modern-day observations and data is not scientifically valid or justified. Harbor and Warburton (1992), MacGregor et al. (2000), and others documented these temporal scale dependencies that regulate relief production and topographic evolution.

It should be noted that our glacier-erosion simulations did not account for scale-dependent ablation variations that are controlled by altitude and the spatiotemporal dynamics of the irradiant flux. We also did not account for ice-velocity variations caused by diurnal and seasonal variations in surface irradiance, ablation, and meltwater production. These scale-dependent parameters and processes and precipitation rate variations would regulate the depth of the ice, basal-sliding velocity, and basal water pressure, which govern the magnitude of erosion over time. Clearly, there should be strong diurnal and seasonal variations to erosion that are not accounted for in our simulation. Furthermore, the magnitude of erosion and temporal latency that is associated with the reduction and removal of lithospheric mass and its influence on isostatic uplift were not accurately represented. The collective operational scale over which input mass fluxes and output mass fluxes are in equilibrium represents a systems perspective of scale that governs relief production, although tectonic structural spatial constraints and crustal thickening also contribute to uplift and system response (Willett et al. 2001; Zeitler et al. 2001). Therefore, spatial data and spatial analysis and modeling efforts need to deterministically account for numerous scale linkages from local to global scales.

Discussion and Conclusions

A multitude of scale concepts are needed to characterize and understand mountain landscape properties, structure, processes, feedbacks, and system couplings. The rapid proliferation of geospatial data and technologies have greatly facilitated our ability to characterize and assess numerous aspects of mountain landscapes such as land cover and lithological variation, topographic properties and structure, landscape hypsometry, surface deformation, and landform distribution (Bishop et al. 2012). Nevertheless, such progress has been limited based upon the availability of new sensors, new spatial data, and traditional spatial analysis and modeling efforts that do not formalize concepts of scale or account for the variety of concepts that are needed to characterize complex mountain systems. The simple concepts of distance, direction, and extent are not so simple when they need to be integrated into the multiple levels of characterization that are required in order for us to represent, analyze, understand, and simulate properties, spatial entities, structural organization, processes, feedbacks, and system response. Effectively, we do not have a space–time explicit framework to represent the inherent linkages of concepts and information to address many scale issues.

In our work, we have identified important concepts that need to be semantically defined and mathematically formalized with respect to topographic characterization, information extraction and integration, mapping, and prediction. This is a notoriously difficult problem and complicated by a relatively large number of concepts that appear to be unique but are functionally interrelated (e.g., distance, direction, and extent). Similarly, our quantitative treatments of scale-related issues (e.g., multiresolution, texture, semivariogram, and fractal dimension) attempt to address representation and computational scale issues without incorporating other required concepts of scale that serve to limit spatial extent and account for the geometric and topological organization of the topography. Furthermore, analysis of geospatial data has primarily been concerned with a limited number of landscape and topographic parameters and is overly dependent upon statistical concepts of stationarity, homogeneity of variance, feature space, self-similarity, and autocorrelation. It is absolutely critical that we account for the structural, functional, and forcing properties of the topography and ultimately integrate information to assess morphometric, morphographic, process–form, process–pattern, and spatiotemporal relationships.

Based upon our analysis results, it is clear that 3D data structures are required to address representation, spatial analysis, and spatial topological relationships accurately. This issue has been addressed by those interested in 3D GIS and others attempting to characterize topographic variation (Goodchild 1992; Rogers et al. 2012; Ying et al. 2014). Similarly, scale-dependent geomorphometric analysis is required to characterize semantically modeled definitions of landforms and features that depict the spatial structural organization of the topography that confines process domains. Difficult issues here include semantic modeling, use of scale-dependent parameters, integration of scale-dependent information, and the need to characterize and represent indeterminate boundaries. This latter issue requires using concepts of scale such as discontinuity, gradient, diffusion, adjacency, heterogeneity, containment, and perhaps other concepts related to surface biophysical properties, morphometry, composition, and chronology, as well as subsurface vertical structure and composition. An ongoing problem has always been that of acquiring data at the appropriate spatial and temporal scales, as well as having access to geophysical data over large geographic areas. Nevertheless, progress on functional morphographic mapping is required to define such spatial constraints and boundaries.

Our results also demonstrate the need for new spatial data structures and algorithms. They must permit the formalization and integration of concepts that provide the framework for analyzing geomorphometric parameters and permit information synthesis to enable the exploration of patterns. Hierarchical and dynamic structures are also required to permit characterization of higher-order variation and concepts of scale that may be diagnostic of process–form relationships. This could potentially make process domain mapping mathematically tractable. Such algorithms could be computationally constrained based upon the spatial structure of the topography.

Finally, formalized spatial frameworks might be able to be incorporated into numerical modeling efforts, where process domain information and knowledge of temporal scale dependencies can be accounted for. Rates of surface processes are governed by a multitude of forcing factors that include radiation, precipitation, morphometry, and lithology. Representation schemes of process mechanics dictate the magnitude and spatial variation of a multitude of energy and matter fluxes. Therefore, scale dependencies in process representational schemes need to be accurately accounted for along with feedback mechanisms that govern topographic forcings. The concept of topographic lability may be related to matter–process–system scale linkage mechanisms, such that the number of scale-dependent parameters and couplings, process–form linkages, and overprinting of surface processes may govern the nature of topographic complexity and evolution. Perhaps the only way to examine such scale-related complexities is to permit the mapping of scale-dependent characteristics of properties, processes, structure, and systems *via* coupled numerical models, as suggested by Bishop et al. (2012). Such space–time representations and simulations should provide for new findings, ideas, and explanations of how to best formalize a scale framework that can serve as the foundation for spatial and topological analyses of mountain topography.

References

Arrell, K., and S. Carver. 2009. Surface Roughness Scaling Trends. In *Proceedings of Geomorphometry*, Zurich, Switzerland, 31 August–2 September, pp. 120–123. Zurich, Switzerland: University of Zurich.

Atkinson, P. M., and N. J. Tate. 2000. Spatial scale problems and geostatistical solutions: a review. *The Professional Geographer* 52, no. 4: 607–623.

Band, L. E. 1989. Spatial aggregation of complex terrain. *Geographical Analysis* 21, no. 4: 1538–4632.

Baudo, R., G. Tartari, and E. Vuillermoz, eds. 2007. *Mountains Witnesses of Global Changes Research in the Himalaya and Karakoram: Share-Asia Project.* Edited by J. F. Shroder Jr. Vol. 10, *Developments in Earth Surface Processes.* Amsterdam: Elsevier.

Benz, U. C., P. Hofmann, G. Willhauck, I. Lingenfelder, and M. Heynen. 2004. Multi-resolution, object-oriented fuzzy analysis of remote sensing data for GIS-ready information. *ISPRS Journal of Photogrammetry and Remote Sensing* 58, no. 3–4: 239–258.

Berger, A. 1977. Long-term variations of the Earth's orbital elements. *Celestial Mechanics* 15, no. 1: 53–74.

Berger, A. L. 1976. Obliquity and precession for last 5000000 years. *Astronomy & Astrophysics* 51, no. 1: 127–135.

Beven, K. 1995. Linking parameters across scales: subgrid parameterizations and scale dependent hydrological models. *Hydrological Processes* 9, no. 5–6: 507–525.

Bian, L., and S. J. Walsh. 1993. Scale dependencies of vegetation and topography in a mountainous environment of Montana. *The Professional Geographer* 45, no. 1: 1–11.

Bishop, M. P. 2013. Remote sensing and GIScience in geomorphology: introduction and overview. In *Treatise on Geomorphology*, edited by J. F. Shroder, pp. 1–24. San Diego, CA: Academic Press.

Bishop, M. P., A. B. G. Bush, L. Copland, U. Kamp, L. A. Owen, Y. B. Seong, and J. F. Shroder, Jr. 2010. Climate change and mountain topographic evolution in the Central Karakoram, Pakistan. *Annals of the Association of American Geographers* 100, no. 4: 772–793.

Bishop, M. P., L. A. James, J. F. Shroder Jr, and S. J. Walsh. 2012. Geospatial technologies and digital geomorphological mapping: concepts, issues and research. *Geomorphology* 137, no. 1: 5–26.

Bishop, M. P., and J. F. Shroder, Jr. 2000. Unroofing of the Nanga Parbat Himalaya. *The Geological society of London* 163–179.

Bishop, M. P., and J. F. Shroder. 2004. GIScience and mountain geomorphology: overview, feedbacks, and research directions. In *Geographic Information Science and Mountain Geomorphology*, edited by P. M. Bishop and J. F. Shroder. Springer, New York.

Bishop, M. P., J. F. Shroder Jr, and J. D. Colby. 2003. Remote sensing and geomorphometry for studying relief production in high mountains. *Geomorphology* 55, no. 1–4: 345–361.

Bishop, M. P., J. F. Shroder Jr, B. L. Hickman, and L. Copland. 1998. Scale-dependent analysis of satellite imagery for characterization of glacier surfaces in the Karakoram Himalaya. *Geomorphology* 21, no. 3–4: 217–232.

Bishop, P. M., I. D. Dobreva, and C. Houser. 2015. Geospatial science and technology for understanding the complexities of the Critical Zone. In Developments in Earth Surface Processes, *Principles and Dynamics of the Critical Zone*, edited by J. R. Giardino and C. Houser, Vol. 19, pp. 523–561. Amsterdam, the Netherlands: Elsevier.

Bolongaro-Crevenna, A., V. Torres-Rodríguez, V. Sorani, D. Frame, and M. A. Ortiz. 2005. Geomorphometric analysis for characterizing landforms in Morelos State, Mexico. *Geomorphology* 67, no. 3–4: 407–422.

Bragato, G. 2004. Fuzzy continuous classification and spatial interpolation in conventional soil survey for soil mapping of the lower Piave plain. *Geoderma* 118, no. 1–2: 1–16.

Bunyavejchewin, S., J. V. Lafranki, P. Pattapong, M. Kanzaki, A. Itoh, T. Yamakura, and P. S. Ashton. 1998. Topographic analysis of a large-scale research plot in seasonal dry evergreen forest at Huai Kha Khaeng Wildlife Sanctuary, Thailand. *Tropics* 8, no. 1–2: 45–60.

Burbank, D. W., J. Leland, E. Fielding, R. S. Anderson, N. Brozovic, M. R. Reid, and C. Duncan. 1996. Bedrock incision, rock uplift and threshold hillslopes in the northwestern Himalayas. *Nature* 379: 505–510.

Burnett, C., and T. Blaschke. 2003. A multi-scale segmentation/object relationship modelling methodology for landscape analysis. *Ecological Modelling* 168, no. 3: 233–249.

Burrough, P. A. 1996. Opportunities and limitations of GIS-based modeling of solute transport at the regional scale. In *Applications of GIS to the Modeling of Non-Point Source Pollutants in the Vadose Zone*, edited by D. L. Corwin and K. Loague, pp. 19–38. Madison, WI: Soil Science Society of America.

Burrough, P. A., and A. U. Frank. 1996. *Geographic Objects with Indeterminate Boundaries*. London: CRC Press.

Chou, Y.-H., Pin-Shuo Liu, and R. J. Dezzani. 1999. Terrain complexity and reduction of topographic data. *Journal of Geographical Systems* 1, no. 2: 179–198.

Cova, T. J., and M. F. Goodchild. 2002. Extending geographical representation to include fields of spatial objects. *International Journal of Geographical Information Science* 16, no. 6: 509–532.

Dehn, M., H. Gärtner, and R. Dikau. 2001. Principles of semantic modeling of landform structures. *Computers & Geosciences* 27, no. 8: 1005–1010.

Deng, Y., and J. P. Wilson. 2008. Multi-scale and multi-criteria mapping of mountain peaks as fuzzy entities. *International Journal of Geographical Information Science* 22, no. 2: 205–218.

Dorsaz, Jean-Marc, J. Gironás, C. Escauriaza, and A. Rinaldo. 2013. The geomorphometry of endorheic drainage basins: implications for interpreting and modelling their evolution. *Earth Surface Processes and Landforms* 38, no. 15: 1881–1896.

Drăguţ, L., and C. Eisank. 2011. Object representations at multiple scales from digital elevation models. *Geomorphology* 129, no. 3–4: 183–189.

Drăguţ, L., C. Eisank, and T. Strasser. 2011. Local variance for multi-scale analysis in geomorphometry. *Geomorphology* 130, no. 3–4: 162–172.

Dungan, J. L., J. N. Perry, M. R. T. Dale, P. Legendre, S. Citron-Pousty, M. J. Fortin, A. Jakomulska, M. Miriti, and M. S. Rosenberg. 2002. A balanced view of scale in spatial statistical analysis. *Ecography* 25, no. 5: 626–640.

Eckert, S., T. Kellenberger, and K. Itten. 2005. Accuracy assessment of automatically derived digital elevation models from aster data in mountainous terrain. *International Journal of Remote Sensing* 26, no. 9: 1943–1957.

Finlayson, D. P., D. R. Montgomery, and B. Hallet. 2002. Spatial coincidence of rapid inferred erosion with young metamorphic massifs in the Himalayas. *Geology* 30, no. 3: 219–222.

Firpo, G., R. Salvini, M. Francioni, and P. G. Ranjith. 2011. Use of digital terrestrial photogrammetry in rocky slope stability analysis by distinct elements numerical methods. *International Journal of Rock Mechanics and Mining Sciences* 48, no. 7: 1045–1054.

Fisher, P., J. Wood, and T. Cheng, 2004. Where is Helvellyn? Fuzziness of multi-scale landscape morphometry. *Transactions of the Institute of British Geographers*, 29: 106–128.

Fisher, P., J. Wood, and T. Cheng, 2005. Fuzziness and ambiguity in multi-scale analysis of landscape morphometry. In *Fuzzy Modeling with Spatial Information for Geographic Problems*, edited by F. E. Petry, V. B. Robinson, and M. A. Cobb, pp. 209–232. Berlin: Springer.

Gallant, J. C., and M. F. Hutchinson. 1997. Scale dependence in terrain analysis. *Mathematics and Computers in Simulation* 43, no. 3–6: 313–321.

Gerçek, D., V. Toprak, and J. Strobl. 2011. Object-based classification of landforms based on their local geometry and geomorphometric context. *International Journal of Geographical Information Science* 25, no. 6: 1011–1023.

Gilchrist, A. R., M. A. Summerfield, and H. A. P. Cockburn. 1994. Landscape dissection, isostatic uplift, and the morphologic development of orogens. *Geology* 22, no. 11: 963–966.

Glenn, N. F., D. R. Streutker, D. J. Chadwick, G. D. Thackray, and S. J. Dorsch. 2006. Analysis of LiDAR-derived topographic information for characterizing and differentiating landslide morphology and activity. *Geomorphology* 73, no. 1–2: 131–148.

Goodchild, M. F. 1992. Geographical data modeling. *Computers & Geosciences* 18, no. 4: 401–408.

Goodchild, M. F. 2011. Scale in GIS: an overview. *Geomorphology* 130, no. 1–2: 5–9.

Grimm, N. B., D. Foster, P. Groffman, J. M. Grove, C. S. Hopkinson, K. J. Nadelhoffer, D. E. Pataki, and D. P. C. Peters. 2008. The changing landscape: ecosystem responses to urbanization and pollution across climatic and societal gradients. *Frontiers in Ecology and the Environment* 6, no. 5: 264–272.

Grohmann, C. H., M. J. Smith, and C. Riccomini. 2011. Multiscale analysis of topographic surface roughness in the Midland Valley, Scotland. *IEEE Transactions on Geoscience and Remote Sensing* 49, no. 4: 1200–1213.

Gueymard, C. 1995. *SMARTS2, A Simple Model of the Atmospheric Radiative Transfer of Sunshine: Algorithms and Performance Assessment.* Cocoa, FL: Florida Solar Energy Center.

Haboudane, D., F. Bonn, A. Royer, S. Sommer, and W. Mehl. 2002. Land degradation and erosion risk mapping by fusion of spectrally-based information and digital geomorphometric attributes. *International Journal of Remote Sensing* 23, no. 18: 3795–3820.

Hájek, F. 2008. Process-based approach to automated classification of forest structures using medium format digital aerial photos and ancillary GIS information. *European Journal of Forest Research* 127, no. 2: 115–124.

Harbor, J., and J. Warburton. 1992. Glaciation and denudation rates. *Nature* 356: 751–751.

Harbor, J., and J. Warburton. 1993. Relative rates of glacial and nonglacial erosion in alpine environments. *Arctic and Alpine Research* 25, no. 1: 1–7.

Hengl, T., and H. I. Reuter, eds. 2009. *Geomorphometry: Concepts, Software, Applications.* Vol. 33. Amsterdam: Elsevier.

Hensley, S., and T. Farr. 2013. 3.3 Microwave Remote sensing and surface characterization. In *Treatise on Geomorphology*, edited by John F. Shroder, pp. 43–79. San Diego, CA: Academic Press.

Höhle, J., and M. Höhle. 2009. Accuracy assessment of digital elevation models by means of robust statistical methods. *ISPRS Journal of Photogrammetry and Remote Sensing* 64, no. 4: 398–406.

Ibanez, D. M., F. Pellon de Miranda, and C. Riccomini. 2014. Geomorphometric pattern recognition of SRTM data applied to the tectonic interpretation of the Amazonian landscape. *ISPRS Journal of Photogrammetry and Remote Sensing* 87, no. 0: 192–204.

Iliffee, J., and R. Lott. 2008. *Datums and Map Projections for Remote Sensing, GIS and Surveying.* 2nd ed. Scotland, UK: Whittles Publishing.

Immerzeel, W. W., P. D. A. Kraaijenbrink, J. M. Shea, A. B. Shrestha, F. Pellicciotti, M. F. P. Bierkens, and S. M. de Jong. 2014. High-resolution monitoring of Himalayan glacier dynamics using unmanned aerial vehicles. *Remote Sensing of Environment* 150, no. 0: 93–103.

Jamieson, S. S. R., H. D. Sinclair, L. A. Kirstein, and R. S. Purves. 2004. Tectonic forcing of longitudinal valleys in the Himalaya: morphological analysis of the Ladakh Batholith, North India. *Geomorphology* 58, no. 1–4: 49–65.

Jantz, C. A., and S. J. Goetz. 2005. Analysis of scale dependencies in an urban land-use-change model. *International Journal of Geographical Information Science* 19, no. 2: 217–241.

Kellogg, L.-K., D. McKenzie, D. L. Peterson, and A. E. Hessl. 2008. Spatial models for inferring topographic controls on historical low-severity fire in the eastern Cascade Range of Washington, USA. *Landscape Ecology* 23, no. 2: 227–240.

Koons, P. O. 1995. Modeling the topographic evolution of collisional belts. *Annual Review of Earth and Planetary Sciences* 23, no. 1: 375–408.

Koons, P. O., P. K. Zeitler, C. P. Chamberlain, D. Craw, and A. S. Meltzer. 2002. Mechanical links between erosion and metamorphism in Nanga Parbat, Pakistan Himalaya. *American Journal of Science* 302, no. 9: 749–773.

Lam, N. S., D. Quattrochi, H. Qiu, and W. Zhao. 1998. Environmental assessment and monitoring with image characterization and modeling system using multiscale remote sensing data. *Applied Geographic Studies* 2, no. 2: 77–93.

Lausch, A., and F. Herzog. 2002. Applicability of landscape metrics for the monitoring of landscape change: issues of scale, resolution and interpretability. *Ecological Indicators* 2, no. 1–2: 3–15.

MacGregor, K. R., R. S. Anderson, S. P. Anderson, and E. D. Waddington. 2000. Numerical simulations of glacial-valley longitudinal profile evolution. *Geology* 28, no. 11: 10311034.

MacMillan, R. A., W. W. Pettapiece, S. C. Nolan, and T. W. Goddard. 2000. A generic procedure for automatically segmenting landforms into landform elements using DEMs, heuristic rules and fuzzy logic. *Fuzzy Sets and Systems* 113, no. 1: 81–109.

Meentemeyer, V., and E. O. Box. 1987. Scale effects in landscape studies. In *Landscape Heterogeneity and Disturbance*, edited by M. G. Turner, pp. 15–34. New York, NY: Springer.

Miliaresis, G. C. 2001. Geomorphometric mapping of Zagros Ranges at regional scale. *Computers & Geosciences* 27, no. 7: 775–786.

Minár, J., and I. S. Evans. 2008. Elementary forms for land surface segmentation: the theoretical basis of terrain analysis and geomorphological mapping. *Geomorphology* 95, no. 3–4: 236–259.

Montgomery, D. R. 1994. Valley incision and the uplift of mountain peaks. *Journal of Geophysical Research: Solid Earth* 99, no. B7: 13913–13921.

Montgomery, D. R. 2001. Slope distributions, threshold hillslopes, and steady-state topography. *American Journal of Science* 301, no. 4–5: 432–454.

Montgomery, D. R. 2003. Predicting landscape-scale erosion rates using digital elevation models. *Comptes Rendus Geoscience* 335, no. 16: 1121–1130.

Montgomery, D. R., G. Balco, and S. D. Willett. 2001. Climate, tectonics, and the morphology of the Andes. *Geology* 29, no. 7: 579–582.

Montgomery, D. R., and M. T. Brandon. 2002. Topographic controls on erosion rates in tectonically active mountain ranges. *Earth and Planetary Science Letters* 201, no. 3–4: 481–489.

Napieralski, J., J. Harbor, and Y. Li. 2007. Glacial geomorphology and geographic information systems. *Earth-Science Reviews* 85, no. 1–2: 1–22.

Niethammer, U., M. R. James, S. Rothmund, J. Travelletti, and M. Joswig. 2012. UAV-based remote sensing of the Super-Sauze landslide: evaluation and results. *Engineering Geology* 128, no. 0: 2–11.

Nuth, C., and A. Kääb. 2011. Co-registration and bias corrections of satellite elevation data sets for quantifying glacier thickness change. *The Cryosphere* 5, no. 1: 271–290.

Ode, Å., C. M. Hagerhall, and N. Sang. 2010. Analysing visual landscape complexity: theory and application. *Landscape Research* 35, no. 1: 111–131.

Owen, L. A., U. Kamp, J. Q. Spencer, and K. Haserodt. 2002. Timing and style of late quaternary glaciation in the eastern Hindu Kush, Chitral, northern Pakistan: a review and revision of the glacial chronology based on new optically stimulated luminescence dating. *Quaternary International* 97–98, no. 0: 41–55.

Paul, F., C. Huggel, and A. Kääb. 2004. Combining satellite multispectral image data and a digital elevation model for mapping debris-covered glaciers. *Remote Sensing of Environment* 89, no. 4: 510–518.

Peuquet, D. J. 2002. *Representations of Space and Time*. New York, NY: The Guilford Press.

Phillips, J. D. 2004. *Independence, Contingency, and Scale Linkage in Physical Geography*. In *Scale and Geographic Inquiry*, edited by, Sheppard and R.B. McMaster, pp. 86–100. Malden, MA: Blackwell Publishing.

Pike, R. J. 2000. Geomorphology - Diversity in quantitative surface analysis. *Progress in Physical Geography* 24, no. 1: 1–20.

Quattrochi, D. A., and M. F. Goodchild, eds. 1997. *Scale in Remote Sensing and GIS*. Boca Raton, FL: CRC Press.

Ramanathan, V. 2007. Global and regional climate change: the next few decades. In *Developments in Earth Surface Processes*, edited by R. Baudo, G. Tartari, and E. Vuillermoz, pp. 9–11. Amsterdam, the Netherlands: Elsevier.

Rasemann, S., J. Schmidt, L. Schrott, and R. Dikau. 2004. Geomorphometry in mountain terrain. In *Geographic Information Science and Mountain Geomorphology*, edited by M. P. Bishop and J. F. Shroder Jr., pp. 101–145. Berlin: Springer.

Raymo, M. E., and W. F. Ruddiman. 1992. Tectonic forcing of late Cenozoic climate. *Nature* 359, no. 6391: 117–122.

Reiners, P. W., T. A. Ehlers, S. G. Mitchell, and D. R. Montgomery. 2003. Coupled spatial variations in precipitation and long-term erosion rates across the Washington Cascades. *Nature* 426, no. 6967: 645–647.

Rogers, D., A. Cooper, P. McKenzie, and T. McCann. 2012. Assessing regional scale habitat area with a three dimensional measure. *Ecological Informatics* 7, no. 1: 1–6.

Romstad, B., and B. Etzelmüller. 2012. Mean-curvature watersheds: a simple method for segmentation of a digital elevation model into terrain units. *Geomorphology* 139–140, no. 0: 293–302.

Roy, A., E. Perfect, W. M. Dunne, and L. D. McKay. 2014. A technique for revealing scale-dependent patterns in fracture spacing data. *Journal of Geophysical Research: Solid Earth* 119, no. 7: 5979–5986.

Safran, E. B., P. R. Bierman, R. Aalto, T. Dunne, K. X. Whipple, and M. Caffee. 2005. Erosion rates driven by channel network incision in the Bolivian Andes. *Earth Surface Processes and Landforms* 30, no. 8: 1007–1024.

Schildgen, T., D. P. Dethier, P. Bierman, and M. Caffee. 2002. 26Al and 10Be dating of late pleistocene and holocene fill terraces: a record of fluvial deposition and incision, Colorado front range. *Earth Surface Processes and Landforms* 27, no. 7: 773–787.

Schmidt, J., and A. Hewitt. 2004. Fuzzy land element classification from DTMs based on geometry and terrain position. *Geoderma* 121, no. 3–4: 243–256.

Selby, M. J. 1982. Controls on the stability and inclinations of hillslopes formed on hard rock. *Earth Surface Processes and Landforms* 7, no. 5: 449–467.

Shaw, Shih-Lung, H. Yu, and L. S. Bombom. 2008. A space-time GIS approach to exploring large individual-based spatiotemporal datasets. *Transactions in GIS* 12, no. 4: 425–441.

Sheppard, E., and R. B. McMaster, eds. 2004. *Scale and Geographic Inquiry: Nature, Society, and Method*. Malden, MA: Blackwell.

Shroder, J. F., and M. P. Bishop. 2000. Unroofing of the Nanga Parbat Himalaya. *Geological Society, London, Special Publications* 170, no. 1: 163–179.

Sklar, L. S., and W. E. Dietrich. 2001. Sediment and rock strength controls on river incision into bedrock. *Geology* 29, no. 12: 1087–1090.

Small, E. E., R. S. Anderson, and G. S. Hancock. 1999. Estimates of the rate of regolith production using 10Be and 26Al from an alpine hillslope. *Geomorphology* 27, no. 1–2: 131–150.

Snyder, N. P., K. X. Whipple, G. E. Tucker, and D. J. Merritts. 2000. Landscape response to tectonic forcing: digital elevation model analysis of stream profiles in the Mendocino triple junction region, northern California. *Geological Society of America Bulletin* 112, no. 8: 1250–1263.

Sumfleth, K., and R. Duttmann. 2008. Prediction of soil property distribution in paddy soil landscapes using terrain data and satellite information as indicators. *Ecological Indicators* 8, no. 5: 485–501.

Tate, N., and J. Wood. 2001. Fractals and scale dependencies in topography. In *Modelling Scale in Geographical Information Science*, edited by N. Tate and P. M. Atkinson, pp. 35–51, Wiley: West Sussex, England.

Uuemaa, E., J. Roosaare, and Ü. Mander. 2005. Scale dependence of landscape metrics and their indicatory value for nutrient and organic matter losses from catchments. *Ecological Indicators* 5, no. 4: 350–369.

Veldkamp, A., P. H. Verburg, K. Kok, G. H. J. de Koning, J. Priess, and A. R. Bergsma. 2001. The need for scale sensitive approaches in spatially explicit land use change modeling. *Environmental Modeling & Assessment* 6, no. 2: 111–121.

Weng, Q., ed. 2014. *Scale Issues in Remote Sensing*. Hoboken, NJ: John Wiley & Sons.

Willett, S. D., R. Slingerland, and N. Hovius. 2001. Uplift, shortening, and steady state topography in active mountain belts. *American Journal of Science* 301, no. 4–5: 455–485.

Wilson, J. P., and M. P. Bishop. 2013. Geomorphometry. In *Treatise on Geomorphology*, edited by J. F. Shroder Jr., pp. 162–186. San Diego, CA: Academic Press.

Wu, F. 1999. GIS-based simulation as an exploratory analysis for space-time processes. *Journal of Geographical Systems* 1, no. 3: 199–218.

Wu, J. 2004. Effects of changing scale on landscape pattern analysis: scaling relations. *Landscape Ecology* 19, no. 2: 125–138.

Wu, J., D. E. Jelinski, M. Luck, and P. T. Tueller. 2000. Multiscale analysis of landscape heterogeneity: scale variance and pattern metrics. *Geographic Information Sciences* 6, no. 1: 6–19.

Ying, Ling-Xiao, Ze-Hao Shen, Shi-Long Piao, Y. Liu, and G. P. Malanson. 2014. Terrestrial surface-area increment: the effects of topography, DEM resolution, and algorithm. *Physical Geography* 35, no. 4: 297–312.

Zahno, C., N. Akçar, V. Yavuz, P. W. Kubik, and C. Schlüchter. 2009. Surface exposure dating of Late Pleistocene glaciations at the Dedegöl Mountains (Lake Beyşehir, SW Turkey). *Journal of Quaternary Science* 24, no. 8: 1016–1028.

Zeitler, P. K., A. S. Meltzer, P. O. Koons, D. Craw, B. Hallet, C. P. Chamberlain, W. S. F. Kidd, et al. 2001. Erosion, Himalayan geodynamics, and the geomorphology of metamorphism. *GSA Today* 11: 4–9.

8

Downscaling on Demand: Examples in Forest Canopy Mapping

Gordon M. Green, Sean C. Ahearn, and Wenge Ni-Meister

CONTENTS

Introduction

Distributed processes dominate the computational landscape, from cloud computing to mobile apps, and on-demand access to spatial data has become routine. Web mapping services allow users to efficiently view large data sets by accessing small subsets on demand, and dynamic services are a part of familiar applications like route navigation, weather forecasting, and traffic monitoring. Most products based on remotely sensed data are published in static form, but on-demand processing offers some promising new possibilities, particularly related to problems of scale. In this chapter, we explore some of the potential of on-demand processing, using downscaling of forest canopy structure as an example.

Mapping the three-dimensional structure of forest canopies has been an area of active research in recent years, driven by the need to quantify the role of forests in the carbon cycle and other biosphere/atmosphere processes. Local and regional applications may suffer from a lack of sufficiently granular data. Airborne lidar, which can measure the height of the forest canopy at high resolution, is widely used to quantify three-dimensional canopy structure at the individual tree level (see, e.g., Popescu et al. 2003), but it is only available for limited areas. Most currently available global canopy height data are at a coarse resolution (≥500 m, especially Lefsky 2010; Simard et al. 2011), and Landsat-scale canopy height data remain limited to national coverage (e.g., Kellndorfer et al. 2010). Improved methods of quantifying the horizontal and vertical distribution of vegetation at higher resolution are needed in fields as diverse as meteorology, climate science, and urban planning (see Jones and Vaughn 2010, for an overview of applications). The recent publication of global Landsat-scale canopy cover data by Sexton et al. (2013) and Hanson et al. (2013) gives us a clearer picture of local canopy cover, but higher resolution canopy height data are not yet available globally, and processing higher resolution remote sensing data at a global scale remains an engineering challenge. Flexible methods of rescaling existing data may help fill the gaps in currently available high-resolution forest canopy height measurements.

The functional resolution of remotely sensed data can be increased by fusing together multiple sources of data. For example, pan-sharpening consists of using a high-resolution monochromatic band to estimate the values of multiple bands acquired at lower resolution. Fusion can take many forms with varying levels of complexity (for a review, see Pohl and van Genderen 1998). As standardized methods of accessing spatial data on demand become widespread, the options available for running fusion processes on demand are increasing as well. Standards like Web mapping services (Beaujardiere 2006), Web coverage services (Baumann 2012), and Web processing services (Schut 2007), as well as proprietary protocols, make this practical. Invoking data fusion processes on demand has several benefits when applied to remotely sensed data:

- Unused data are not processed.
- High-resolution data can be made available with global or continental coverage by only processing those regions of interest that are actually requested.
- The underlying data can also be accessed on demand, reducing the overhead of processing precursor data.
- New data can be incorporated as they become available.

Researchers are finding applications for on-demand processing, in areas such as feature extraction (e.g., Mansourian et al. 2008), where the underlying process is complex and time-consuming; disaster management (e.g., Joyce et al. 2009), where up-to-the-minute data are required; or distributed geoprocessing (e.g., Shi et al. 2012), which requires the integration of heterogeneous data sources. High-resolution canopy height estimation is a good candidate for on-demand processing because it requires the integration of heterogeneous data; its output is too large to preprocess; and the underlying data processing can be time-consuming.

The basic architecture of an on-demand fusion system is shown in Figure 8.1. A client process makes a request to a server process, that request primarily consisting of a region of interest. The server application acquires the data it needs for the fusion process at the time of the request, from local resources such as relational databases or file systems, and from remote resources such as Web map services. After carrying out the fusion process, the Web application returns the results to the client process. The client process may also access the remote services directly, as would be the case if the client process were a Web map.

The different resolutions in question are shown in Figure 8.2. The 1-km resolution height data do not capture the high degree of fragmentation within the sample scene. The 30-m data bring us much closer to capturing the variability in canopy cover, without height. The high-resolution airborne lidar captures canopy height at the scale of individual trees, offering a picture of canopy structure that can be readily integrated with field measurements. By combining data from all of these sources—the large selection of coarse-resolution

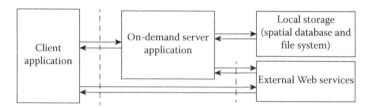

FIGURE 8.1
On-demand architecture diagram.

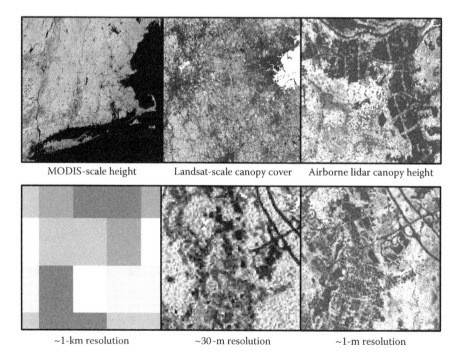

FIGURE 8.2
Source and target scales of downscaling.

biophysical and climatological data at moderate-resolution imaging spectro-radiometer (MODIS) resolution, newly available Landsat-scale canopy cover with global coverage, and the increasing body of high-resolution data from airborne lidar—we can begin to form a more complete picture of forest canopy cover at high resolution, by downscaling on demand.

Methodology

The downscaling process we are using involves simulating individual trees and constraining the simulation by global canopy cover data [see Dungan (1999) for a discussion of conditional simulation in vegetation mapping]. It follows a method outlined in Green et al. (2013) but uses simple geometrical models of individual trees instead of lidar-based samples. Figure 8.3 shows a part of Central Park in New York City processed using this method. The left-hand column shows an aerial photograph, the middle column shows a canopy height model (CHM) derived from airborne lidar, and the third column shows a synthetic CHM created by simulating individual trees, independent of the lidar data and within the constraints of 30-m canopy cover data. The individual tree data are rendered with a 1-m pixel size, but the spatial accuracy is limited to that of the 30-m data.

We looked at four methods that demonstrate some of the possibilities of downscaling on demand. In the simplest case, the downscaling consists of placing individual trees within 30-m pixels based on a single constant height value and 30-m canopy cover data. The second uses a similar method, but instead of using a single constant height estimate, it uses the context-specific

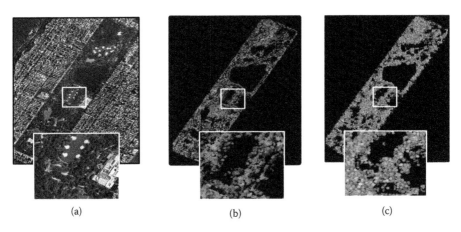

(a)   (b)   (c)

FIGURE 8.3
(a) Central Park in New York City with a detail inset. (b) DSM of the same area derived from airborne lidar. (c) Simulated canopy height model generated independently from the lidar data.

height value from the 1-km resolution global canopy height by Simard et al. (2011). The third estimates canopy height at high resolution by sampling from a database of lidar-based CHMs using a nearest-neighbor approach. The last builds a random forest predictive model from a database of Geoscience Laser Altimeter (GLAS) waveforms (Zwally et al. 2011) and uses this to estimate the maximum height within each 30-m pixel [see, e.g., Pal (2005) for a more detailed description of random forest classification of remotely sensed data]. The four methods are referred to as *constant, simple, nearest neighbor,* and *random forest* in the discussion below.

The constant model serves as a baseline against which to compare the other methods. It involves the fewest remote data accesses and fastest calculations. The simple downscaling model scales the preexisting height data at coarse resolution by associating a height value with each 30-m pixel based on canopy cover. This has the advantage of simplicity and efficiency, as well as global availability, but its accuracy is limited to that of the 1-km height data. The nearest-neighbor approach, in contrast, makes use of the full resolution lidar data, locating the nearest samples using simple Euclidean distance among normalized ancillary values. This approach has the advantage of retaining the realism of the original measurements and can be easily be extended using additional reference data. This feature is particularly relevant in this example, because the amount of available lidar data is increasing at a rapid pace. However, the underlying lidar samples are only available for small areas within the contiguous United States. The last approach uses a random forest classifier built from GLAS waveforms. One-kilometer resolution land cover and climatological data together with 30-m canopy cover data are used to build the model, which predicts canopy height at 30-m resolution. This classifier is then invoked on demand for each region of interest, and a 1-m synthetic CHM is built from the resulting estimate. The data underlying this method are also available globally.

The downscaling process is shown in Figure 8.4. The predictor variables are accessed *via* Web services, fed into a downscaling model that simulates individual trees, and the output is rendered into a raster containing a simulated CHM for the given region of interest.

The constant model uses only the canopy cover and tree shape and downscales a constant maximum height value (25 m) using a linear distribution of tree heights within each 30-m pixel. The simple model replaces the constant height value with the height value from the 1-km height data by Simard et al. (2011). To fill in the gaps where 1-km pixels are ascribed a zero height but nonetheless have a nonzero canopy coverage according to the canopy cover data, an average value of the nonzero pixels within the region of interest is used.

The nearest-neighbor model samples randomly from the five airborne lidar samples that are nearest to the region of interest, each sample consisting of a height distribution, canopy cover, location, and tree type. These reference samples were selected randomly from among ~8,000 (30 m × 30 m) sample sites spaced evenly among all of the lidar sites. Distance is calculated as the simple Euclidean distance between each pixel within the region of interest

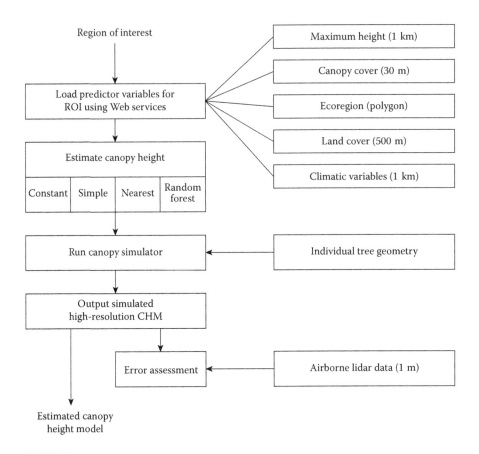

FIGURE 8.4
Overview of the general on-demand downscaling process. The four models are shown in the context of the overall process.

and the reference pixels, using a combination of geographic distance, difference in 1-km height value, difference in canopy cover, and tree type.

The random forest model uses location, 1-km height, land cover type, 30-m canopy cover, and the four 1-km climatic variables as predictors of canopy height. The classifier was created by first loading the predictor values using Web services for each of the GLAS samples from across North America. Test classifiers were then generated using the Orange machine learning environment (Demšar et al. 2013). The random forest model was chosen because it performed slightly better than other classifiers based on comparing the root mean square error (RMSE) using a threefold cross-validation procedure. It was also selected because the random forest method helps reduce the chance of overfitting the training data, compared to other methods such as regression trees. The generated classifier was then saved to a file for the test system to load and run on demand.

We implemented the test system using the Python programming language and a selection of libraries that manage the component parts of the system.

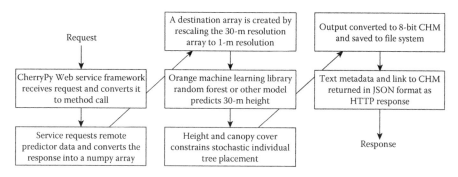

FIGURE 8.5
System implementation using the Python programming language.

The building blocks include NumPy for handling array data, the CherryPy lightweight Web server application framework, the Orange machine learning library (Demšar et al. 2013), and ArcGIS Server (Esri, Relands, CA, USA) for map services. The Python processing sequence is shown in Figure 8.5.

Source Data

For the on-demand model, we needed to either locate or create Web services that publish the required data. While on-demand subsetters and query interfaces are available for some data sets, many are difficult to access programmatically. For the test application, we created Web map services using ArcGIS Server 10.1, so that the precursor data could be accessed with a point query or as a gridded surface, using the Identify and Export Map operations, respectively. The data used here include global 30-m canopy cover from Sexton et al. (2013); global 1-km canopy height from Simard et al. (2011); airborne lidar gathered from nine locations from within the contiguous United States; full-waveform lidar data from the GLAS sensor from within North America; polygonal ecoregion information from Olson et al. (2001); MODIS-based 2005 land cover type by Friedl et al. (2010); and WorldClim climatological data from Hijmans et al. (2005). These data sources are listed in Table 8.1.

The four climatological variables were loaded into a four-band image mapping service so that they could be acquired with a single Web service call for any region of interest. The maximum 1-km canopy height was also loaded into a Web mapping service, as was the MODIS land cover type data.

The 30-m canopy cover data from Sexton et al. (2013) contain percent tree cover downscaled from the MOD44B product (Hansen et al. 2006) using Landsat data and validated using lidar. It does not include detailed canopy height at the individual tree level, which is the target variable for the downscaling described here. It was downloaded and loaded into a Web mapping service

TABLE 8.1

Source Data Used for Downscaling

Source	Resolution	Description	Coverage
MODIS	1 km	Land cover type (MCD12Q1, 2005)	Global
MODIS and ICESat	1 km	Estimated maximum canopy height (Simard et al. 2011)	Global
ICESat	~65 m	GLAS waveforms	North America
Landsat	30 m	Downscaled global canopy cover from MOD44B (Sexton et al. 2013)	Global
Airborne lidar	1 m	Canopy height models (OpenTopography.org and NCALM)	Locations within contiguous United States
WorldClim	1 km	Mean annual temperature (Hijmans et al. 2005)	Global
WorldClim	1 km	Temperature standard deviation	Global
WorldClim	1 km	Annual precipitation	Global
WorldClim	1 km	Precipitation coefficient of variation	Global
WWF	Vector	Ecoregion, used to determine tree shape (Olson et al. 2001)	Global

Note: GLAS, Geoscience Laser Altimeter; MODIS, moderate-resolution imaging spectroradiometer; NCALM, National Center for Airborne Laser Mapping.

for on-demand access. These 30-m data are the bridge between the MODIS-scale land cover data and the airborne lidar data. The product defines *canopy cover* as vegetation over 5 m and, like MOD44B, saturates at approximately 80% cover.

Airborne lidar data at 1-m resolution lidar data were acquired from OpenTopography.org and National Center for Airborne Laser Mapping (NCALM); converted to CHMs, consisting of the surface created by subtracting a bare earth digital elevation model from a digital surface model; mosaicked into nine sample sites; and published *via* Web services. For the present purposes, the lidar CHMs were assumed to be accurate and were treated as reference data. The locations and areas of the source sites are shown in Figure 8.6.

GLAS data were downloaded for North America and processed using the method described in Selkowitz et al. (2012), which included discarding waveforms that failed various threshold tests related to cloud cover, geolocation error, noise thresholds, and others. Waveforms from terrain over 5 degrees in slope as derived from 90-m resolution data from the Shuttle Radar Topography Mission were discarded, and the remaining waveforms were slope-corrected by applying the slope correction formula from Yang et al. (2011). The waveforms were loaded into a spatial database, and a random subset of 50,000 was selected for use in training the random forest classifier.

Each model requires an estimate of tree shape. We chose a simple and computationally efficient method of estimating the size and shape of each tree crown by associating the 14 primary ecoregion types from Olson et al. (2001)

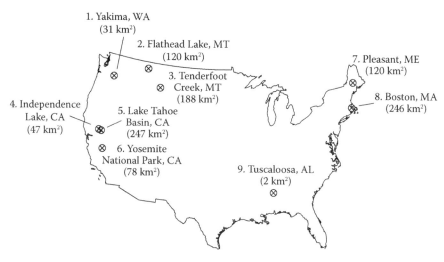

FIGURE 8.6
Location of lidar sites.

Central Park, NYC

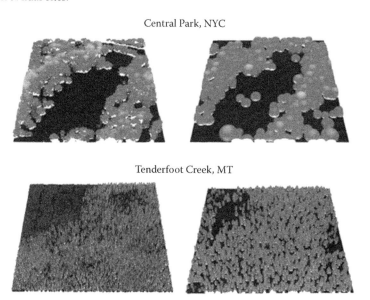

Tenderfoot Creek, MT

FIGURE 8.7
Conical *versus* rounded tree geometries.

with a dominant tree shape, either conical or rounded, assuming conical shapes for boreal, montane, and temperate conifer forests and rounded for all others. Crown diameter was estimated as a function of height and canopy cover, derived empirically from the airborne lidar data. Figure 8.7 shows the effect of the different tree shapes relative to reference lidar CHMs. The effects of crown shape on canopy volume can be large. For example, using a rounded tree shape for a sample coniferous site resulted in a 40% overestimate of canopy volume. Nelson (1997) included a detailed discussion of the effect of tree shape on canopy height distribution.

To evaluate the on-demand approach, we selected a random set of sample sites drawn from a regularly spaced set of 500 × 500 m candidate sites spread across the nine lidar reference sites (n = 446). The metrics used to compare the results to the reference data were 98th percentile height, canopy volume, and forested area calculated using a 1-m height threshold. These values were extracted from each sample site from the lidar DSM and the model output and compared for each site and each model.

Results and Discussion

Sample output for a single site is shown in Figure 8.8. The first row shows an aerial photograph and lidar CHM of the site, and the images below show the outputs of the four models for the same location. The airborne lidar data in the upper right panel show a high degree of variability in the size and height of individual trees, including localized clusters of tall trees and patches of lower trees. The output of the constant model, in the middle left, shows a range of sizes, but an unrealistically uniform height. The simple model shows a spatial distribution of heights similar to the reference data, but underestimates the height of the taller trees. The nearest neighbor model has a similar level of fragmentation but a diverging spatial arrangement, while the random forest model is close to the constant model in this particular case.

Across all of the sample sites and the three values being compared—98th percentile height, canopy volume, and forested area—the picture is varied. Figure 8.9 compares the error among the different approaches. RMSE was calculated per site, then averaged, to avoid a disproportionate influence from the larger sites, with error bars showing the standard deviation among the RMSE values. Generally, the results showed a lower RMSE among the nearest-neighbor and random forest models. The simple model, which is directly dependent on the data from Simard et al. (2011), shows an RMSE close to the 6-m RMSE those authors described. The least error was found in the nearest-neighbor model. The random forest model, which is a model similar to that used by Simard, shows a similar RMSE. The improved accuracy of the nearest-neighbor model over the random forest is likely due to the fact that the nearest-neighbor model uses airborne lidar data drawn from sites similar to the test sites. The random forest model uses GLAS data, which tend to be less accurate and are drawn from locations scattered across North America.

Figure 8.10 shows the per-site scatter plots of reference *versus* estimated volume for the random forest model within each sample site, in 10^6 m³; Figure 8.11 shows reference *versus* estimated 98th percentile height in meters for each site.

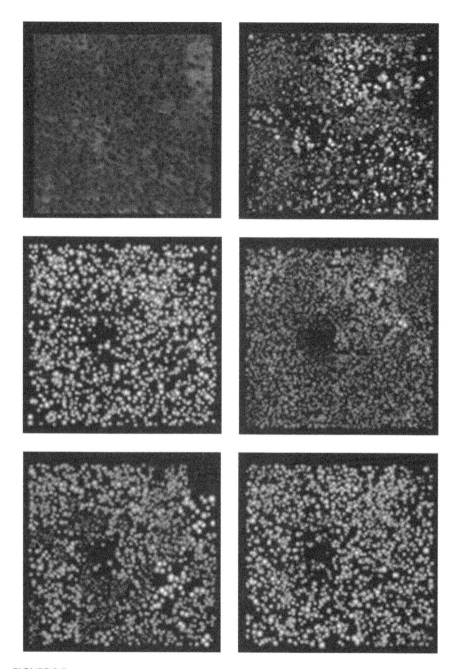

FIGURE 8.8

Sample results using each downscaling method. Upper left is an aerial photograph; upper right is a reference canopy height model derived from airborne lidar; middle left is the constant model; middle right is the simple model; lower left is nearest neighbor model; lower right is the random forest. Grey scale in height models is linear from 0 m (black) to 35 m and above (white).

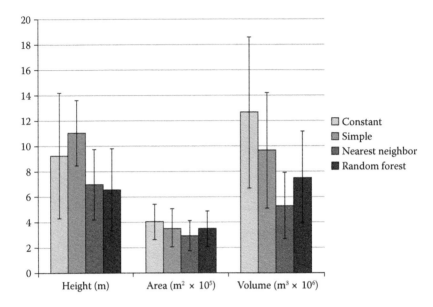

FIGURE 8.9
Average root mean square error of the different downscaling models.

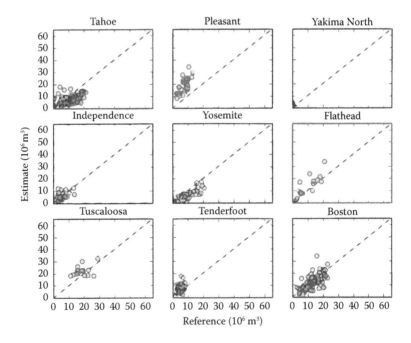

FIGURE 8.10
Per-site scatterplots showing reference and estimated canopy volume: in 10^6 m^3.

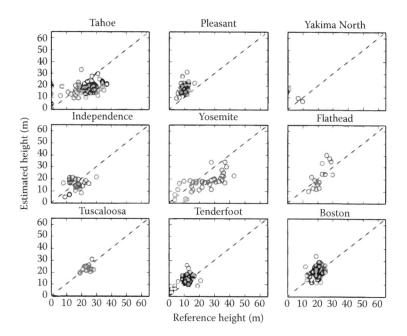

FIGURE 8.11
Per-site scatterplots showing reference and estimated canopy heights in meters.

Although the models show a high degree of error relative to the measured values, it is in the same range of the Simard data, upon which the simple model was based and which follows a similar approach as the random forest model. The most pronounced overestimates are in the Pleasant, ME, location, which is also overestimated by the Simard data. This area includes wetlands and multiple bodies of water, and one hypothesis is that the reference DSM may have been affected by water. Overall, although the downscaling does not improve accuracy, it does offer an approximation of increased level of detail, as the available global coarse-resolution height data have no information about the vertical or horizontal distribution within each pixel. Each simulated CHM in the data generated by the on-demand service, in contrast, gives a first approximation of the same level of detail as CHM derived from airborne lidar.

Conclusion

This sample application demonstrates a few of the many possible approaches to using on-demand processing to address issues of scale in remote sensing. The on-demand model reduces the computational burden of processing high-resolution data. It also allows new data to be incorporated easily. The constant and simple models can be updated with new versions of the land cover and canopy cover data sets without modifications to the code or database. Expanding the nearest-neighbor model to include new airborne lidar data only requires adding new data to the Web service and spatial database, without modifications to the code. Similarly, updating or replacing the GLAS data underlying the random forest model requires updating the database and generating a new random forest classifier. This highlights a key advantage of the on-demand architecture: the system can be updated without any changes to the code that performs the on-demand fusion.

The test version of this system implemented the constant and simple downscaling methods globally, but the nearest-neighbor and random forest models were only implemented for the contiguous United States and North America, respectively. This is due to the lack of airborne lidar data outside of the United States and the data processing requirements for global GLAS data. In any case, a production system would likely entail selecting and developing only one of these models and optimizing it further to improve both its accuracy and performance. Currently, the system takes 10 seconds to process a 1-km^2 area; therefore, it can only be used for a few square kilometers at a time, making it unsuitable for display in Web maps, and artifacts at pixel boundaries can make the display inconsistent. We anticipate that the more practical use of this particular on-demand system would consist of special-purpose use in third-party applications that need simulated forest canopy data at high resolution. This or similar on-demand architectures may be useful for other applications with similar requirements. We were encouraged to find that the Python implementation offered a straightforward path from data-processing scripts to Web services. As more researchers publish data in a programmatically accessible format, the need to download and preprocess existing data sets should decrease. On-demand access to existing remote sensing data sets may reduce the logistical barriers to extending the resolution of existing data and working with data at different scales.

Acknowledgments

The authors thank the anonymous reviewer for the very helpful and thorough comments and suggestions. Special thanks to Dr. Shihyan Lee for the Integrated Data Language (IDL) code used for part of the GLAS data processing. New York City lidar data are from the Center for the Advanced Research of Spatial Information (Sean C. Ahearn, Director). This material is based on data services provided by the OpenTopography Facility with support from the National

Science Foundation (NSF) under award numbers 0930731 and 0930643. Yosemite National Park, CA, Tuscaloosa, AL, Pleasant, ME, Flathead Lake, MT, Independence Lake, CA, and Tenderfoot Creek, MT, lidar data acquisition and processing completed by NCALM (http://www.ncalm.org). NCALM funding provided by the NSF Division of Earth Sciences, Instrumentation, and Facilities Program (EAR-1043051). This material is based on services provided to the Plate Boundary Observatory (PBO) by NCALM. PBO is operated by UNAVCO for EarthScope (http://www.earthscope.org) and supported by the NSF (EAR-0350028 and EAR-0732947), from the EarthScope Yakima Lidar Project, WA. Lake Tahoe Basin lidar data provided by the Tahoe Regional Planning Agency. Boston-area airborne lidar data were made available by the Office of Geographic Information (MassGIS), Commonwealth of Massachusetts, Information Technology Division. MODIS data products were obtained through the online data pool at the National Aeronautics and Space Administration Land Processes Distributed Active Archive Center, United States Geological Survey (USGS)/Earth Resources Observation and Science Center, Sioux Falls, SD (https://lpdaac.usgs.gov/data_access).

References

Baumann, P. Ed., 2012. *WCS 2.0 Interface Standard (WCS) Implementation Specification OGC 09-110r4*, Open Geospatial Consortium, Wayland, MA.

Beaujardiere, J. Ed., 2006. *OpenGIS Web Map Service (WMS) Implementation Specification, OGC 06-942*, Open Geospatial Consortium, Wayland, MA.

Demšar, J., Curk, T., and Erjavec, A., 2013. Orange: Data mining toolbox in python, *Journal of Machine Learning Research*, 14: 2349–2353.

Dungan, J., 1999. Conditional simulation: An alternative to estimation for achieving mapping objectives, in *Spatial Statistics for Remote Sensing*, Stein et al., Eds., Berlin, Germany: Kluwer Academic Publishers.

Friedl, M. A., Sulla-Menashe, D., Tan, B., Schneider, A., Ramankutty, N., Sibley, A., and Huang, X., 2010. MODIS collection 5 global land cover: Algorithm refinements and characterization of new datasets, *Remote Sensing of Environment*, 114: 168–182.

Green, G., Ahearn, S., and Ni-Meister, W., 2013. A multi-scale approach to mapping canopy height, *Photogrammetric Engineering and Remote Sensing*, 79(2): 185–194.

Hansen, M., DeFries, R., Townshend, J. R., Carroll, M., Dimiceli, C., and Sohlberg, R., 2006. *Vegetation Continuous Fields MOD44B, 2005 Percent Tree Cover, Collection 4*, University of Maryland, College Park, MD.

Hansen, M. C., Potapov, P. V., Moore, R., Hancher, M., Turubanova, S. A., Tyukavina, A., Thau, D., et al., 2013. High-resolution global maps of 21st-century forest cover change, *Science*, 342: 850–853.

Hijmans, R. J., Cameron, S. E., Parra, J. L., Jones, P. G., Jarvis, A., 2005. Very high resolution interpolated climate surfaces for global land areas, *International Journal of Climatology*, 25: 1965–1978.

Jones, H., and Vaughan, R., 2010. *Remote Sensing of Vegetation, Principles, Techniques, and Applications.* Oxford, United Kingdom: Oxford University Press.

Joyce, K. E., Belliss, S. E., Samsonov, S. V., McNeill, S. J., Glassey, P. J., 2009. A review of the status of satellite remote sensing and image processing techniques for mapping natural hazards and disasters. *Progress in Physical Geography,* 33(2): 183-207.

Kellndorfer, J., Walker, W., LaPoint, E., and Kirsch, K., 2010. Statistical fusion of Lidar, InSAR, and optical remote sensing data for forest stand height characterization: A regional-scale method based on LVIS, SRTM, Landsat ETM+, and Ancillary data sets. *Geophysical Research Letters,* 115: 1–10.

Lefsky, M. A., 2010. A global forest canopy height map from the moderate resolution imaging spectroradiometer and the geoscience laser altimeter system, *Geophysical Research Letters,* 37: L15401.

Mansourian, A., Valadan Zoje, M. J., Mohammadzadeh, A., Farnaghi, M., 2008. Design and implementation of an on-demand feature extraction web service to facilitate development of spatial data infrastructures, *Computers, Environment and Urban Systems,* 32(5): 377–385.

Nelson, R., 1997. Modeling forest canopy heights: The effects of canopy shape, *Remote Sensing of Environment,* 60(3), 327–334, ISSN 0034-4257, http://dx.doi.org/10.1016/S0034-4257(96)00214-3.

Olson, D. M., Dinerstein, E., Wikramanayake, E. D., Burgess, N. D., Powell, G. V. N., Underwood, E. C., D'Amico, J. A., et al., 2001. Terrestrial ecoregions of the world: A new map of life on Earth, *Bioscience* 51(11): 933–938.

Pal, M., 2005. Random forest classifier for remote sensing classification, *International Journal of Remote Sensing,* 26(1): 823–854.

Pohl, C., and Van Genderen, J. L., 1998. Review article multisensor image fusion in remote sensing: Concepts, methods and applications, *International Journal of Remote Sensing,* 19(5): 823–854.

Popescu, S. C., Wynne, R. H., and Nelson, R. F., 2003. Measuring individual tree crown diameter with lidar and assessing its influence on estimating forest volume and biomass, *Canadian Journal of Remote Sensing,* 29(5): 564–577.

Schut, P. Ed., 2007. *OpenGIS Web Processing Service,* Open Geospatial Consortium, Wayland, MA.

Selkowitz, D. J., Green, G., Peterson, B., and Wylie, B., 2012. A multi-sensor lidar, multi-spectral and multi-angular approach for mapping canopy height in boreal forest regions, *Remote Sensing of Environment,* 121: 458–471.

Sexton, J. O., Song, X., Feng, M., Noojipady, P., Anand, A., Huang, C., and Kim, D., 2013. Global 30-m resolution continuous fields of tree cover: Landsat-based rescaling of MODIS vegetation continuous fields with lidar-based estimates of error, *International Journal of Digital Earth,* 6(5): 427–448.

Shi, S., and Nigel, W., 2012. Automated geoprocessing mechanism, processes and workflow for seamless online integration of geodata services and creating geoprocessing services, Selected topics in applied earth observations and remote sensing, *IEEE Journal,* 5(6): 1659–1664.

Simard, M., Pinto, M. J. F., Fisher, J., and Baccini, A., 2011. Mapping forest canopy height globally with spaceborne lidar, *Geophysical Research Letters,* 116: G04021.

Yang, W., Ni-Meister, W., and Lee, S., 2011. Assessment of the impacts of surface topography, off-nadir pointing and vegetation structure on vegetation lidar waveforms using an extended geometric optical and radiative transfer model, *Remote Sensing of Environment*, 115(11), 2810–2822.

Zwally, H. J., Schutz, R., Bentley, C., Bufton, J., Herring, T., Minster, J., Spinhirne, J., et al., 2011. *GLAS/ICESat L2 Antarctic and Greenland Ice Sheet Altimetry Data V002*, National Snow and Ice Data Center. Digital media, Boulder, CO.

9

Multiscale Analysis of Urban Areas Using Mixing Models

Dar Roberts, Michael Alonzo, Erin B. Wetherley, Kenneth L. Dudley, and Phillip E. Dennison

CONTENTS

Introduction

Globally, more than 50% of the world's population lives in urban areas, with the percentage forecasted to increase to 66% by 2050 (UN-DESA 2014). Although urban areas constitute less than 3% of the Earth's terrestrial surface, they have a significant ecological footprint, accounting for an estimated three-fourths of global carbon emissions and 60% of residential water use (Grimm et al. 2008). They are also major local and regional sources of airborne pollutants and waste products (Grimm et al. 2008) and can have a strong impact on local environmental conditions. For example, elevated air temperatures in urban areas, known as the *urban heat island effect* (Voogt and Oke 2003; Weng et al. 2004; Weng 2009), can negatively impact human health (Patz et al. 2005) and can modify urban phenology by extending the growth season in cold environments (Zhang et al. 2004). Urban form, such as the distribution of urban green space, impacts urban climate (cooling by tree shading and advection; Middel et al. 2014), whereas road structure impacts transportation flows and associated emissions (Hoek et al. 2008). Of particular interest is the relative proportion of green cover and impervious surfaces, which impacts urban air temperatures (Weng et al. 2004; Lu and Weng 2006; Myint et al. 2010), surface runoff (Leopold 1968; Cuo et al. 2008), and avian diversity (Fontana et al. 2011).

The clear relationship between urban surface cover and environmental response has prompted considerable research on the use of remote sensing to map vegetated and impervious surface fractions (e.g., Ridd 1995; Phinn et al. 2002; Wu and Murray 2003; Weng and Lu 2008). However, urban environments are challenging in that they are composed of extremely diverse surface types (Herold et al. 2003) localized into objects that are smaller than the spatial resolution of most spaceborne, and some airborne, systems (Jensen and Cowen 1999). Airborne systems can provide data at scales from several centimeters to a few meters and thus can resolve fine-scale objects. However, high spatial resolution imagery's spectral range and resolution is frequently insufficient to discriminate urban materials such as soils from senesced grass (Roberts et al. 1993) or tile roofs from other roof types (Herold et al. 2003). Imaging spectrometers, which sample many wavelengths, can discriminate a wide range of materials in the urban environment, but have been deployed over only a small subset of cities, primarily in North America and Europe. Spaceborne sensors, such as WorldView-2 (1.84 m multispectral, 0.46 m panchromatic), have fine enough spatial resolution to resolve most urban objects but also lack the spectral resolution needed for accurate classification. Furthermore, data from commercially available sensors are relatively expensive and typically have a small swath (16.4 km at nadir for WorldView-2). Landsat data (Thematic Mapper [TM], Enhanced Thematic Mapper, and Operational Land Imager), with a 30-m spatial resolution, lack the spatial resolution to resolve all but the largest objects in the urban environment but does provide global, repeat sampling over most of the world's urban areas at no cost. For this reason, Landsat data are often used to map green cover and impervious surface area in urban areas

To illustrate the impact of spatial scale, we show images acquired by the Classic and Next Generation (NG) Airborne Visible Infrared Imaging Spectrometer (AVIRIS) sensors over a portion of downtown Santa Barbara, California (Figure 9.1). In Figure 9.1a through d, we show AVIRIS data acquired at different spatial resolutions including 3.9 m (AVIRIS-NG, 2014), 7.5 m (AVIRIS-Classic, 2011), 18 m (AVIRIS-Classic, 2013), and 30 m simulated (AVIRIS-Classic, 2013). The finest resolution of 3.9 m is similar to resolutions provided by sensors such as RapidEye, 18 m is similar to the multispectral bands of SPOT-5, and 30 m corresponds to the spatial resolution of the Landsat series. At 3.9 and 7.5 m, most of the main urban features such as roads, red tile, composite shingle, and commercial roofs are readily apparent (Figure 9.1a and b). At 18 m, dark, linear road features can still be resolved, while at 30 m they are almost entirely lost (Figure 9.1c and d). At the finest spatial resolution, individual cars are discernable (Figure 9.1e). Large features, such as city parks and green sports fields, are apparent at all scales, as are the largest commercial roofs (bright white in all images).

The effect of spatial scale can also be observed in spectra by comparing 3.9 m AVIRIS-NG spectra to AVIRIS-Classic at 30 m (Figure 9.1f). Large, low stature areas such as a green sports field (Location 3 on Figure 9.1e)

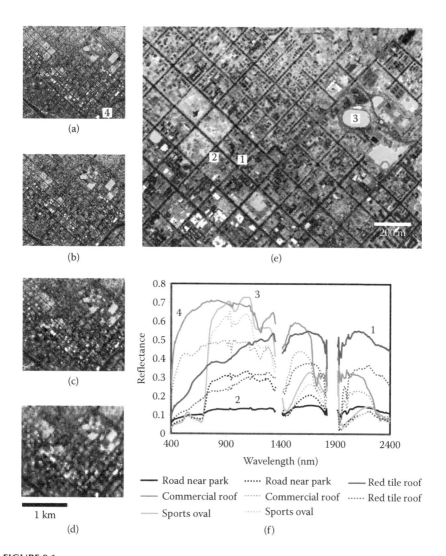

FIGURE 9.1
Example of the impact of spatial scale on the spectra and pattern of surface features in an urban environment (downtown Santa Barbara). Images acquired at (a) 3.9 m (Airborne Visible Infrared Imaging Spectrometer, Next Generation [AVIRIS-NG], 2014-06-03), (b) 7.5 m (AVIRIS-Classic, 2011-07-19), (c) 18 m (AVIRIS-Classic), and (d) simulated 30 m (2013-04-11). Wavelengths displayed include 1650, 830, and 650 nm as red, green, and blue. (e) An expanded view of AVIRIS-NG data that include the Santa Barbara Courthouse. (f) Spectra at 3.9 m (solid) and simulated 30 m (dashed) for red tile roof (1: red), asphalt road (2: black), a green sports field (3: green), and commercial roof (4 on 1a: gray). Reflectance spectra were supplied by JPL as standard reflectance products (From Thompson et al. *Remote Sensing of Environment*, 2015).

are similar at both spatial scales, showing minor brightness differences due to differences in seasonality and lighting (June 2014 compared to April 2013). Large, tall objects (red tile roofs and commercial roofs, Location 1 on Figure 9.1e and Location 4 on Figure 9.1a) also show similar spectral shapes, as well as more significant differences in brightness due to a high solar zenith in the April imagery. Fine-scale features, such as roads, are well resolved spectrally at 3.9 m but become mixed with adjacent surfaces at 30 m (Location 2 on Figure 9.1e). At 30 m, a mixed spectrum consisting of road and adjacent vegetation appears similar to that of dark vegetation (black dashed line in Figure 9.1f).

Spectral diversity is also a challenge in urban remote sensing. Spectral variability among surfaces is a product of absorption by specific chemical compounds such as iron oxides, hydrocarbons, clays, carbonates, and pigmented paints (Herold et al. 2004; Roberts and Herold 2004; Heiden et al. 2007). Spectral variation may be enhanced due to aging, weathering, vertical structure, and illumination (Herold and Roberts 2005; Van der Linden and Hostert 2009). For example, roofs can be made of composite shingle, metal, tile, asphalt, wood, concrete, slate, brick, or several other materials, all of which can vary spectrally as a product of manufacturing, age, weathering, color, slope, and aspect. Similarly, transportation surfaces can be composed of asphalt, concrete, tile, cobble, gravel, or several other materials, all of which may vary based on coating, age, and weathering. Considerable variation can also be present in vegetation. For example, over 450 tree species are present in the city of Santa Barbara, although more than 80% of the crown area is represented by fewer than 30 tree species (Alonzo et al. 2014). The addition of urban lawns, hedges, and shrubs adds further spectral variability at a range of spatial scales.

Spectral variability significantly impacts the ability to discriminate materials in the urban environment. To illustrate, spectra acquired using a field spectrometer and high spatial resolution AVIRIS are shown for six urban material classes: nonphotosynthetic vegetation (NPV), green vegetation (GV), rocks, soils and sand, impervious ground, and roofs (Figure 9.2). Some classes, such as NPV and soils, show subtle spectral differences, while others show very pronounced differences. For example, within the impervious surface category (roofs and impervious ground), over 25 unique spectra are shown, although these represent less than 10% of the spectra used to map Santa Barbara (see case study). The GV is nearly as diverse, showing considerable variation even within a subclass, such as low vegetation (Figure 9.2b). Only a fraction of the spectral diversity present in urban trees in the Santa Barbara is shown here (see Alonzo et al. 2014).

The combination of high spectral diversity and coarse spatial resolution relative to object size has led to a suite of specialized analysis approaches designed to address the limitations of remotely sensed imagery and provide the necessary information on urban form and composition, in particular green cover and impervious surface fractions. Many of these approaches

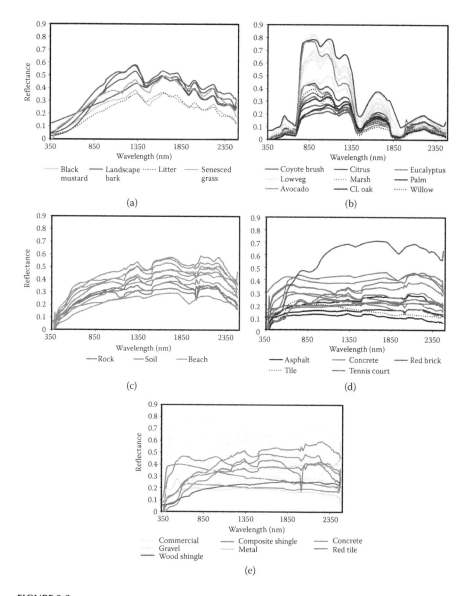

FIGURE 9.2
Example spectra of (a) nonphotosynthetic vegetation (NPV), (b) green vegetation (GV), (c) rock and soil, (d) impervious ground surfaces, and (e) roofs.

follow the universal vegetation-impervious surface-soils (VIS) framework proposed by Ridd (1995), in which urban materials are categorized into these three broad classes. However, they differ in their general approach and include two broad classes of methodology: classification, in which a single pixel or group of pixels is placed into a specific class based on its spatial or spectral attributes, and mixture analysis, in which a pixel consisting of two or more materials is decomposed into an estimate of the fractional abundances of each material. Classification, especially object-based image analysis, has been widely used in urban areas and is particularly well suited to fine spatial resolution data where urban objects can be identified based on their form and spectral properties (for a review see Blaschke 2010). At coarser spatial scales, where most objects are smaller than the spatial resolution of the sensor, the more common approach is spectral mixture analysis (SMA), in which a mixed pixel is decomposed into fractional estimates of pure, sub-pixel components, often called *endmembers* (Roberts et al. 1993; Settle and Drake 1993; Keshava and Mustard 2002). In this chapter, we focus on SMA in the urban environment, including a case study using multiple endmember spectral mixture analysis (MESMA; Roberts et al. 1998) applied to AVIRIS data acquired over the city of Santa Barbara to illustrate the potential of mixing models. Note that fine spatial scale does not eliminate the presence of mixed pixels. Even at submeter scales, a pixel imaging a part of a tree can include branches, leaves, and shadows and a pixel imaging a roof often includes multiple tiles and shadows. Fine-scale variation within a pixel often leads to misclassification within urban objects (i.e., a tree crown classified as multiple species) that can be mitigated at object scales using, for example, a pixel majority filter (e.g., Alonzo et al. 2014). Both classification approaches and SMA are further complicated by high within-class spectral variability, where a single class may have very high spectral variance leading to classification errors. In these cases, a single spectrum may be insufficient to represent the spectral variability within that class (Figure 9.2).

SMA in the Urban Environment

SMA is highly appropriate as a tool for estimating VIS fractions in the urban environment because it is designed to retrieve subpixel fractions from images where the spatial resolution is coarser than objects in the environment. SMA is a technique in which a mixed pixel is modeled as a combination of pure spectral components (endmembers) (Keshava and Mustard 2002). The most common approach used in the urban environment is linear mixture analysis, in which each pixel is modeled as a linear combination of several classes of spectra, most often GV, impervious surface, and soil. Typically, a single set of endmembers is applied to an image. Endmembers can be derived from field or laboratory spectra, images (Keshava and Mustard 2002), estimated spectra (e.g., virtual endmembers; Tompkins et al. 1997), or even radiative transfer models (Sonnentag et al. 2007). Fractional abundances are typically estimated using a variety of inversion approaches, including least-squares regression (Keshava and Mustard 2002), Gram–Schmidt orthogonalization

(Adams et al. 1993), and singular value decomposition (Boardman 1989). The solution space can be constrained to produce physically reasonable fractions or left unconstrained (Keshava and Mustard 2002). Regression approaches and nonlinear models also exist for estimating surface fractions, using quantitative training samples (i.e., pixels that include a range of known surface fractions) to train artificial neural networks (e.g., Pu et al. 2008) and support vector machines (SVMs) (Okujeni et al. 2015).

One of the first applications of SMA in an urban environment image was presented by Small (2001), who employed a three-endmember model consisting of low albedo (e.g., asphalt), high albedo (e.g., concrete), and vegetation to estimate GV cover in the city of New York using Landsat TM data. This approach was extended by Wu and Murray (2003) to include soil as a fourth endmember, thus incorporating all three key elements of the original VIS model. Wu and Murray (2003) applied this model to Landsat TM data acquired over Columbus, Ohio. In their study, a single set of endmembers was used, identified as extremes in minimum noise fraction transforms plotted in *n*-dimensional space. Modeled estimates of impervious fraction were highly accurate, with minor underestimates in densely urbanized areas and slight overestimates in less developed areas. Pu et al. (2008) also applied a similar approach in Yokohama, Japan, using ASTER data, but found that an artificial neural network provided more accurate measures of fractional cover than standard SMA using unconstrained least squares regression and low albedo, high albedo, soil, and vegetation as endmembers.

The potential adverse impact of within-class spectral variability (Figure 9.2) on fractional cover estimates has been recognized, and several researchers have developed alternative approaches. For example, Roessner et al. (2001) proposed an iterative approach, in which pure endmembers are identified in an urban scene and used as seeds in SMA to unmix their neighbors. The use of local sources enabled them to account for within-class spectral variability and urban spatial structure using automation. Song (2005) proposed Bayesian spectral mixture analysis, in which the impact of endmember variability is captured in a probability distribution function for that endmember and fractions are then estimated using Bayes' theorem. The most common approach employed for analysis of urban environments has been MESMA. MESMA accounts for endmember variability by allowing the number and types of endmembers to vary on a per-pixel basis. A spectral library is developed, which includes more than one spectrum for each endmember class (e.g., several spectrally distinct types of vegetation within the GV class). All possible combinations of two, three, four, or more endmembers are tested in a linear mixture for a pixel, and the model that produces physically reasonable fractions and the lowest error in fit (as expressed by a root mean square error [RMSE]) is selected.

MESMA can be used as a classifier, in which a class is assigned to a pixel based on the spectrum selected as the best endmember for that pixel for the two-endmember case (one bright endmember and shade, e.g., Dennison and

Roberts 2003). At higher levels of complexity (three endmembers or more), MESMA is often used only to estimate fractional cover. These two applications represent very different goals. As a classifier, the objective is to include all spectra needed to quantify the spectral variability of a class, including a class that itself may be mixed. For example, when mapping orchards, orchard spectra may include pure canopy or a mixture of canopy and the soil between rows. At higher levels of complexity, where the goal is to map fractional cover of GV, soil, or impervious, a pure canopy spectrum would be a viable endmember, but a mixture of canopy and soil would not. In addition, as the number of endmembers included in a model increases, the number of combinations required increases as the product of the number of endmembers within each class (i.e., 10 soils, 10 GV, and 10 impervious endmembers would result in 1,000 possible combinations) and the ability to discriminate materials decreases (Roberts et al. 1998). Thus, it is often necessary to subselect candidate spectra from a spectral library to improve both computational efficiency and discrimination. For this reason, numerous techniques have been developed for improved endmember selection (see the Santa Barbara case study).

MESMA was employed by Rashed et al. (2003) to map vegetation, impervious, soil, and shade fractions in the city of Los Angeles using Landsat TM data. Eight endmembers, including two vegetation, three impervious surfaces, two soils, and shade, were identified using the pixel-purity index (Boardman et al. 1995). These spectra were subsequently used to generate a total of 63 models including all possible combinations of two, three, or four endmembers and combinations with and without a shade endmember. Powell et al. (2007) expanded the use of MESMA to analyze Landsat data acquired over the city of Manaus, Brazil. In their study, endmembers were derived from a combination of image and field spectra, and endmember selection was accomplished using three-endmember subselection tools: count-based endmember selection (COB; Roberts et al. 2003), endmember average RMSE (EAR; Dennison and Roberts 2003), and minimum average spectral angle (MASA; Dennison et al. 2004). Potential endmembers were further subselected to match actual cover fractions derived from digital videography, then refined to discard endmembers that performed poorly in the image. Final models included a complexity term (in which the choice between a two, three, or four endmember model was based on an RMSE change threshold) where the more complex model (the model that used the higher number of endmembers) was only selected if the RMSE decrease between complexity levels exceeded a given threshold. Urban composition was shown to vary with the age of each neighborhood, with the highest soil fractions found in the youngest parts of Manaus, the highest impervious fraction in intermediate aged neighborhoods, and increased vegetation cover in the oldest neighborhoods. Powell and Roberts (2008) subsequently ported this approach to 10 cities in the Brazilian state of Rondonia, demonstrating that the approach could be applied over large geographic areas and multiple dates to observe the growth and evolution of a frontier city. A significant finding in that study was that the complexity term (number of models used in a pixel) provided a quantitative measure of urban extent, in which urbanized areas frequently required four endmembers (GV, impervious, soil, and shade), disturbed landscapes three, and natural landscapes two. There was a clear linear relationship between the natural log of population (x) and the natural log of urban area (y), defined as all areas requiring four endmembers, matching the classic relationship between built-up area and population first described by Nordbeck (1965) and Tobler (1969).

Numerous variants of MESMA have been subsequently applied in urban remote sensing (Franke et al. 2009; Myint and Okin 2009; Demarchi et al. 2012; Michishita et al. 2012; Liu and Yang 2013; Okujeni et al. 2013; Fan and Deng 2014; Wu et al. 2014; Okujeni et al. 2015). For example, Wu et al. (2014) developed spatially constrained MESMA, in which candidate endmembers for each class were drawn from a predefined neighborhood. Endmember candidates were identified within the search window using k-means clustering; thus, the endmember selection process could be automated. Spatially constrained MESMA was tested using Landsat ETM+ data acquired over Franklin County, Ohio. Comparable accuracy was achieved compared to endmember sets applied to the entire image using MESMA. Franke et al. (2009) proposed hierarchical unmixing, in which the mixing process was divided into stages with early stage analysis restricting the types of endmembers used in the later stages of analysis. Hierarchical unmixing was applied to 4-m HyMap data acquired over the city of Bonn, Germany. First, the image was modeled as a two-endmember case with only two categories, impervious and pervious. Optimal endmembers for each category were identified from a comprehensive spectral library developed from the HyMap Bonn data set using EAR, MASA, and COB and used to unmix the image. The impervious/pervious map (a classification based on which endmember was selected) was estimated at 97% accuracy and was subsequently used to spatially restrict the types of endmembers that could be selected at the second level. Thus, soils, water, and vegetation were restricted to areas previously modeled as pervious, while built-up represented the impervious class. At the third level, seven classes were mapped, including trees and grass (vegetation); soil, river, and lakes (water); and roads and roofs/buildings (built-up areas), generating both a classified product and an estimate of fractional cover. The classified product was generated by assigning a pixel to the class of the endmember with the largest fractional cover. The final level of complexity included 20 classes, including several tree species and multiple roof types, each spatially constrained by the previous level of complexity. Through this approach, the higher accuracy achieved at lower levels of complexity could be used to reduce confusion between spectra at higher levels of complexity and to improve accuracy.

Recently, several studies have evaluated the potential of accurate estimates of vegetation and impervious fractional cover using imaging spectrometry at relatively coarse (18–30 m) spatial resolutions. Fan and Deng (2014) applied MESMA to map forest, grass, impervious, and soil fractions using 30-m Hyperion data acquired over Guangzhou City, China. In that study, the authors employed a spectral angle, rather than RMSE criteria, to select between models. Demarchi et al. (2012) applied MESMA to map the fractional cover of vegetation, soil, and impervious surfaces in Leuven, Belgium, using CHRIS/Proba data. Similar to Powell et al. (2007), the authors selected different levels of complexity based on RMSE differences but used a relative measure of RMSE change rather than an absolute measure based on the percentage decline in RMSE at one level compared to the previous level of complexity. Roberts et al. (2012) analyzed combined AVIRIS and MODIS/ASTER airborne simulator data to evaluate potential synergies between the visible-shortwave infrared (VSWIR) and thermal infrared (TIR) parts of the spectrum. Mixing models were produced at spatial resolutions ranging from 7.5 to 60 m to evaluate the impact of the proposed 60-m spatial resolution of the Hyperspectral Infrared Imager (HyspIRI) on cover fraction estimates. While they found that material classification would be difficult for small objects at 60 m, accurate estimates of GV, impervious, NPV, and soil fractions could be derived at all spatial resolutions because spectral variability of surface materials was incorporated within MESMA. In this case, spectra were combined from two sources: field spectra of very small surfaces such as sidewalks and composite shingle roofs, and AVIRIS 7.5-m spectra of vegetation, commercial roofs, and other large surfaces. Okujeni et al. (2013) further extended the MESMA concept, in this case generating preformulated mixtures of the main surface components (roof, pavement, tree, grass, other), then using an SVM approach to classify each pixel as one of the predefined mixtures. They found this approach produced higher accuracies than conventional MESMA. Most recently, Okujeni et al. (2015) applied SVM classification of predefined mixtures to simulated ENMAP data for the city of Berlin, demonstrating that subpixel fractions of tree, grass, roof, roads, and soils could be generated even at the proposed 30-m resolution of ENMAP.

Case Study from Santa Barbara

Here, we highlight the potential value of MESMA for mapping an urban environment by analyzing AVIRIS-Classic data acquired at 7.5-m spatial resolution on July 19, 2011 (Figure 9.3). We produced a classification based on a two-endmember model and estimates of fractional cover for GV, NPV, paved, rock, roof, and soil based on two-, three-, and four-endmember models. We subdivided the impervious class into *paved* and *roof* endmembers to evaluate the potential of using MESMA to discriminate the height classes of impervious surface, similar to Okujeni et al. (2015). We further discriminated low vegetation from high vegetation in the two-endmember classified product, based on which type of endmember was selected. To evaluate

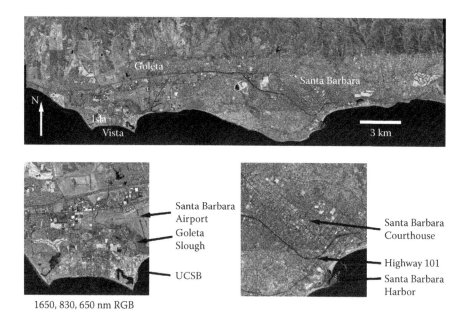

1650, 830, 650 nm RGB

FIGURE 9.3
Study site. Subsets include a section of the image in the west (left), which includes Isla Vista, the airport, and portions of Goleta, and east (right), which shows downtown Santa Barbara.

the effectiveness of urban-height classification using MESMA, we compared MESMA-derived classified height classes to lidar-based classes using broad classes of low vegetation, high vegetation, ground, and building.

The Santa Barbara urban environment represents an ideal test bed for remote sensing in that it includes considerable spectral variability in vegetated and nonvegetated materials. Furthermore, this area has been extensively studied, including studies by Herold et al. (2003; 2004) and Herold and Roberts (2005), establishing some of the first urban spectral libraries and more recent studies evaluating the potential of MESMA in urban (Roberts et al. 2012) and mixed natural/anthropogenic covers (Roth et al. 2012). At very high spatial resolutions, AVIRIS (Alonzo et al. 2013) and combined AVIRIS and waveform lidar (Alonzo et al. 2014) have been used to map 15 and 29 common urban tree species, respectively. In this chapter, we used both very high spatial resolution digital orthophotos (1 m) and waveform lidar to validate some of the MESMA products.

Study Site Description

The study site is a 34 × 10 km subset of three east–west AVIRIS flights flown on June 19, 2011. The subset extends from the city of Goleta, CA, on the west to the border between Santa Barbara and Montecito, CA, on the east. In 2010, the cities of Goleta and Santa Barbara had a combined population in excess of 170,000, including the unincorporated areas between the two cities and Isla Vista (US Census 2010). Based on census tract data, mean household income varies greatly from a mean below $40,000 to $200,000 (U.S. Department of Commerce, U.S. Census Bureau, Geography Division. 2010). Urban land cover is diverse, ranging from low- to high-density residential housing, apartments, industrial areas (in Goleta), commercial businesses (clustered along Highway 101 in Goleta and downtown Santa Barbara), parks, golf courses, open fields, and less frequent reservoirs and small lakes. Small-scale agriculture is practiced along the fringe of urbanized areas, particularly in Goleta. Citrus and avocado orchards are prevalent north of each city, located between the more urbanized coastal plain and south-facing slope of the Santa Ynez Mountains. The coastal plain is dominated by human-modified "natural cover" including European introduced grasslands, Eucalyptus groves, and extensive invasions by black mustard. Native vegetation in the plains consists of small-stature shrublands dominated by coyote bush and extensive marshes south of the Santa Barbara Airport and east and west of the University of California, Santa Barbara (UCSB; Figure 9.3). Native vegetation is primarily restricted to riparian corridors, which are dominated by willow, sycamore, and coast live oak. Native vegetation also occurs in infrequent forested slopes within the urban boundaries (mostly coast live oak) and extensive evergreen shrublands on the south flank of the Santa Ynez range.

Methods

Our primary analysis tool was MESMA, which was used to classify land cover at 7.5-m spatial resolution based on spectra selected using only two endmembers and to map the fractional cover of GV, NPV, paved, rock, roof, and soil using a fusion of two-, three-, and four-endmember models. The basic procedure for selecting endmembers, classifying the image, and mapping fraction cover is shown schematically in Figure 9.4. The left side of the figure shows the procedure for building the initial spectral library using iterative endmember selection (IES; Schaaf et al. 2011; Roth et al. 2012). A master library was extracted from the AVIRIS data using image and field-assessed polygons. Here polygons either consisted of >70% of a single species, a single land-use class (e.g., orchards), or a single urban material identified in Google Earth imagery (e.g., asphalt road or commercial roof). Polygon and image information were used to generate metadata for each spectrum (Table 9.1). This library was then randomly sampled to generate training and validation libraries as in Roth et al. (2012). Image spectra were not suitable for some surfaces that were poorly resolved at 7.5-m spatial resolution, such as sidewalks and composite shingle residential roofs. Field spectra derived from Herold et al. (2004) were used to augment the training library to create a combined field/image library (Powell et al. 2007; Roberts et al. 2012).

MESMA can be used as a classifier using a single bright endmember and shade and can be used to map fractional cover as mixtures of two-, three-,

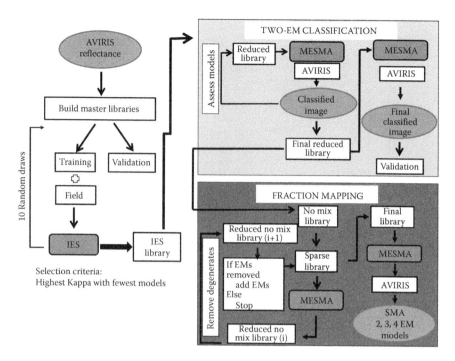

FIGURE 9.4
Flow chart showing steps for spectral library development, endmember selection using iterative endmember selection, endmember reduction to improve efficiency and accuracy, classification, further library reduction, and final mixture modeling.

or four-endmember models plus shade. IES uses two endmember models to model a spectral library, where potential endmembers are taken from the library itself to maximize the classification accuracy of MESMA (Schaaf et al. 2011). The combined field/image library was input into IES using the finest level of spectral categorization (L7; Table 9.1) to determine classification accuracy. IES was run 10 times using input spectral libraries containing randomly selected training spectra, and the IES run that produced the highest kappa (Cohen 1960) with the lowest number of endmembers was selected as the final IES library. For this case study, IES generated libraries consisting of 378–651 models with a kappa between 0.753 and 0.911. By regressing the number of endmembers in a model (x) against kappa (y) and fitting a line, we identified the best IES library as one that plots above the line, yielding a kappa of 0.855 using 497 endmembers.

The right side of Figure 9.4 shows the image analysis steps and further model refinement. IES generates very large endmember libraries, including spectra that may be unique in an endmember library but poorly represented in an image. In addition, IES may select endmembers that can be confused with other materials in the image that were not well represented

TABLE 9.1

Subset of Metadata Fields

Name	L1	L2	L3	L4	L5	L7	Source
DTR_19	per	GV	shrub	evergreen	needleleaf	ADFA	AVIRIS_r22
BAPI_I05	per	GV	shrub	evergreen	broadleaf	BAPI	AVIRIS_r21
EUSP_I04	per	GV	tree	evergreen	broadleaf	EUSP	AVIRIS_r20
golfcourse18	per	GV	herb	annual	grass	IRGR	AVIRIS_r21
MARSH07	per	GV	herb	perennial	NA	MARSH	AVIRIS_r20
PEAM_I08	per	GV	tree	evergreen	broadleaf	PEAM	AVIRIS_r22
closed_canopy_oak	per	GV	tree	evergreen	broadleaf	QUAG	AVIRIS_r21
ndbnyg.002-	per	NPV	bark	new	NA	BARK	ASD
BRNI_I03	per	NPV	herb	annual	forb	BRNI	AVIRIS_r20
basketball_asphalt_field_n1	imp	Paved	parking_lot	asphalt	unknown	ASPHALT_PARKING_LOT	AVIRIS_r21
asphalt_road_surface_1	imp	Paved	road	asphalt	101	ASPHALT_ROAD	AVIRIS_r21
airport_asphalt_n2	imp	Paved	airport_rway	asphalt	old	ASPHALT_RUNWAY	AVIRIS_r21
parking_structure_concrete	imp	Paved	parking_lot	NA	NA	CONCRETE_PARKING_LOT	AVIRIS_r20
rpcsof.005-	imp	Paved	road	concrete	old	CONCRETE_ROAD	ASD
spcsye.003-	imp	Paved	sidewalk	concrete	new	CONCRETE_SIDEWALK	ASD
rotrye.006-	imp	Paved	road	tile	new	TILE_ROAD	ASD
RDP_115	imp	Rock	NA	NA	NA	ROCK	AVIRIS_r22
public_storage_n1	imp	Roof	commercial	shingle	white	COMMERCIAL_ROOF	AVIRIS_r21
fsctmg.032-	imp	Roof	residential	comp_shingle	lt_tan	COMP_SHINGLE_ROOF	ASD
rompero_respervoir_n2	imp	Roof	concrete	grey	unknown	CONCRETE_ROOF	AVIRIS_r21

(Continued)

TABLE 9.1 (Continued)

Subset of Metadata Fields

Name	L1	L2	L3	L4	L5	L7	Source
phelps_asphalt&grey_gravel	imp	Roof	asphalt/gravel	NA	old	GRAVEL_ROOF	AVIRIS_r21
mission_red_tile_1	imp	Roof	church	tile	old	RED_TILE	AVIRIS_r21
fsfnof.002-	imp	Roof	residential	wood_shingle	old	WOOD_SHINGLE_ROOF	ASD
lrxnxx.002-	per	Soil	soil	NA	NA	SOIL	ASD

Note: Metadata are required because techniques such as iterative endmember selection require knowledge of which spectra fall into each search category. Here we present a hierarchical approach in which spectra are labeled first under very broad classes, then further subdivided into finer classes. The name is the name of a spectrum based on the original polygon, or field spectrum used as a spectrum source. At the broadest level (L1), we include two classes: pervious (per) and impervious (imp). At L2 we further divide this into seven categories: GV, NPV, soil, rock, paved, roof, and water (not shown). At L3, we divide each L2 class into the next finer level, such as shrub or tree (GV), bark or herb (NPV), road or parking lot (paved), or residential or commercial (roof). Within each L3 class we divide by phenology (e.g., evergreen, annual) to create L4, then by leaf form (e.g., needleleaf, broadleaf) to create L5, and finally by species to create L7. Impervious surfaces have no equivalents at L4 or higher, so we use material at L4 (asphalt, concrete), followed by color, age (e.g., light tan = lt_tan), or other characteristic. The highest metadata field (L7) divides plants by species or genus while defining the finest descriptor for an urban surface, such as asphalt road, concrete roof, or red tile. Source describes the source of spectra (AVIRIS or ASD spectrometer). Several metadata fields have been removed for clarity.

in the training library. The final objective was twofold: to generate the most accurate classification using a subset of the best performing spectra (two-EM classification, Figure 9.4) and to produce fraction maps using a subset of pure, spectrally distinct endmembers (fraction mapping, Figure 9.4). Two-endmember classification began by applying MESMA to the AVIRIS images using the 497-endmember IES library. Rare (<100 pixels modeled in the image) and overmapped (e.g., an avocado orchard spectrum that models riparian areas or native forests more often than avocado orchard) endmembers were removed (Assess Models on Figure 9.4), to produce a reduced library. This process was repeated iteratively, producing a new reduced library with each iteration, until a satisfactory classified map was produced that excluded rare spectra or spectra that mapped the wrong material more often than the right one based on visual assessment (final reduced library, Figure 9.4). In this study, the final classified map was generated using a final reduced library of 376 endmembers applied to the AVIRIS image.

We used a different procedure to map fractional cover (fraction mapping: Figure 9.4). The goal of this step was to identify a subset of the purest end-members for each class that accounted for the greatest within-class variance and minimized within-class spectral degeneracy. Fraction mapping started with the final reduced library used for classification. However, because a land use/land cover class may consist of a mixture of materials, a library developed for classification may not be suitable for mapping fractional cover. An example in an urban environment would be an orchard, where tree cano-pies may be separated by rows of bare soil or plant residues. In this case, only the pure canopy spectra or bare soil between rows of trees would be suit-able for a mixing model. Mixed spectra were removed by visual assessment of each spectrum. Water was also removed from the endmember library, because water is readily confused with dark impervious surfaces in a mixing model. Water is often masked out when using SMA in urban areas (e.g., Wu and Murray 2003; Powell et al. 2007; Pu et al. 2008). The resulting endmember library, termed the *no-mix library*, consisted of 258 spectra.

While greatly reduced in size, an endmember library with 258 spectra still allows for a large number of potential endmember combinations and thus a large number of models to run and compare for each pixel. There are 923,349 unique models for a 258-endmember library if all possible two-, three-, and four-endmember models are used. Furthermore, many of the endmembers contained within the library are only subtly different and thus represent unnecessary added complexity and increased run time. These "degenerate" spectra were removed from the no-mix library using the following process. First, the highest reflectance endmember was selected from each L2 class (one GV, NPV, rock, soil, roof, and paved), generating a sparse library. This endmember subset was then used to model the reduced no-mix library. In the first pass ($i = 0$, where i is the iteration), the no-mix library and reduced no-mix libraries were the same. Any spectrum that was modeled by the sparse library without requiring an additional nonshade endmember was then discarded to produce a new iteration of the reduced no-mix library ($i + 1$). The next candidate spectrum for each L2 class was then selected from this library and added to the previous subset using two criteria: (1) it had high reflectance and (2) it had been modeled as highly mixed in the previous step. These highly mixed spectra were considered to be poorly modeled by the previous endmember subset and thus potentially unique. The augmented subset of endmembers was again used to unmix the next iteration of the reduced no-mix library ($i = 1$), and well-modeled spectra from this library were again discarded to generate another iteration of the reduced no-mix library ($i = 2$). A new set of highly mixed spectra from each L2 class was selected from this library and added to the endmember subset, and the process continued until no more degenerate spectra were present in any endmember class, generating the final library. Through this process, the original no-mix library of 258 spectra was reduced to a final subset of 90 endmembers, resulting in 90; 3,113; and 47,980 potential two-, three-, and four-endmember models.

The final library was applied to the AVIRIS images to generate all possible two-, three-, and four-endmember models. A single model was determined for each pixel by using an RMSE threshold of 0.007 to select between the best-fit two-, three-, or four-endmember models, similar to Powell et al. (2007) and Roberts et al. (2012). For example, if the RMSE decreased by more than 0.007 between the best-fit two- and three-endmember models, the three-endmember model was selected.

Results and Discussion

Using MESMA to separately model NPV, GV, soils, rocks, roof, and paved as distinct classes offers a powerful alternate framework for describing urban surfaces. In our first example (Figure 9.5), we show a VIS, SMA, and two-endmember classification product (based on 376 endmembers) for the western subset (see Figure 9.3 for subset location). The class was determined based on which endmember was selected by the two-endmember MESMA model, with each spectrum placed into one of 24 land use/land cover classes based on its metadata. This portion of Goleta is composed of orchards in the north, Highway 101 and commercial businesses and industry in the center, the UCSB campus in the southeast, and high-density residential Isla Vista in the southwest. The Santa Barbara Airport is also a dominant feature. Vegetated areas include marshes south of the airport, mixtures of native and introduced European grasslands, and eucalyptus groves (Figure 9.5d).

In the first frame, we show a modified form of a VIS model based on fractions modeled in MESMA (Figure 9.5a). The VIS model was generated from the MESMA result by combining the GV and NPV fractions as vegetation, the rock, paved, and roof fractions for impervious, and using the soil fraction for soil. Fractions could result from any complexity level (two, three, or four endmembers). In Figure 9.5a, the impervious fractions are shown in red, the vegetation fractions are shown in green, and the soil fractions are shown in blue.

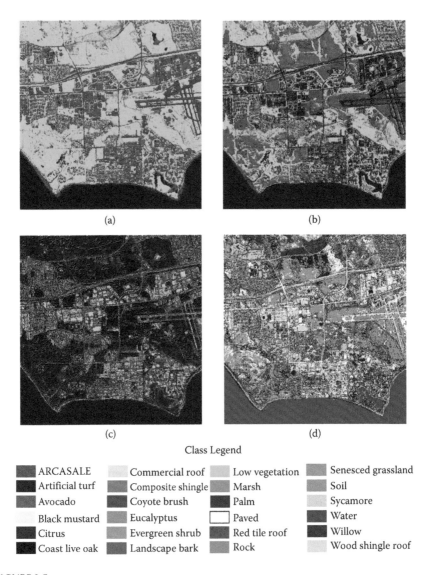

Class Legend

■ ARCASALE	Commercial roof	Low vegetation	Senesced grassland
■ Artificial turf	Composite shingle	Marsh	Soil
Avocado	■ Coyote brush	■ Palm	Sycamore
Black mustard	Eucalyptus	□ Paved	■ Water
Citrus	Evergreen shrub	Red tile roof	Willow
■ Coast live oak	Landscape bark	Rock	Wood shingle roof

FIGURE 9.5
(a) Modified VIS model, (b) NPV-GV soil, (c) impervious ground-roof-rock, and (d) multiple endmember spectral mixture analysis (MESMA) classified map. Colors for (a) through (c) are loaded as RGB, and the legend refers to colors shown in (d). The region shown is the western subset (Figure 9.3). Black areas are unclassified pixels.

This VIS model differs from the traditional VIS model, which does not include NPV as a category and most likely places NPV into the *soil* category because broad-band sensors lack the multiple shortwave infrared (SWIR) bands needed to discriminate them (Numata et al. 2008). In this urban environment, NPV is common and it would be inappropriate to treat it as soil.

Bare soil surfaces are prone to wind and water erosion, whereas NPV protects surfaces from these effects, which is a primary motivation behind retaining plant residues on agricultural surfaces (Daughtry et al. 2004). Furthermore, earlier in the growth season senesced areas would have been classified as GV, contributing ecosystem services, such as photosynthesis and cooling by transpiration. Even when fully senesced, vertically oriented senesced plant materials cast shadows, cooling the surface (Roberts et al. 2012). NPV and soils do have some common elements; however, both are pervious surfaces and neither photosynthesizes.

In Figure 9.5b, we show the classic NPV, GV, soil model (loaded as RGB) that is commonly used to describe natural landscapes. Here, NPV and GV are separately modeled, enabling us to show which parts of the landscape are senesced and which may be actively photosynthesizing. We further discriminate vegetation classes (Figure 9.5d) based on the specific model that was selected using the reduced library from IES (Figure 9.4). Here we discriminate NPV into two classes, black mustard and senesced grassland. The former is an invasive species, while the latter includes a mix of invasive and native species. Black mustard forms tall stands of senesced plant material, in contrast to the shorter-stature plants in senesced grassland. Orchards, such as citrus (Figure 9.5d), also stand out as vegetated areas that have a high soil fraction, capturing the exposed soil between rows of trees (Figure 9.5b). By contrast, avocado orchards form dense stands with closed canopies and show little soil. Low stature vegetation, such as green sports fields and golf courses, are discriminated from all other vegetated surfaces (Figure 9.5d).

The use of multiple classes of materials in MESMA allows different classification schemes that extend beyond traditional VIS or SMA models (Figures 9.5c and 9.5d). In Figure 9.5c, ground-level impervious surfaces are shown in red, roofs in green, and rock in blue. Road networks and the airport runway appear as linear features displayed in red, while most roofs are displayed as regularly shaped objects in green. In this case, MESMA did not map surface height, but rather mapped spectrally unique impervious surfaces that occurred either at ground level or were most often used as roofing material. The ability to discriminate roofs from roads, even at the coarse 30-m spatial resolution of the proposed ENMAP sensor, has been shown (Okujeni et al. 2015).

Our next example shows the same basic products but focuses on the eastern subset (see Figure 9.3) centered over the central business district of Santa Barbara (Figure 9.6). In this area, vegetation in the modified VIS model (Figure 9.6a) is nearly identical to vegetation mapped as GV in an NPV, GV, soil RGB image (Figure 9.6b). In the highly urbanized urban core, very little open space remains and large areas of senesced plant material are uncommon. This scenario is likely more representative of a typical large urban city, where space is at a premium. Bare soil surfaces are present but are mostly concentrated along the beach (Figure 9.6b).

The urban core in this subset is dominated by impervious surfaces (Figure 9.6a). Impervious surfaces are nearly equally represented by

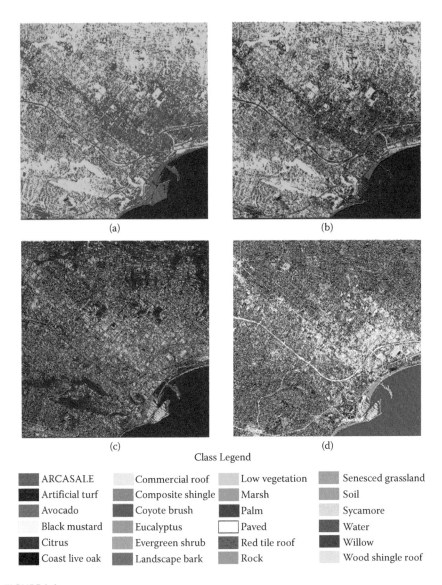

FIGURE 9.6
(a) Modified VIS model, (b) NPV-GV soil, (c) impervious ground-roof-rock, and (d) MESMA classified map. Colors for (a) through (c) are loaded as RGB, and the legend refers to colors shown in (d). The region shown is the eastern subset (Figure 9.3). Black areas are unclassified pixels.

ground-level surfaces (shown in red) and roofs (shown in green; Figure 9.6c). Roads are well represented as linear features, and Highway 101 is particularly well mapped as consisting of asphalt surfaces considerably wider than the 7.5-m spatial resolution (Figure 9.6c and d). Commercial roofs dominate downtown Santa Barbara, displayed in cyan on the classified map (Figure 9.6d).

Of the smaller-scale roofs, red tile roofs are also prominent, both on public buildings and on apartments. Residential roofs consisting of composite shingle are not obvious and, in fact, are often modeled as commercial roofs north of the city. This result is not surprising given that many commercial roofs in the Santa Barbara area are made of the same material as composite shingles. Vegetation classified in the urban area is mostly incorrect at Level 7, mapped either as avocado or citrus rather than the actual species present. Given that the model did not include most of the tree species in the Santa Barbara area, MESMA classified the vegetation as the next most similar class. However, it is also notable that native vegetation (evergreen shrub in Figure 9.6d) was not mapped in the urban core, suggesting that native vegetation and urbanized vegetation are distinct enough to be discriminated. It is possible to classify urban vegetation to the species level (e.g., Alonzo et al. 2013, 2014); however, the 7.5-m spatial resolution of this data set would result in lower accuracy and the segmentation used by Alonzo et al. may not be suitable for smaller canopies at 7.5-m spatial resolution.

The potential to map height classes using imaging spectrometry and MESMA is exciting. However, is it accurate? We evaluated this question by comparing MESMA-derived height classes, created by assigning specific materials to each class, against a height class map generated using waveform lidar (Figure 9.7). In this figure, roads, sidewalks, parking lots, green sports fields, golf courses, and irrigated lawns were assigned to ground; shrubs and chaparral to low vegetation; citrus, avocado, oak, sycamore, evergreen shrubs, and any other tall vegetated surface to high vegetation; and all roofs to buildings.

Visually, the two maps correspond well. The impact of the 7.5-m spatial resolution of AVIRIS is obvious by the pixelated nature of most surfaces. By comparison, the waveform lidar (with approximately 22 pulses per m^{-2}) provided clearly defined shapes (Figure 9.7, Polygon 8, lower left frame). MESMA also included a much larger number of unclassified pixels. These represent pixels that could not be classified using a single endmember and shade (Figure 9.4) but could still be modeled as a mixture of NPV, GV, soil, rock, paved, or roof.

To provide a more detailed comparison, height class fractions derived from lidar (proportion of a polygon within a derived height class) and MESMA height-class fractions were compared in 13 polygons distributed across the city (Figures 9.7 and 9.8). Here we show fractional cover for high vegetation (Figure 9.8a), buildings (Figure 9.8b), and ground (Figure 9.8c). Polygon descriptions are provided on the lower right.

Based on this analysis, the MESMA classification appears to consistently underestimate high vegetation cover (Figure 9.8a). Underestimates varied from only a few percent (Polygon 3, oak stand and orchards) to close to 20% in industrial or commercial areas (Polygons 1 and 2). High vegetation cover is most similar in residential areas (Polygons 4, 5, 7, and 8), although the difference still approached 10% in some areas (Polygons 9, 11, and 12). Errors in the MESMA vegetation height classification tended to be compensated for

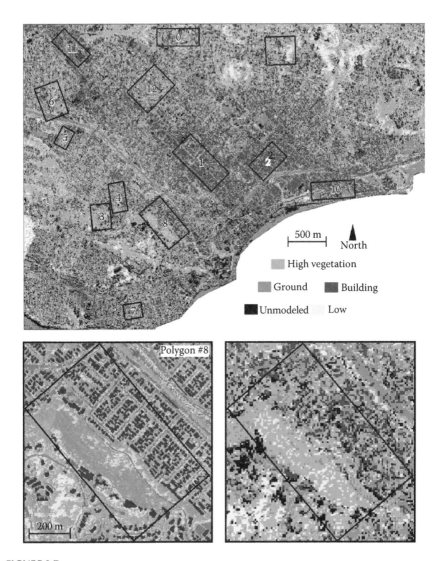

FIGURE 9.7
MESMA-derived height classes (top and lower right) showing building, ground, high vegetation, and low vegetation. Polygons on the upper map correspond to 13 locations that cover a range of urban surface covers. The figure on the lower left was generated from classification of waveform lidar. The figure on the lower right was generated from the MESMA classified map by placing different surface classes into their corresponding height class.

either as an error in mapping buildings or ground surfaces. For example, in Polygons 1, 2, and 11, an underestimate of high vegetation by MESMA is complemented by an overestimate of ground. This error may have two origins. First, MESMA may be classifying vegetation shadows as vegetation with a high shade fraction, whereas lidar would correctly assign these pixels

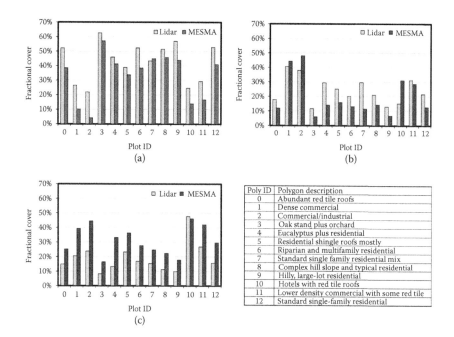

FIGURE 9.8
Surface composition for the 13 polygons including (a) high vegetation, (b) building, and (c) ground. Ground included soils, low vegetation, and impervious ground surfaces

to ground. Second, the MESMA classification likely has a cover threshold below which a pixel will either be left unclassified or assigned to ground or building. The lidar, with finer spatial resolution, can correctly detect the presence of small trees or shrubs. Given that smaller trees and shrubs are common in the more commercial or industrial parts of Santa Barbara, it is not surprising that many of them were missed by the MESMA classifier.

MESMA classification of buildings is more mixed (Figure 9.8c). In Polygons 3 to 9, MESMA consistently underestimated the proportion of building area. With the exception of Polygon 3, these polygons can all be defined as some variant of residential. Thus, it appears that MESMA is underestimating roof area in residential zones. We suspect that this result is a product of spectral ambiguity between roofs and roads (Table 9.1; Herold et al. 2003; Roberts et al. 2012). A composite shingle roof is made up of many of the same materials as an asphalt road surface. In residential zones, composite shingle roofs erroneously classified as asphalt roads would result in an overestimate of ground at the expense of building. We find it notable that Polygon 11, in which a dominant roof material is red tile, shows only a minor difference in the area of building mapped compared to lidar. Thus, the ability of MESMA to discriminate roofs and roads will vary between roof materials (Okujeni et al. 2015). In Polygons 1, 2, and 10, which are described as commercial, industrial, and hotels,

MESMA tended to overclassify building area. Again, we suspect material composition is playing a role. For example, a red-tiled courtyard would be classified by MESMA as red-tile roof, yet it is actually a ground object. Similarly, a concrete parking structure, bridge, or sidewalk modeled by MESMA using the spectrum from a concrete roof would be called *building*, but it would be classified as ground based on lidar. Concrete parking structures are particularly challenging—based on height they operate more like a building, but functionally they are used to park cars.

In Figures 9.7 and 9.8, we compared lidar-based results to MESMA results to evaluate the quality of the MESMA height class map. We can also assess the accuracy of fraction maps (Figure 9.9). Here, we compared high spatial resolution data to MESMA-derived fractions for (a) GV, (b) roof, (c) paved ground, (d) impervious, (e) NPV, and (f) soil. The reference fractions were generated in eCognition (v. 6.4, Trimble, Munich, Germany) based on object-based image analysis of 0.3-m digital aerial imagery combined with the aforementioned data set. Buildings, paved ground, and high vegetation were discriminated automatically using the lidar; grass was distinguished from impervious surface using a greenness threshold on the aerial imagery; and NPV and soil were discriminated manually based on visual assessment of the imagery. MESMA and reference fractions were compared in 37 100×100 m polygons distributed randomly across the urban core.

Overall, the MESMA fraction accuracies were high, especially for GV (Figure 9.9a). The relationship between reference and MESMA GV had an r^2 of 0.92. However, contrary to the two-endmember classified product, MESMA consistently overestimated GV cover, with a bias of 3% and slope of 1.15. The roof fraction was also mapped with a relatively high r^2 (0.73) and showed a linear relationship between the model and reference. Still, MESMA appeared to underestimate the roof fraction, with a slope of 0.75.

MESMA estimates of paved ground also are highly correlated with reference data, but the best-fit model is not linear (Figure 9.9c: $r^2 = 0.84$ polynomial, $r^2 = 0.76$ for a linear equation with a slope of 0.731 and intercept of –0.058). When compared to reference fractions, MESMA underestimated the paved-ground fraction at low surface cover but estimated the paved-ground fraction accurately at high surface cover. This result is consistent with the hypothesis that shadows cast by vegetation are being modeled as containing a GV fraction. When vegetation cover is low, the shadows cast by plants would be modeled as vegetation at the expense of ground, resulting in an underestimate of the ground fraction. Notably, roofs, which would be expected to be less shadowed than roads or sidewalk, show a more linear relationship. In the MESMA classified image, many of these same dark pixels likely went unmodeled, leading to an underestimate of vegetation cover due to shadows. The poorest correlations were observed for NPV and soil (e and f), but these materials were also uncommon in much of this part of the study site (see Figure 9.6b). In addition, these materials are the most difficult to discriminate in digital imagery, making the reference data suspect.

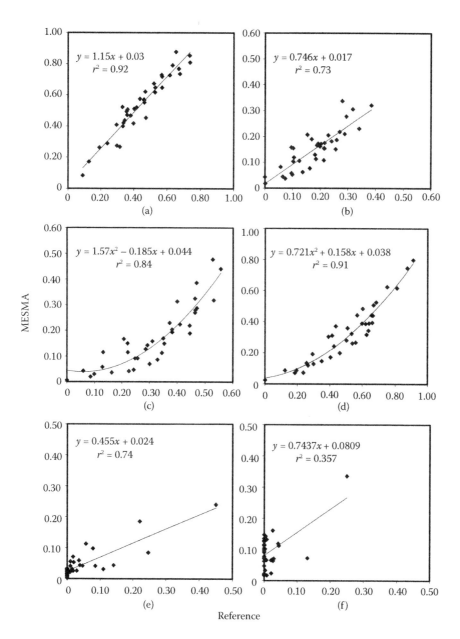

FIGURE 9.9
Validation of MESMA fractions (*y*) compared to reference fractions (*x*) derived from high resolution digital photography using 37,100 m^2 polygons. Surfaces include (a) GV, (b) roof, (c) paved
ground, (d) impervious (roof + paved), (e) NPV, and (f) soil.

The products generated in this case study have considerable potential. For example, accurate estimates of impervious fraction generated by MESMA could be used as input layers into a hydrological model to estimate storm runoff (see USDA 1986). Alternatively, the modified VIS model presented here could be used to evaluate socioeconomic drivers of urban land cover. Here, we compare MESMA-derived measures of vegetation cover and impervious surface fraction to mean income by census tract (Figure 9.10). In many cities in North America, there is a strong relationship between vegetation cover, impervious cover, and wealth (e.g., Iverson and Cook 2001; Jenerette et al. 2013), although this relationship does not hold true in some of the larger cities such as New York City, where some of the wealthiest residents live in apartments, or Baltimore, where the forested urban landscape more accurately represents past rather than present demographics (Boone et al. 2010).

In Santa Barbara, we see the classic relationship between wealth and fractional cover. Wealthier census tracts can be characterized as having the lowest impervious cover and highest vegetation cover. In comparison, less wealthy neighborhoods have lower vegetation cover and higher impervious cover.

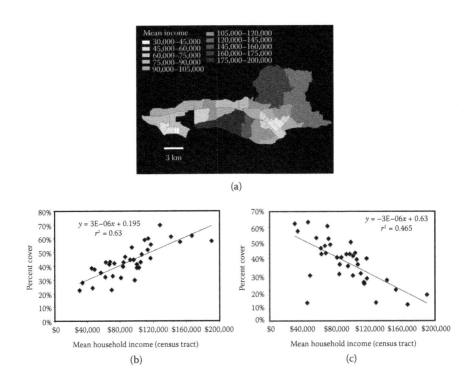

FIGURE 9.10
(a) Comparison of income by census tract, (b) against GV, and (c) impervious fractions. Outliers in Figure 9.10c are census tracts in the west that are sparsely populated.

This result is not surprising given the high cost of land in Santa Barbara. Areas with more parks and sports fields tend to have higher income, while areas with higher-density housing tend to have lower income.

Limitations and Future Research

In this chapter, we presented an overview of the use of SMA to map urban composition, emphasizing the use of SMA and its variants. We illustrated some of the potential of MESMA with a case study, applying MESMA to 7.5-m AVIRIS data acquired over the cities of Santa Barbara and Goleta. We used MESMA both as a classifier and a tool for mapping fractional cover. MESMA proved to be extremely versatile in mapping urban land cover. Use of multiple endmembers and multiple levels of complexity provides a high level of detail, enabling mapping of a variety of impervious surface types. This high level of detail can also be easily generalized using metadata, allowing mapping of broader land cover classes. This flexibility, combined with MESMA's ability to provide subpixel fractional cover, makes MESMA well suited to urban applications. Comparisons of mapped land cover to high spatial resolution digital imagery showed high accuracies, although relationships between reference and MESMA fractions were not always linear. We also demonstrated that MESMA can be useful for mapping height classes, which may be useful where lidar data are not available.

However, the analysis presented in this case study was initially only conducted at 7.5-m spatial resolution. Based on prior work (Schaaf et al. 2011; Roberts et al. 2012; Roth 2014), we know spatial scale can significantly impact the performance of MESMA classifications. For example, Roberts et al. (2012) evaluated the performance of MESMA at spatial resolutions of 7.5, 15, and 60 m. In that study, MESMA could not classify small-scale objects such as residential roofs and roads at the coarser resolutions. A comparison of spectra at 3.9 and 30 m (Figure 9.1e) illustrates how spatial scale might impact classification. Large-scale objects, such as commercial roofs and sports fields, could be classified accurately at all spatial scales from 3.9 to 30 m. By contrast, asphalt road surfaces could be classified at resolutions as coarse as 7.5 m, but not 30 m. The 7.5-m spatial resolution used here also has its limits for classification. For example, many urban trees have relatively small crowns. These trees could be classified at 3.5 m (Alonzo et al. 2013) but likely would not have been accurately classified at 7.5 m. Natural vegetation, however, often forms patches that are significantly larger than 60 m and can be classified accurately even at relatively coarse spatial resolutions; accuracies may actually improve at coarser spatial scales (Schaaf et al. 2011; Roth 2014).

To evaluate the impact of spatial scale on classification and fractional cover mapping, we spatially degraded the original 7.5-m AVIRIS to 30 m, similar to the spatial resolution of Landsat. We then applied the 376-endmember final reduced library to the 30-m AVIRIS to generate a two-endmember classified product and the 90-endmember final library to the same image to generate estimates of fractional cover (Figure 9.4). To compare classified products, we then took the 30-m classified product and resampled it back down to 7.5 m using nearest-neighbor resampling. Thus, a 30-m pixel consisting of a single class was split into 16 7.5-m pixels consisting of a single class. This enabled direct, pixel–pixel comparison of the 30- and 7.5-m classified products. Fraction images were treated using the reverse scaling operation, aggregating fraction maps at 7.5 to 30 m based on pixel average, thus converting 16 different estimates of fractional cover to a single fraction for each endmember class and permitting direct comparison of one 30-m product to another. Based on prior work, we would predict fractions to scale well between 7.5 and 30 m, but agreement between classifications to be low.

As predicted, agreement between spatial scales was relatively low. At Level 7, overall agreement was only 30%, with only one class (senesced grassland) showing greater than 50% agreement. It is notable that this class typically forms large, homogeneous blocks that are resolved well at 30 m. Citrus orchards were a close second, showing 47% agreement, followed by low vegetation at 34%. When individual pixels were placed into the broader classes of high vegetation, low vegetation, NPV, paved, roof, soil, rock, and water (Table 9.2), the agreement between scales was much higher, exceeding 60%. However, much of the increased accuracy is due to three classes: high vegetation, NPV, and inland waters (lakes). Assuming 7.5 m represents the correct class, high vegetation had a 79.6% producer's accuracy and 70.7% user accuracy. NPV had lower accuracies at 57% producer's accuracy and 65.2% user accuracy, whereas inland water had a 50.6% producer's accuracy but the higher user accuracy at 74.9%. Thus, while two-endmember MESMA could

TABLE 9.2

Comparison of Two Endmember Classification at 7.5 and 30 m for Seven Broad Classes

Class	7.5 m	30 m	Difference	Producer's Accuracy	User Accuracy
All	3718200	3798126	79926	0.615	0.602
High veg	2208982	2488032	279050	0.796	0.707
Low veg	454908	515460	60552	0.293	0.259
NPV	444689	388584	−56105	0.57	0.652
Paved	168758	80505	−88253	0.174	0.365
Roof	264610	207991	−56619	0.267	0.339
Soil	74061	40533	−33528	0.256	0.468
Rock	79670	61822	−17848	0.153	0.197
Inland water	22522	15199	−7323	0.506	0.749

Note: The numbers below the column heads 7.5 and 30 m report the number of pixels in each class at 7.5 m. Veg, vegetation; NPV, nonphotosynthetic vegetation.

not discriminate one type of L7 class from another at 30 m, it did select high vegetation. NPV and water, which form some of the largest, most homogeneous surfaces in the region, also cross scales well. Both high vegetation and low vegetation tended to be overmapped at 30 m, as shown by an increase in the area mapped from 7.5 to 30 m. Accuracies for low vegetation, paved, roof, soil, and rock also tended to be higher when aggregated but were below 30% in all cases, with paved mapped with a 17.4% producer's accuracy and a 36.5% user accuracy. This suggests that paved at 30 m was undermapped, most likely at the expense of high and low vegetation. This is reflected in the area mapped, in which the number of pixels classified as paved at 7.5 m dropped by more than a factor of 2 at 30 m.

As predicted, fractions scaled well between 7.5 and 30 m for GV and NPV, producing r^2 values of 0.77 and 0.74, respectively, slopes near 1.0, and intercepts near zero (Figure 9.11). Similar to Roberts et al. (2012), impervious fractions did not scale as well but still resulted in a reasonably high r^2 value of 0.61. When subdivided into roof and paved fractions, r^2 values were significantly lower, suggesting that impervious fractions can be scaled between 7.5 and 30 m, but the ability to discriminate roofs from paved surfaces degrades at 30-m resolution. We conclude that MESMA used to map fractional cover is more appropriate than classification at 30 m.

TIR remote sensing, potentially available from sensors such as the proposed HyspIRI mission, may also contribute significantly to urban analysis (Roberts et al. 2012). TIR sensors provide a direct measurement of the urban heat island effect (Lo et al. 1997; Lu and Weng 2006; Weng 2009). The TIR complements the SWIR because asphalt, which may be difficult to discriminate from soil in the VSWIR, can be spectrally distinct from soil in the TIR (Roberts et al. 2012). Furthermore, dark surfaces such as asphalt, which can be confused with water or modeled as shade, would be hot in a daytime TIR image, potentially discriminating shadows from dark surfaces. In this study, we found MESMA to overestimate the GV fraction at low cover, likely due to tree shadows. The use of TIR data may reduce this error.

Potentially one of the most exciting new developments in urban remote sensing is the potential of newer, push-broom imaging spectrometers, such as AVIRIS-NG. In principle, AVIRIS-NG can acquire data at spatial resolutions as fine as 0.5 m with a significantly higher signal-to-noise ratio than can be achieved using the whisk broom design used by sensors such as AVIRIS-Classic or HyMap (Hamlin et al. 2011). Not only do sensors such as AVIRIS-NG offer the potential of improved, fine spatial resolution mapping of urban surfaces and urban tree species discrimination, they will also allow for the construction of spectral libraries that can be used to improve classification accuracy and fraction mapping at coarser spatial scales (Schaaf et al. 2011; Roberts et al. 2012). In this study, field spectra of composite shingle roofs and sidewalks were used since they could not be resolved as pure at 7.5-m spatial resolution. In the future, fine spatial resolution AVIRIS-NG data could be used in a similar fashion, providing representative endmembers that can be applied

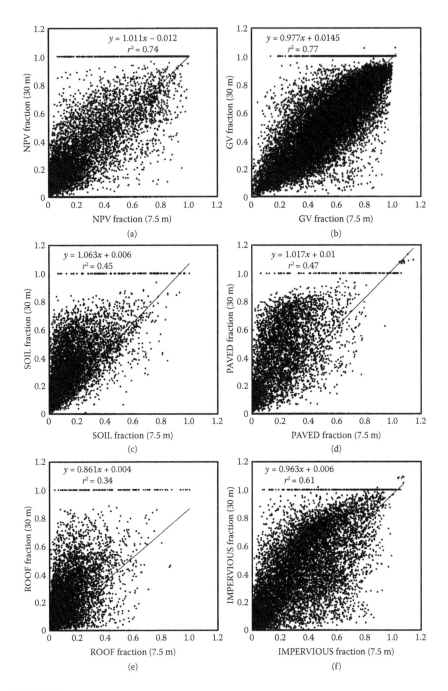

FIGURE 9.11
Scatterplots showing fractions for (a) NPV, (b) GV, (c) soil, (d) paved, (e) roof, and (f) impervious (combined roof and paved) with 7.5 m plotted on (*x*) and 30 m (on *y*).

to coarser resolution data. The use of MESMA with image spectra collected at multiple spatial scales could supplant the need for most field spectra and provide a robust means for mapping urban land cover across spatial scales.

Acknowledgments

The authors wish to thank William Ouellette and Keely Roth for assistance in developing the reference polygons used in this research. AVIRIS imagery at all spatial scales was kindly supplied by the Jet Propulsion Laboratory, which also supplied georectified reflectance at 3.9, 18, and 30 m. This research was supported in part by National Aeronautics and Space Administration Grants NNX11AE44G, "Evaluation of Synergies between VNIR-SWIR and TIR Imagery in a Mediterranean-Climate Ecosystem," and NNX12AP08G, "HyspIRI Discrimination of Plant Species and Functional Types along a Strong Environmental–Temperature Gradient." In this chapter, MESMA was implemented using Viper Tools 2, an update to Viper Tools partially supported by the Belgian Science Policy Office in the framework of the STEREO III program, Project VIPER2 (SR/XX/171).

References

Adams, J.B., Smith, M.O., & Gillespie, A.R. 1993. Imaging spectrometry: Interpretation based on spectral mixture analysis. In *Remote geochemical analysis: Elemental and mineralogical composition* (eds. C.M. Pieters and P. Englert). pp. 145–166. New York, NY: Cambridge University Press.

Alonzo, M., Bookhagen, B., & Roberts, D.A. 2014. Urban tree species mapping using hyperspectral and lidar data fusion. *Remote Sensing of Environment* 148: 70–83.

Alonzo, M., Roth, K., & Roberts, D. 2013. Identifying Santa Barbara's urban tree species from AVIRIS imagery using canonical discriminant analysis. *Remote Sensing Letters* 4(5): 513–521.

Blaschke, T. 2010. Object based image analysis for remote sensing. *ISPRS Journal of Photogrammetry and Remote Sensing* 65: 2–16.

Boardman, J.W. 1989. Inversion of imaging spectrometry data using singular value decomposition. In *Quantitative remote sensing: An economic tool for the Nineties; Proceedings of IGARSS '89 and 12th Canadian Symposium on Remote Sensing,* Vancouver, Canada, July 10–14, 1989. Volume 4 (A91-15476 04-43). pp. 2069–2072 New York, NY: Institute of Electrical and Electronics Engineers.

Boardman, J.W., Kruse, F.A., & Green, R.O. 1995. Mapping target signatures via partial unmixing of AVIRIS data. In *Summaries of JPL Air- borne Earth Science Workshop* (ed. R.O. Green), vol. 1, pp. 23–26. Pasadena, CA: JPL Publication, 95–1, v. 1.

Boone, C.G., Cadenasso, M.L., Grove, J.M., Schwarz, K., & Buckley, G.L. 2010. Landscape, vegetation characteristics, and group identity in an urban and suburban watershed: Why the 60s matter. *Urban Ecosystems* 13(3): 255–271.

Cohen, J. 1960. A coefficient of agreement for nominal scales. *Educational and Psychological Measurement* 20(1): 37–46.

Cuo, L., Lettenmaier, D.P., Mattheussen, B.V., Storck, P., & Wiley, M. 2008. Hydrologic prediction for urban watersheds with the distributed hydrology—Soil-Vegetation Model. *Hydrological Processes* 22: 4205–4213.

Daughtry, C.S.T., Hunt Jr., E.R., & McMurtrey III, J.E. 2004. Assessing crop residue cover using shortwave infrared reflectance. *Remote Sensing of Environment* 90: 126–134.

Demarchi, L., Canters, F., Chan, J.C-W., & Van de Voorde, T. 2012. Multiple endmember unmixing of CHRIS/Proba imagery for mapping impervious surfaces in urban and suburban environments. *IEEE Transactions on Geoscience and Remote Sensing* 50(9): 3409–3424.

Dennison, P.E., Halligan, K.Q. & Roberts, D.A. 2004. A comparison of error metrics and constraints for multiple endmember spectral mixture analysis and spectral angle mapper. *Remote Sensing of Environment* 93: 359–367.

Dennison, P.E., & Roberts, D.A. 2003. Endmember selection for multiple endmember spectral mixture analysis using endmember average RSME. *Remote Sensing of Environment* 87(2–3): 123–135.

Fan, G., & Deng, Y. 2014. Enhancing endmember selection in multiple endmember spectral mixture analysis (MESMA) for urban impervious surface area mapping using spectral and spectral distance parameters. *International Journal of Applied Earth Observation and Geoinformation* 33: 290–301.

Fontana, S., Sattler, T., Bontadina, F., & Moretti, M. 2011. How to manage the urban green to improve bird diversity and community structure. *Landscape and Urban Planning* 101: 278–285.

Franke, J., Roberts, D., Halligan, K., & Menz, G. 2009. Hierarchical multiple endmember spectral mixture analysis (MESMA) of hyperspectral imagery for urban environments. *Remote Sensing of Environment* 113: 1712–1723.

Grimm, N.B., Faeth, S.H., Golubiewski, N.E., Redman, C.L., Wu, J., Bai, X., & Briggs, J.J. 2008. Global change and the ecology of cities. *Science* 319: 756–760.

Hamlin, L., Green, R.O., Mouroulis, P., Eastwood, M., Wilson, D., Dudik, M., & Paine, C. 2011. Imaging spectrometer science measurements for terrestrial ecology: AVIRIS and new developments. *IEE Aerospace Conference*, 1–7.

Heiden, U., Segl, K., Roessner, S., & Kaufmann, H. 2007. Determination of robust spectral features for identification of urban surface materials in hyperspectral remote sensing. *Remote Sensing of Environment* 111: 537–552.

Herold, M., Gardner, M., & Roberts, D.A. 2003. Spectral resolution requirements for mapping urban areas. *IEEE Transactions on Geoscience and Remote Sensing* 41(9 Part 1): 1907–1919.

Herold, M., & Roberts, D. 2005. Spectral characteristics of asphalt road aging and deterioration: Implications for remote-sensing applications. *Applied Optics* 44(20): 4327–4334.

Herold, M., Roberts, D., Gardner, M., & Dennison, P. 2004. Spectrometry for urban area remote sensing—Development and analysis of a spectral library from 350 to 2400 μm. *Remote Sensing of Environment* 91: 304–319.

Hoek, G., Beelen, R., de Hoogh, K., Vienneau, D., Gulliver, J., Fischer, P., & Briggs, D.J. 2008. A review of land-use regression models to assess spatial variation of outdoor air pollution. *Atmospheric Environment* 42(33): 7561–7578.

Iverson, L.R., & Cook, E.A. 2001. Urban forest cover of the Chicago region and its relation to household density and income. *Urban Ecosystems* 4: 105–124.

Jenerette, G.D., Miller, G., Buyantuev, A., Pataki, D.E., Gillespie, T.W., & Pincell, S. 2013. Urban vegetation and income segregation in drylands: A synthesis of seven metropolitan regions in the southwestern United States *Environmental Research Letters* 8: 1–9.

Jensen, J.R., & Cowen, D.C. 1999. Remote sensing of urban/suburban infrastructure and socioeconomic attributes. *Photogrammetric Engineering and Remote Sensing* 65(9): 611–622.

Keshava, N., & Mustard, J.F. 2002. Spectral unmixing. *IEEE Signal Processing Magazine* 19(1): 44–57.

Leopold, L.B., 1968. Hydrology for urban land planning—A guidebook on the hydrologic effects of urban land use. *US Geological Survey, Geological Survey Circular* 554: 18.

Liu, T., & Yang, X. 2013. Mapping vegetation in an urban area with stratified classification and multiple endmember spectral mixture analysis. *Remote Sensing of Environment* 133: 251–264.

Lo, C.P., Quattrochi, D.A., & Luvall, J.C. 1997. Application of high-resolution thermal infrared remote sensing and GIS to assess the urban heat island effect. *International Journal of Remote Sensing* 18(2): 287–304.

Lu, D. & Weng, Q. 2006. Spectral mixture analysis of ASTER images for examining the relationship between urban thermal features and biophysical descriptors in Indianapolis, Indiana, USA. *Remote Sensing of Environment* 104(2): 157–167.

Michishita, R., Jiang, Z., & Xu, B. 2012. Monitoring two decades of urbanization in the Poyang Lake area, China through spectral unmixing. *Remote Sensing of Environment* 117: 3–18.

Middel, A., Hab, K.I., Brazel, A.J., Martin, C.A., & Guhathakurta, S. 2014. Impact of urban form and design on mid-afternoon microclimate in Phoenix Local Climate Zones. *Landscape and Urban Planning* 122: 16–28.

Myint, S.W., Brazel, A., Okin, G., & Buyantuyev, A. 2010. Combined effects of impervious surface and vegetation cover on air temperature variations in a rapidly expanding desert city. *GIScience & Remote Sensing* 47(3): 301–320.

Myint, S., & Okin, G. 2009. Modelling land-cover types using multiple endmember spectral mixture analysis in a desert city. *International Journal of Remote Sensing* 30(9): 2237–2257.

Nordbeck, S. 1965. The law of allometric growth. *Michigan Inter-University Community of Mathematical Geographers,* Discussion Paper 7, p. 28, Department of Geography, University of Michigan, Ann Arbor, MI.

Numata, I., Roberts, D.A., Chadwick, O.A., Schimel, J.P., Galvao, L.S., & Soares, J.V. 2008. Evaluation of hyperspectral data for pasture estimate in the Brazilian Amazon using field and imaging spectrometers. *Remote Sensing of Environment* 112: 1569–1583.

Okujeni, A., van der Linden, S., & Hostert, P. 2015. Extending the vegetation-impervious-soil model using simulated EnMAP data and machine learning. *Remote Sensing of Environment* 158: 69–80.

Okujeni, A., van der Linden, S., Tits, L., Somers, B., Somers, B., & Hostert, P. 2013. Support vector regression and synthetically mixed training data for quantifying urban land cover. *Remote Sensing of Environment* 137: 184–197.

Patz, J.A., Campbell-Lendrum, D., Holloway, T., & Foley, J.A. 2005. Impact of regional climate change on human health. *Nature* 438(7066): 310–317.

Phinn, S., Stanford, M., Scarth, P., Murray, A.T. & Shyy, P.T. 2002. Monitoring the composition of urban environments based on the vegetation-impervious surface-soil model by subpixel analysis techniques. *International Journal of Remote Sensing* 23 (20): 4131–4153.

Powell, R.L., & Roberts, D.A. 2008. Characterizing variability of the urban physical environment for a suite of cities in Rondonia, Brazil. *Earth Interactions* 12(13): 32.

Powell, R., Roberts, D.A., Hess, L., & Dennison, P. 2007. Sub-pixel mapping of urban land cover using multiple endmember spectral mixture analysis: Manaus, Brazil. *Remote Sensing of Environment* 106: 253–267.

Pu, R., Gong, P., Michishita, R., & Sasagawa, T. 2008. Spectral mixture analysis for mapping abundance of urban surface components from the Terra/ASTER data. *Remote Sensing of Environment* 112: 939–954.

Rashed, T., Weeks, J.R., Roberts, D.A., Rogan, J., & Powell, R.L. 2003. Measuring the physical composition of urban morphology using multiple endmember spectral mixture models. *Photogrammetric Engineering and Remote Sensing* 69(9): 1011–1020.

Ridd, M.K. 1995. Exploring a V-I-S (vegetation-impervious surface-soil) model for urban ecosystem analysis through remote sensing: Comparative anatomy for cities. *International Journal of Remote Sensing* 16(12): 2165–2185.

Roberts, D.A., Adams, J.B., & Smith, M.O. 1993. Green vegetation, non-photosynthetic vegetation and soils in AVIRIS Data. *Remote Sensing of Environment* 44(2/3): 255–269.

Roberts, D.A., Dennison, P.E., Gardner, M., Hetzel, Y., Ustin, S.L., & Lee, C. 2003. Evaluation of the potential of Hyperion for fire danger assessment by comparison to the airborne visible/infrared imaging spectrometer. *IEEE Transactions on Geoscience and Remote Sensing* 41(6 Part 1): 1297–1310.

Roberts, D.A., Gardner, M., Church, R., Ustin, S., Scheer, G., & Green, R.O. 1998. Mapping chaparral in the Santa Monica Mountains using multiple endmember spectral mixture models. *Remote Sensing of Environment* 65: 267–279.

Roberts, D.A., & Herold, M. 2004. Imaging spectrometry of urban materials, in molecules to planets. In *Infrared Spectroscopy in Geochemistry, Exploration Geochemistry and Remote Sensing* (eds. P. King, M. Ramsey and G. Swayze). pp. 155–183. Québec, Canada: Mineral Association of Canada, London, Ontario, Canada.

Roberts, D.A., Quattrochi, D.A., Hulley, G.C., Hook, S.J., & Green, R.O. 2012. Synergies between VSWIR and TIR data for the urban environment: An evaluation of the potential for the hyperspectral infrared imager (HyspIRI) decadal survey mission. *Remote Sensing of Environment* 117, 83–101.

Roessner, S., Segl, K., Heiden, U., & Kaufmann, H. 2001. Automated differentiation of urban surfaces based on airborne hyperspectral imagery. *IEEE Transactions on Geoscience and Remote Sensing* 39(7): 1525–1532.

Roth, K.L. 2014. Discriminating Among Plant Species and Functional Types Using Spectroscopy Data: Evaluating Capabilities Within and Across Ecosystems, Across Spatial Scales and Through Seasons, UC Santa Barbara PhD Dissertation.

Roth, K.L., Dennison, P.E., & Roberts, D.A. 2012. Comparing endmember selection techniques for accurate mapping of plant species and land cover using imaging spectrometer data. *Remote Sensing of Environment* 127: 139–152.

Schaaf, A.N., Dennison, P.E., Fryer, G.K., Roth, K.L., & Roberts, D.A. 2011. Mapping plant functional types at multiple spatial resolutions using imaging spectrometer data. *GIScience and Remote Sensing* 48(3): 324–344.

Settle, J.J., & Drake, N.A. 1993. Linear mixing and the estimation of ground cover proportions. *International Journal of Remote Sensing* 14(6): 1159–1177.

Small, C. 2001. Estimation of urban vegetation abundance by spectral mixture analysis. *International Journal of Remote Sensing* 22(7): 1305–1334.

Song, C. 2005. Spectral mixture analysis for subpixel vegetation fractions in the urban environment: How to incorporate endmember variability? *Remote Sensing of Environment* 95: 248–263.

Sonnentag, O., Chen, J.M., Roberts, D.A., Talbot, J., Halligan, K.Q., & Govind, A. 2007. Mapping tree and shrub leaf area indices in an ombrotrophic peatland through multiple endmember spectral unmixing. *Remote Sensing of Environment* 109: 342–360.

Thompson, D.R., Gao, B-C, Green, R.O., Roberts, D.A., Dennison, P.E., & Lundeen, S. 2015. Atmospheric correction for global mapping spectroscopy: ATREM advances for the HysPIRI preparatory campaign. *Remote Sensing of Environment*, 167: 64–77.

Tobler, W.R. 1969. Satellite confirmation of settlement size coefficients. *Area* 52: 167–175.

Tompkins, S., Musard, J.F. Pieters, C.M., & Forsyth, D.W. 1997. Optimization of endmembers for spectral mixture anaylsis. *Remote Sensing of Environment* 59(3): 472–489.

United States Department of Agriculture 1986. *Urban hydrology for small watersheds.* Technical Release 55 (TR-55) (Second Edition). Natural Resources Conservation Service, Conservation Engineering Division.

U.S. Department of Commerce, U.S. Census Bureau, Geography Division. 2010. *2010 U.S. Census Demographic Profile Data (vector digital data).* http://www.census.gov/geo/maps-data/data/tiger.html

United Nations, Department of Economic and Social Affairs, Population Division 2014. *World Urbanization Prospects: The 2014 Revision, Highlights (ST/ESA/SER.A/352).*

Van der Linden, S., & Hostert, P. 2009. The influence of urban structures on impervious surface maps from airborne hyperspectral data. *Remote Sensing of Environment* 113: 2298–2305.

Voogt, J.A., & Oke, T.R. 2003. Thermal remote sensing of urban climates. *Remote Sensing of Environment* 86: 370–384.

Weng, Q. 2009. Thermal infrared remote sensing for urban climate and environmental studies: Methods, applications, and trends. *ISPRS Journal of Photogrammetry and Remote Sensing* 64: 335–344.

Weng, Q., & Lu, D. 2008. A sub-pixel analysis of urbanization effect on land surface temperature and its interplay with impervious surface and vegetation cover in Indianapolis, United States. *International Journal of Applied Earth Observations and Geoinformation* 10: 68–83.

Weng, Q., Lu, D., & Schubring, J. 2004. Estimation of land surface temperature–vegetation abundance relationship for urban heat island studies. *Remote Sensing of Environment* 89(4): 467–483.

Wu, C., Deng, C., & Jia, X. 2014. Spatially constrained multiple endmember spectral mixture analysis for quantifying subpixel urban impervious surfaces. *IEEE Journal of Selected Topics in Applied Earth Observations and Remote Sensing* 7(6): 1976–1984.

Wu, C., & Murray, A. 2003. Estimating impervious surface distribution by spectral mixture analysis. *Remote Sensing of Environment* 84: 493–505.

Zhang, X., Friedl, M.A., Schaaf, C.B., Strahler, A.H., & Schneider, A. 2004. The footprint of urban climates on vegetation phenology. *Geophysical Research Letters* 31: L12209.

10

Urban Road Extraction from Combined Data Sets of High-Resolution Satellite Imagery and Lidar Data Using GEOBIA

Minjuan Cheng and Qihao Weng

CONTENTS

Introduction

Remotely sensed data have been widely used in urban road extraction. Low resolution data cannot provide adequately detailed information for accurate estimation of road surface conditions. In low-resolution images, road extraction is often regarded as linear feature extraction because roads are considered to be "continuous and smooth lines" (Amini et al. 2002). Most traditional classification methods are based on statistical analysis of individual pixels. Those classifiers are well suited to images with relatively low spatial resolution (Wang et al. 2004). As high-resolution satellite images become available, a large amount of detailed information of ground features may be readily available to extract. On high-resolution images, a road is no longer a linear network of a few pixels, but a ribbon of features with a certain width. Therefore, many factors, such as lanes, cars, pedestrians, and shadows of trees and buildings, may affect the extraction of roads from the satellite images.

Because of large data volume, high-resolution imagery is often difficult to process and analyze. Some researchers tried to apply traditional feature extraction and classification methods to high-resolution imagery for road network detection (Shi and Zhu 2002; Long and Zhao 2005; Zhu et al. 2005; Gautama et al. 2006; Hu et al. 2007; Péteri and Ranchin 2007). These studies faced difficulties due to two reasons. First, the traditional road extraction methods considered roads as linear features, whereas in high-resolution images roads are no longer linear features but "continuous and elongated homogeneous regions" along a certain direction (Long and Zhao 2005). Second, increased spectral variance makes it difficult to separate a section of road from spectrally mixed urban land-cover types by using traditional pixel-based methods (Shaban and Dikshit 2001). The high-resolution imagery can reveal very fine details of the Earth's surface, and geometric information on ground features becomes clearly visible. Consequently, the advent of these images benefits the recognition and understanding of roads and expands the possibilities of the extraction methods. In contrast, the highly complex land-cover details brought by the high-resolution images cause an increase in geometric noise. For instance, a section of road may include pixels with various spectral values due to differences in materials, shading, or detailed conditions (Zhou and Troy 2008). Roads are often highly complex feature-packed information with geometric parameters that are critical in road recognition. Road extraction from high-resolution imagery also faces the problem that sharply increased noise such as trees, lanes, vehicles, and shadows would influence the data analysis. Numerous studies have been conducted to explore more effective and efficient methods for road extraction from high-resolution images as a surface. For example, Yang and Wang (2007) presented an improved model for road detection based on the principles of perceptual organization and classification fusion in the human vision system. The model consisted of four levels (pixel, primitives, structures, and objects) and two additional subprocesses (automatic classification of road scenes and global integration of multiform roads). Teng and Fairbairn (2002) used a GIS data set to assist the extraction of shape descriptors, reflectance, and height above ground characteristics of the road using a fuzzy expert system and an adaptive neuro-fuzzy system. Some approaches are based on edge- and line-detecting algorithms (Long and Zhao 2005), whereas Mourad et al. (2010) proposed a rule-based classification method using multispectral segmentation with inputs of digital maps and spectral data.

Instead of pixel, geographic object–based image analysis (GEOBIA) methods have been developed to process and analyze "meaningful objects" formed by grouping neighboring pixels according to certain rules (Blaschke 2010). Thus, these methods can take into account of not only spectral but also spatial information, such as size, texture, and contexture. Research has shown that GEOBIA methods have great potential over traditional pixel-based methods to improve the accuracy of feature extraction from high-resolution satellite images (Ivits et al. 2005; Im et al. 2007; Platt and Rapoza 2008;

Zhou and Troy 2008). Roads are difficult target features in an urban environment due to the complexity and packed information of ground features. Considering the advantages of GEOBIA methods, they are a good choice for road extraction. GEOBIA methods may overcome the limitations of pixel-based methods in processing high-resolution imagery, which show high spectral similarity among different land cover types and high spectral variation within the same land cover type (Zhou and Troy 2008). The key to these spectral characteristics is optimal segmentation for further processing (Baatz and Schape 2000), where scale is the most decisive parameter. Most previous studies attempted to identify a single optimal scale, which can lead to minimal over- and undersegmentation. However, Ke et al. (2010) demonstrated that there should be a range of scales, which would produce statistically similar classification accuracy. In addition, a multiscale approach has been advocated (Blaschke 2003; Arbiol et al. 2006).

Theoretically, roads can be distinguished from other features by exploiting spectral and spatial information. However, due to complex and sharp land cover change as well as undesired information included with the vast data content, some parts of roads might be missing while noise is included in the extraction result. Satellite images can only provide spectral and spatial information of the Earth's surface, whereas lidar data, which contains elevation information, can provide a three-dimensional description of ground features. Moreover, lidar data can provide intensity data, which is determined by the materials and is not influenced by shadow or other environmental factors. Previous studies had tried to find suitable means to make use of such combined data sets. Clode et al. (2007) presented a road extraction technique using lidar data in which a hierarchical technique was used to classify the lidar points into road and nonroad points, and the classification result was then vectorized to provide centerline and direction information. Other studies applied a multiclassifier system to lidar data to detect roads in the complex urban areas (Clode et al. 2005; Samadzadegan et al. 2009). In addition, lidar data have been widely used in urban classification (Charaniya et al. 2004; Brennan and Webster 2006; Miliaresis and Kokkas 2007; Im et al. 2008).

Fewer studies have tried to use a combined data set of high-resolution satellite imagery and lidar data for urban road extraction. Nevertheless, with useful information provided by lidar data, as well as the multiple types of information that can be integrated with GEOBIA, urban road extraction using high-resolution satellite imagery has potential for great improvement. Sander and George (2009) used airborne laser data fused with 2D topographic data to build an automated method for 3D modelling of highway interchanges. Tiwari et al. (2009) explored an integrated approach to road extraction using airborne laser scanning altimetry and high-resolution data. This method was applied to Amsterdam, the Netherlands, and yielded an accuracy of over 90% for automatic road extraction. It has also been demonstrated that the accuracy of classification gained great improvement by integrating high-resolution multispectral imagery and lidar data in urban environments (Hodgson et al. 2003). Zhou and Troy (2008) demonstrated that the usage of lidar data was very helpful for the separation of buildings and pavements. Hu et al. (2004) integrated the processing of high-resolution imagery with lidar data for automatic extraction of an urban road network in Toronto. They demonstrated that the integration of contexture information greatly improved the correctness and accuracy of the extraction result. Cai and Rasdorf (2008) used geographic information systems (GIS) and lidar point cloud data in a 3D modeling environment to obtain road centerline lengths. They concluded that use of lidar data greatly helped to obtain 3D roads with satisfying accuracy.

The objectives of this research are as follows: (1) to explore potential techniques for urban road extraction from combined data sets of IKONOS data and lidar data using an object-based method and (2) to compare extraction results using IKONOS data alone *versus* the combined IKONOS and lidar data set. The study area was Indianapolis, Indiana, USA. Experimental rules and parameters were mainly applied to downtown Indianapolis due to the limited processing time.

Study Area

Indianapolis was selected as the study area (Figure 10.1). It is located in Marion County, Indiana, USA. According to the 2010 US Census, Indianapolis' population was 820,445. The Indianapolis metropolitan area has been experiencing growth in population and urban land. It has 130 total highway miles and 4,004 total roadway miles. The city is known as "the crossroads of America." It is served by more highways, especially interstates, than any other cities in the United States. There are also many state roads, as well as the Indianapolis International Airport, serving this area. The geographic, socioeconomic, and transportation characteristics of Indianapolis indicated that the selected study area was a typical populous and highly developed US city, with complex urban road networks, and thus was suitable for a study of road extraction from remotely sensed data.

Data and Methodology

Figure 10.2 illustrates the analytical procedure of this research. First, road extraction was conducted using an IKONOS image with GEOBIA, and then the extraction was from the combined data set of IKONOS and lidar. It contained three steps: data preprocessing, segmentation, and rule-based extraction.

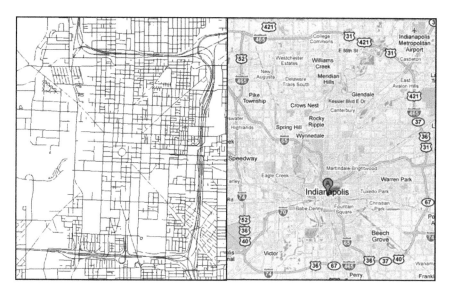

FIGURE 10.1
Study area: Downtown Indianapolis, IN, USA.

Data Sets and Preprocessing

Three types of data were used in this study: IKONOS images, lidar, and digital elevation models (DEMs). An IKONOS image dated October 6, 2003, was secured for research. Radiometric and geometrical distortions of the images were corrected to a quality level of 551 M before data delivery. Lidar data acquired in March and April 2003 were obtained from Indianapolis Mapping and Geographic Infrastructure Services. The lidar data set contained both last return (LST) and first return (FST) data. The accuracy of each lidar point (root mean square error) is 6 inches for the vertical and 1 foot for the horizontal. The DEM data, with a resolution of 5 feet, was downloaded from the Indiana Spatial Data Portal. According to the information provided by the Indiana Spatial Data Portal, the 2005 Indiana digital elevation and surface models were created from the 2005 IndianaMap Color Orthophotography Project data collected during the March and April leaf-off season at a minimum resolution of 1 foot statewide. Google Earth was used as the reference data.

The original 16 FST and 16 LST files contained numeric coordinates, as well as the elevation and intensity of each point. After being converted into various data formats (TIFF, JSP, etc.), they could be directly used in mapping software such as ArcMap, Erdas, ENVI, and ENVI Zoom. The shapefile data was converted into raster data format. Next, the lidar raster data were geometrically corrected so that it would match the IKONOS data. We eliminated abnormal values in the absolute elevation data through certain correction process. One main cause of the abnormal values was the lost points, usually caused by water bodies. To create a combined data set, the final absolute

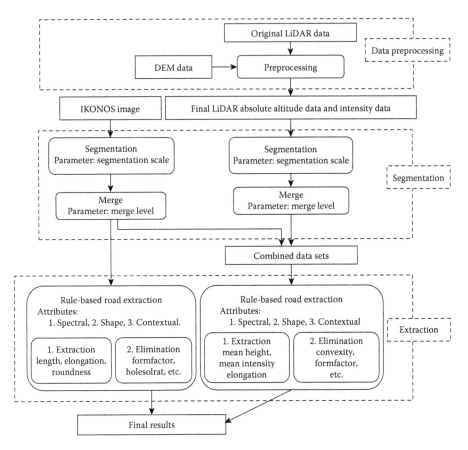

FIGURE 10.2
Road extraction procedures with geographic object–based image analysis.

elevation lidar data and the final intensity data were stacked to IKONOS data as the fifth and sixth bands, respectively. The resolution of the lidar data was set to the same as the IKONOS data (1 m).

Road Extraction

Most man-made objects are characterized by fairly homogeneous areas and by a regular shape. For instance, buildings and many roads are bounded by geometrical edges such as straight lines, so that they can be easily detected using edge-based techniques (Mueller et al. 2003). The ENVI Zoom employs an edge-based segmentation algorithm that is suitable for road extraction due to its fast running speed and simple one-parameter input requirement (scale level). By suppressing weak edges to different levels, the algorithm can yield multiscale segmentation results. The scale level parameter is an abstract term influencing the average object size. If the scale is too large, different features such as roads and buildings might be combined into one single object. If the scale is too small, one feature might be divided into so many pieces that each segment would be less meaningful. In the segmentation step, an analyst should make sure that the feature of interest is not divided into too many small segments. The next step, merging, can be used to slightly adjust the segmentation result and to improve the delineation of feature boundaries.

Rule-based classification was then applied to classify the image. According to the characteristics of urban roads, the following four attributes were selected to create key rules to detect roads:

1. *Length*: The combined length of all boundaries of the road polygons. Because roads are zonal features with long boundaries, the length value may be high so roads appear as a light color in Figure 10.3a.

2. *Roundness*: A shape measure that compares the area of a polygon to the square of the maximum diameter of the polygon (Figure 10.3b).

3. *Elongation*: A shape measure that indicates the ratio of the major axis of a polygon to the minor axis of the polygon (Figure 10.3c).

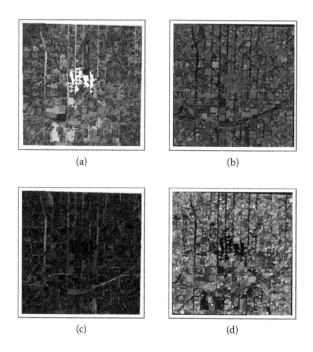

(a)

(b)

(c)

(d)

FIGURE 10.3
Four attributes and the appearance of roads: (a) Length, roads appear as bright colors; (b) roundness, roads appear as dark colors; (c) elongation, roads appear as light colors; (d) formfactor, roads appear as dark colors.

4. *Formfactor*: A shape measure that compares the area of a polygon to the square of the total perimeter; it is computed as follows:

$$\text{Formfactor} = 4\pi \times (\text{area})/(\text{total perimeter})^2 \qquad (10.1)$$

Because roads have a narrow shape, their perimeter value was large; thus the roads appear dark in Figure 10.3d. Additional attributes (including Convexity, Bandratio, Holesolrat, Numholes, Band 3) were selected as supplementary rules for the extraction from the combined data set of IKONOS image and lidar data. For every rule, the parameter fuzzy tolerance was adjusted to better model the uncertainty inherent in the classification. By adjusting Fuzzy Tolerance to a larger value, more objects would appear in the transition area between road and nonroad, and *vice versa*. According to Im et al. (2008), the mean height can distinguish buildings and trees from roads and grass due to their differences in elevation. The mean intensity value was able to distinguish between the grass and roads because of their difference of material. The mean height and the mean intensity derived from the lidar data were used to refine the extraction result from the combined data set.

Accuracy Assessment

To validate the results of the road extraction, measures developed by Jin and Davis (2005) were employed. The extracted roads and the manually delineated roads were compared. All objects in the image were categorized into four types:

1. *Road objects (RD)*: The extracted and manual results both labeled the object as road, that is, the overlapping part of the extracted and the referenced road was RD.
2. *Nonroad objects (NR)*: The extracted and manual results both labeled the object as a nonroad feature.
3. *False road objects (FR)*: The extracted result incorrectly labeled the object as a road.
4. *False nonroad objects (FNR)*: The extracted result incorrectly labeled the object as a nonroad feature.

In order to calculate the four measures, 90 random samples sized 10×10 m were selected. Within each square, the reference road data, extracted road result, and RD were individually digitized, compared, and their areas calculated. Because the total area of each sample was known (100 m²), the rest of the three measures were calculated as follows:

$$\text{NR} = 100 - \text{Reference road data} - \text{Extracted road result} + \text{RD} \qquad (10.2)$$

$$\text{FR} = \text{Extracted road result} - \text{RD} \qquad (10.3)$$

$$\text{FRD} = \text{Reference road data} - \text{RD} \qquad (10.4)$$

Once RD, NR, FR, and FNR were determined, they could be applied to evaluate the accuracy of the road extraction result, and the following measures were computed:

$$\text{Overall accuracy} = \frac{RD + NR}{RD + NR + FR + FNR} \times 100 \tag{10.5}$$

$$\text{Branching factor} = \frac{FR}{RD} \tag{10.6}$$

$$\text{Miss factor} = \frac{FNR}{RD} \tag{10.7}$$

$$\text{Detection percentage} = \frac{RD}{RD + FNR} \times 100 \tag{10.8}$$

$$\text{Quality percentage} = \frac{RD}{RD + FR + FNR} \times 100 \tag{10.9}$$

The branching factor was a measure of the commission error where the system incorrectly labeled nonroad features as roads. The miss factor measured the omission error where the system incorrectly labeled road objects as nonroad features. The detection percentage denoted the percentage of road objects correctly labeled by the extracted result. The quality percentage, the most important measure, indicated the absolute quality of the extraction. To obtain high quality, the extraction result needed to correctly label as many objects as possible with little mislabeling of nonroad objects.

Results

Segmentation Result

Image segmentation was conducted at different spatial scales. At a segmentation scale of 60–64, objects were broken into disaggregated parts. This type of segmentation made it problematic for discrimination between roads and surrounding features. As the scale increased, adjacent independent objects became merged to form meaningful road entities. At a scale of 65, this procedure preserved the best configuration for the roads. When scales larger than 65 were chosen, nonroad features became increasingly integrated with roads, indicating undersegmentation. Accordingly, the scale parameter 65 was chosen as the optimal scale to be applied in generating the final segmentation result (Figure 10.4). The result shows that many fragments were removed and became merged as meaningful entities. For example, a parking lot was divided by lanes and cars formed numerous fragments inside. In the segmentation result, most lanes were removed from the parking lot, and the number of cars was decreased (marked with white circles in Figure 10.4).

Next, merging of objects was conducted. Several merging levels were tested and applied to the segmentation result. More and more neighboring road fragments were merged into an integral object as the merging level increased. This process reduced the road partitioning and made the road shape clearer. However, surrounding features were also combined with roads when the merge level was over 76. For example, a building and a parking lot were connected with the road object at the merging levels of 78 and 79. This issue would ultimately cause errors in the road extraction result, demonstrating that levels higher than 76 were not appropriate for use in generating the final results. A level of 76 generated the most favorable merging result because it preserved the best road configuration and further reduced fragments. For the road crossing marked with white circles in Figure 10.5, the lanes in opposite directions became blurred and formed a square feature in the center of the segmentation image. After merging, this intersection was properly configured and reflected the actual road dimensions.

ENVI Zoom employed an edge-based segmentation algorithm that classified pixels as edge or non edge. The use of lidar-derived height and intensity information increased the variance in pixels for different feature types, making the boundaries of features more discernible. This enabled better discrimination of features and facilitated more rational image partitioning than when a single

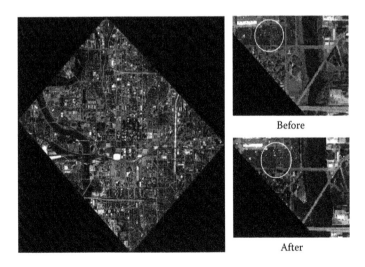

Before

After

FIGURE 10.4
Image segmentation result using the scale parameter of 65.

data source was used. The combined data set resulted in fewer errors than the IKONOS image at the same segmentation scale. Figure 10.6 shows that at a scale of 63, the mixture of roads and nonroad features was clearly visible in the segmentation result of the IKONOS image (marked with red circles in Figure 10.6). As the scale increased to 70, a substantial number of roads were merged with surrounding nonroad features (marked with red circles in Figure 10.6). However, by using the combined data set at the same segmentation scale, a minimum amount of error occurred. Road objects were clearly

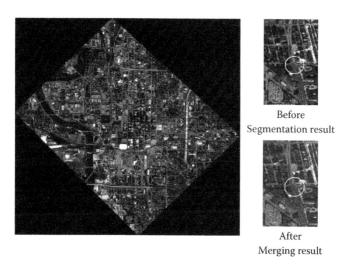

Before
Segmentation result

After
Merging result

FIGURE 10.5
Merging result at the level of 76.

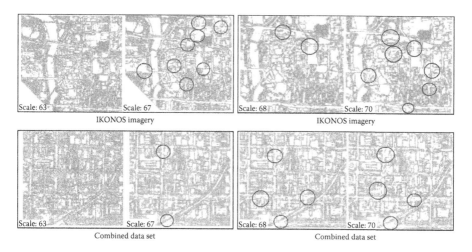

FIGURE 10.6
Segmentation result as a function of the scale parameter.

separable from the surrounding features, although slight oversegmentation existed. The combined data set allowed for a gradual variation in the segmentation results as the scale changed. This facilitated a more accurate estimation of the optimal scale at which to generate road shapes.

Rule-Based Extraction

Figure 10.7 shows all six attributes involved in the road extraction from the combined data set. Among them, roads displayed distinctive colors in mean height, mean intensity, and elongation. These attributes were aimed at recognizing all potential roads, whereas the remaining attributes functioned to filter out noise. Roads and noise were distinguished from each other by determining the divergence of their height, intensity, and shape. For example, buildings were easily distinguishable in mean height and convexity because of their height and convex shape; the circular and rectangular features appeared brighter than roads in formfactor and roundness. Other spectral, textural, or shape attributes, whose function apparently duplicated these three (such as length) or that did not enhance road extraction (such as solidity), were not selected in this study.

Following the selection of attributes for the discrimination, the ranges for each attribute were determined for separating roads from nonroad features.

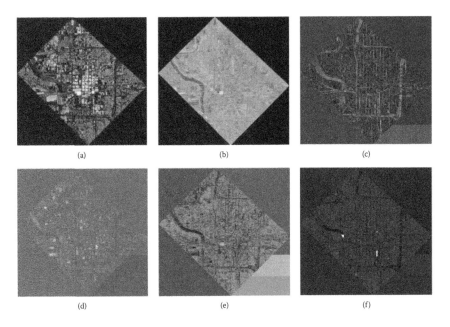

FIGURE 10.7
Attributes used in the rule-based extraction: (a) Avgband_5 (mean height), (b) avgband_6 (mean intensity), (c) elongation, (d) convexity, (e) formfactor, and (f) roundness.

TABLE 10.1

Rules Used in the GEOBIA Method for Combined Data Set

Rule	Appearance of Roads	Fuzzy Tolerance	Function
Mean height <0.1000	Dark	2	To extract objects with low absolute altitude
Mean intensity [29.0000, 65.0000]	Gray	0	To extract objects with similar superficial density and textures as roads
Elongation >1.5000	Bright	5	To further detect objects with zonal shape
Convexity not >1.8000	Bright or light gray	5	To eliminate convex objects, which are primarily buildings
Formfactor <0.3500, roundness <0.2800	Bright or gray	1	To remove objects with nonzonal shapes
Avgband_5 not >8.0000; Avgband_6 not <26.0000	Bright or light gray	1	To further eliminate noise with different elevations and intensities from roads

Note: GEOBIA, geographic object–based image analysis.

The procedure used for distinguishing roads from nonroad features from the combined data set was similar to that from the IKONOS image. An attribute value was iteratively tested until an acceptable threshold was obtained. The following rules were found to be optimal for separating roads from nonroad features (Table 10.1). For the IKONOS data set, three shape attributes were primarily used to extract objects. For the combined data set, lidar-derived height and intensity attributes extracted a larger number of potential road objects, whereas the shape attribute of elongation also assisted in the process. Moreover, Fuzzy Tolerance also proved useful in improving the extractions.

Accuracy Assessment

Figure 10.8 illustrates the final urban road extraction result from the combined data set. This method produced a clearer and more complete road network. It outperformed the method using the IKONOS image alone. First, it effectively detected more roads and successfully extracted most of the east–west roads and many tributary roads, which were absent in the extraction result from the IKONOS data. Second, most of the roads shaded by high-rise buildings in the downtown area of Indianapolis become visible. Finally, it contained fewer nonroad features, such as buildings and parking lots.

The accuracy analysis was implemented using the method discussed above. The total number of the four measurements is shown in Table 10.2. It shows that the result from the combined data set achieved 59.9% higher

W⬧E

N
S

Meters
0 245 490   980

Legend
■ Road

FIGURE 10.8
Object-based urban road extraction result using the combined data set.

TABLE 10.2

RD, NR, FNR, and FR statistics (in m²)

	RD	NR	FNR	FR
Combined data set	1587.35	6346.14	392.25	674.26
IKONOS imagery	992.57	6424.92	987.03	595.48

Note: FNR, false nonroad objects; FR, false road objects; NR, nonroad objects; RD, road objects

RD and 60.3% lower FNR than those from the IKONOS data. This suggests that the combined data set allowed extracting more roads correctly with less nonroad objects mislabeled as roads. However, the result from combined data set obtained 13.2% higher FR and 1.2% lower NR, implying that it was a less effective method compared to the single data source in recognizing nonroad objects. According to Table 10.3, the combined data set approach correctly extracted 7933.5 m² of objects with 1066.5 m² of error, achieving an overall accuracy of 88.15%, which was 5.73% higher than the result using the IKONOS image alone. The FR was 1.7 times that of FNR, indicating that the primary reason influencing the accuracy was mislabeling of nonroad objects.

The accuracy measurements derived from these four indices are presented in Table 10.3. It further demonstrated that with the combined data set we achieved

TABLE 10.3

Overall Accuracy Measurements

	Combined Data Set	IKONOS Data
Branching factor	0.43	0.60
Miss factor	0.25	0.99
Detection percentage	80.19%	50.14%
Quality percentage	59.81%	38.55%
Overall accuracy	88.15%	82.42%

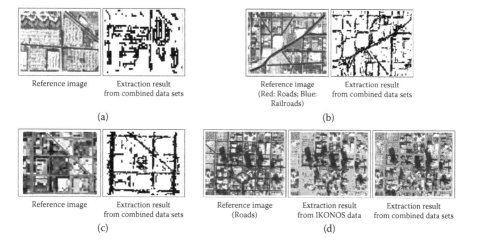

Reference image   Extraction result from combined data sets

(a)

Reference image (Red: Roads; Blue: Railroads)   Extraction result from combined data sets

(b)

Reference image   Extraction result from combined data sets

(c)

Reference image (Roads)   Extraction result from IKONOS data   Extraction result from combined data sets

(d)

FIGURE 10.9
Confusion between roads and other urban objects. (a) Error of parking lots; roads are marked in red lines. (b) Error of railroad; roads are marked in red; railroad in blue. (c) Error of buildings; roads are marked in blue and buildings in red circles. (d) Error of shades; roads are marked in red lines.

a better extraction result. First, it produced a 21.3% higher quality percentage. A higher ratio of correctness to error suggests an increase in the precision of road extraction. Second, it produced a 30% higher detection percentage, indicating an improvement in extracting more roads. Third, the branching factor and miss factor decreased, indicating fewer commission and omission errors. The decline of the miss factor by 0.75 indicated the accuracy in properly distinguishing roads from nonroads. Nevertheless, the branching factor was higher than the miss factor with the combined data set, which implied that the inclusion of nonroad features was the main reason for accuracy reduction.

Mislabeled objects were categorized into four main types: parking lot, railroad, building, and shading. A comparison of extracted results with the reference data allowed us to better understand the sources of errors. Figure 10.9 shows the confusion of roads with other urban objects.

Parking lots (Figure 10.9a) were one of the most difficult features to separate from roads, primarily because they have similar heights and spectral and texture characteristics. The boundaries between parking lots and roads were spectrally blurry. Railroads had almost the same height, similar spectral characteristics, and zonal shape as roads and they may have been wrongly classified (Figure 10.9b). The combined use of lidar and IKONOS data produced a better result in detecting buildings, because the difference in elevation between buildings and roads was easily distinguishable by using lidar-derived height information. However, some irregularly roofed buildings altered the characteristics of buildings, which may have ultimately caused errors in the extraction (Figure 10.9c). Features and artifacts like trees, buildings, and shade also affected the spatial characteristics of road objects and greatly decreased the reliability of spatial features in the road extraction. In Figure 10.9d, a comparison of the downtown extraction results clearly shows that the combined data set yielded a better result in extracting roads in shaded areas, although the improvement was not optimal.

Discussion and Conclusions

Accurate extraction of urban roads from high-resolution satellite imagery remains a challenging task. It requires consideration of the image's spatial resolution, the selection of features, the extraction method, and the complexity of the urban landscape. This chapter provided a case study of the extraction of urban roads from IKONOS images and lidar data with a GEOBIA method. The results demonstrated that road extraction from the combined IKONOS and lidar data set has great potential to improve road extraction by extracting the majority of visible roads and including fewer nonroad features. The accuracy analysis result indicated that it obtained an overall accuracy of 88.15%, which was 5.73% higher than the extraction from the IKONOS image alone. The accuracy was mainly impaired by four classes of mislabeled nonroad objects: parking lots, shades, railroads, and buildings.

A few issues need to be discussed further:

1. *Data sources*: By using IKONOS and lidar images, this study demonstrated that the combined use of two different data sets can greatly improve road extraction results. Employment of two types of images resulted in an improvement in the segmentation process by causing fewer errors at the same segmentation scale and achieving more subtle variations in the segmentation results as the scale changed. It also resulted in improvement in the rule-based extraction by

achieving higher extraction accuracy. The data selected for the extraction should complement each other. In this study, IKONOS images and lidar data were used because the former provides spatial and spectral information, whereas the latter provides elevation information; each provided unique characteristics for the objects under investigation.

The time and season when the specific data were acquired would also influence the feature extraction result. At the time of acquiring the IKONOS images, many roads running east–west were shaded by buildings. Shade can cause spectral mixture of ground features. Moreover, in the season during which the IKONOS imagery was acquired, the leaves were still on the trees for some parts of the roads. Satellite imagery acquired at another time or season may reduce the negative impacts of shadows and leaves.

Finer resolution imagery, such as those from QuickBird and WorldView-2, would potentially enhance the extraction of road objects, with more accurate separation from the surrounding features, such as adjacent buildings and parking lots. Meanwhile, noise on the road (such as cars) would become more visible in ultrahigh-resolution images. This noise may be eliminated based on spectral, geometric, and/or other characteristics.

2. *Features for rule-based extraction*: In this study, the mathematic morphology approach, which emphasizes the description of geometrical shape, helped to further remove nonroad objects with similar spectral characteristics. A comparison of the road extraction results suggests that geometric attributes were critical in the delineation of road objects. Lidar-derived height and intensity attributes were effective because they caused less confusion between roads and nonroad features. The spectral, textural, and contextual attributes can be used as supplementary means to improve road extraction through the elimination of nonroad objects. Each of these attributes may aid in removing different types of noise. It would also be interesting to conduct a comparative analysis of two or more cities by applying the same methods and data sets. This type of study might unearth new findings beyond those revealed in this study.

3. *The selection of algorithm parameters*: Because the GEOBIA required input for several parameters (e.g., segmentation and merging levels), it was a difficult job to specify the ideal values for these parameters. To apply the current methodology to other study areas, these parameters should be modified to reflect the road characteristics and data set used. The characteristics of target features to be extracted should be examined in detail. For example, the parameters specified for rural roads should not be the same as those for urban roads.

References

Amini, J., Saradjian, M.R., Blais, J.A.R., & Azizi, A. (2002). Automatic road-side extraction from large scale imagemaps. *International Journal of Applied Earth Observation and Geoinformation*, 4, 96–98.

Arbiol, R., Zhang, Y., & Paia, V. (2006). Advanced classification techniques: a review. *ISPRS Commission VII Mid-term Symposium "From Pixel to Processes"*, May 8–11, 2002, Enschede, The Netherlands.

Baatz, M., & Schape, A. (2000). Multiresolution segmentation: an optimization approach for high quality multi-scale image segmentation. *ISPRS Journal of Photogrammetry and Remote Sensing*, 58(3–4), 12–23.

Blaschke, T. (2010). Object based image analysis for remote sensing. *ISPRS Journal of Photogrammetry and Remote Sensing*, 65, 2–16.

Brennan, R., & Webster, T.L. (2006). Object-oriented land cover classification of Lidar derived surfaces. *Canadian Journal of Remote Sensing*, 32(2), 62–172.

Cai, H., & Rasdorf, W. (2008). Modeling road centerlines and predicting lengths in 3-D using LiDAR point cloud and planimetric road centerline data. *Computer-Aided Civil and Infrastructure Engineering*, 23, 157–173.

Charaniya, A.P., Manduchi, R., & Lodha, S.K. (2004). Supervised parametric classification of aerial LiDAR data. *IEEE Workshop on Real-Time 3D Sensors*, June 2004, Washington, DC. 25–32.

Clode, S.P., Rottensteiner, F., & Kootsookos, P. (2005). Improving city model determination by using road detection from LiDAR Data. *IAPRS*, XXXVI(3/W24), 159–164.

Clode, S.P., Rottensteiner, F., Kootsookos, P., & E. Zelniker. 2007. Detection and vectorization of roads from Lidar Data. *Photogrammetric Engineering & Remote Sensing*, 73(5), 517–535.

Gautama, S., D'Haeyer, J., & Philips, W. (2006). Graph-based change detection in geographic information using VHR satellite images. *International Journal of Remote Sensing*, 27(9), 1809–1824.

Hodgson, M., Archer, C., Jensen, J., Tullis, J., & Riordan, K. (2003). Synergistic use of lidar and color aerial photography for mapping urban parcel imperviousness. *Photogrammetric Engineering & Remote Sensing*, 69, 973–980.

Hu, J., Razdan, A., Femiani, J., Cui, M., & Wonka, P. (2007). Road network extraction and intersection detection from aerial images by tracking road footprints. *IEEE Transactions on Geoscience and Remote Sensing*, 45(12), 4144–4157.

Hu, X., Tao, C.V., & Hu, Y. (2004). Automatic road extraction from dense urban area by integrated processing of high resolution imagery and LiDAR. *International Archives of the Photogrammetry, Remote Sensing and Spatial Information Sciences*, Istanbul, 35, 320–325.

Im, J., Jensen, J., & Hodgson, M. (2008). Object-based land cover classification using high-posting-density LiDAR data. *GIScience and Remote Sensing*, 45(2), 209–228.

Im, J., Jensen, J.R., & Tullis, J.A. (2007). Object-based change detection using correlation image analysis and image segmentation. *International Journal of Remote Sensing*, 29(2), 399–423.

Ivits, E., Koch, B., Blaschke, T., Jochum, M., & Adler, P. (2005). Landscape structure assessment with image grey-values and object-based classification at three spatial resolutions. *International Journal of Remote Sensing*, 26(4), 2975–2993.

Jin, X., & Davis, C. (2005). An integrated system for automatic road mapping from high resolution multi-spectral satellite imagery by information fusion. *Information Fusion*, 6(4), 257–273.

Ke, Y.H., Quackenbush, L.J., & Im, J. (2010). Synergistic use of QuickBird multispectral imagery and LIDAR data for object-based forest species classification. *Remote Sensing of Environment*, 114, 1141–1154.

Long, H., & Zhao, Z.M. (2005). Urban road extraction from high-resolution optical satellite images. *International Journal of Remote Sensing*, 26(22), 4907–4921.

Miliaresis, G., & Kokkas, N. (2007). Segmentation and object-based classification for the extraction of the building class for LIDAR DEMs. *Computers and Geosciences*, 33(8), 1076–1087.

Mourad, B., Kalifa, G., & He, D. (2010). Rule-based classification of a very high resolution image in an urban environment using multispectral segmentation guided by cartographic data. *IEEE Transactions on Geoscience and Remote Sensing*, 48(8), 3198–3211.

Mueller, M., Segl, K., & Kaufmann, H. (2003). Extracting characteristic segments in high-resolution panchromatic imagery as basic information for object-driven image analysis. *Canada Journal of Remote Sensing*, 29(4), 453–457.

Pe'teri R., & Ranchin, T. (2007). Road networks derived from high spatial resolution satellite remote sensing data. In Q. Weng (ed.), *Remote Sensing of Impervious Surfaces*, pp. 215–236. Boca Raton, FL: CRC Press.

Platt, R., & Rapoza, L. (2008). An evaluation of an object-oriented paradigm for land use/land cover classification. *The Professional Geographer*, 60(1), 87–100.

Samadzadegan, F., Bigdeli, B., & Hahn, M. (2009). Automatic road extraction from LIDAR data based on classifier fusion in urban area. *2009 Joint Urban Remote Sensing Event*, Shanghai, China, pp. 1–6.

Sander J.O.E., & George, V. (2009). 3D information extraction from laser point clouds covering complex road junctions. *The Photogrammetric Record*, 24(125), 23–36.

Shaban, M.A., & Dikshit, O. (2001). Improvement of classification in urban areas by the use of textural features: the case study of Lucknow City, Uttar Pradesh. *International Journal of Remote Sensing*, 4, 565–593.

Shi, W.Z., & Zhu, C.Q. (2002). The line segment match method for extracting road network from high-resolution satellite images. *IEEE Transactions on Geoscience and Remote Sensing*, 40(2), 511–514.

Teng, C.H., & Fairbairn, D. (2002). Comparing expert systems and neural fuzzy systems for object recognition in map dataset revision. *International Journal of Remote Sensing*, 23(3), 555–567.

Tiwari, P.S., Pande, H., & Pandey, A.K. (2009). Automatic urban road extraction using airborne laser scanning/altimetry and high resolution satellite data. *Journal of the Indian Society of Remote Sensing*, 37(2), 223–231.

Wang, L., Sousa, W.P., & Gong, P. (2004). Integration of object-based and pixel-based classification for mapping mangroves with IKONOS imagery. *International Journal of Remote Sensing*, 25(24), 5655–5668.

Yang, J., & Wang, R.S. (2007). Classified road detection from satellite images based on perceptual organization. *International Journal of Remote Sensing*, 28(20), 4653–4669.

Zhou, W., & Troy, A. (2008). An object-oriented approach for analyzing and characterizing urban landscape at the parcel level. *International Journal of Remote Sensing*, 29(11), 3119–3135.

Zhu, C., Shi, W., Pesaresi, M., Liu, L., Chen, X., & King, B. (2005). The recognition of road network from high-resolution satellite remotely sensed data using image morphological characteristics. *International Journal of Remote Sensing*, 26, 5493–5508.

11

Integrating Remotely Sensed Climate and Environmental Information into Public Health

Pietro Ceccato, Stephen Connor, Tufa Dinku, Andrew Kruczkiewicz, Jerrod Lessel, Alexandra Sweeney, and Madeleine C. Thomson

CONTENTS

Introduction

During the last 30 years, the development of geographical information systems (GIS) and satellites for Earth observation has made important progress possible in the monitoring of the weather, climate, and environmental and anthropogenic factors that influence the reduction or the reemergence of vector-borne diseases. Analyses resulting from the combination of GIS and remote sensing have improved knowledge of climatic, environmental, and biodiversity factors (Witt et al. 2011; Al-Hamdan et al. 2014), influencing vector-borne diseases such as malaria (Beck et al. 1997; Omumbo et al. 1998; Ceccato et al. 2005a; Baeza et al. 2013), visceral leishmaniasis (VL) (Bhunia et al. 2010, 2012; Sweeney et al. 2014), dengue (Buczak et al. 2012; Moreno-Madriñán et al. 2014; Machault et al. 2014), Rift Valley fever (Linthicum et al. 1999; Anyamba et al. 2009), schistosomiasis (Simoonga et al. 2009; Walz et al. 2015), Chagas' disease (Kitron et al. 2006; Roux et al. 2011), and leptospirosis (Herbreteau et al. 2006; Skouloudis and Rickerby 2015). This knowledge and products developed using remotely sensed data helped and continue to help decision makers to better allocate limited resources in the fight against vector-borne diseases.

Because vector-borne diseases are linked to climate and environment, we present here our experience over the last 10 years helping Ministries of Health in Africa and the World Health Organization (WHO) to integrate climate and environmental information into decision processes.

Epidemics of vector-borne diseases still cause millions of deaths every year. Malaria remains one of the foremost global health problems, with an estimated 3 billion people at risk of infection in over 109 affected countries, 250 million cases annually, and 1 million deaths. The Roll Back Malaria (RBM) partnership involving the Global Fund to Fight AIDS, Tuberculosis and Malaria, the President's Malaria Initiative (PMI), national governments, nongovernmental organizations, and other donor agencies have supported antimalaria programs in a number of African countries over the last decade. In some countries, such as Ethiopia, malaria is highly sensitive to variations in the climate. In this context, climate information can either be used as a resource, for example in the development of early warning systems (EWS) (DaSilva et al. 2004), or must be accounted for when estimating the impact of interventions (Aregawi et al. 2014).

In Ethiopia, the determinants of malaria transmission are diverse and localized (Yeshiwondim et al. 2009), but altitude (linked to temperature) is certainly a major limiting factor in the highland plateau region and rainfall in the semi arid areas. A devastating epidemic caused by unusual weather conditions was documented in 1958, affecting most of the central highlands between 1,600 and 2,150 m with an estimated 3 million cases and 150,000 deaths (Fontaine et al. 1961). Subsequently, cyclic epidemics of various dimensions have been reported from other highland areas, at intervals of approximately 5–8 years, with the last such epidemic occurring in 2003. Most of these epidemics have been attributed to climatic abnormalities, sometimes associated with El Niño, although other factors such as land-use change may also be important.

Endemic regions of VL exist within East Africa, with a geographic hotspot in the northern states of South Sudan (Seaman et al. 1996). This region is known to experience seasonal fluctuations in cases that typically peak during the months of September, October, November, December, and January (SONDJ) (Seaman et al. 1996; Gerstl et al. 2006; WHO 2013). In the northern states of South Sudan alone, VL epidemics have recently been observed with a reported 28,512 new cases from January 2009 to December 2012 (Ministry of Health – Republic of South Sudan 2013). Without proper treatment, mortality in South Sudan is high, with numbers approaching 100,000 during one multi year epidemic in the late 1980s and early 1990s (Seaman et al. 1996). In the northern states, the sandfly (Figure 11.1) responsible for transmission of VL is *Phlebotomus orientalis* (Quate 1964; Hoogstraal and Heyneman 1969; Schorscher and Goris 1992; Seaman et al. 1996; Kolaczinski et al. 2008).

The habitat of *P. orientalis* is determined by specific ecological conditions, including the presence of black cotton soils, or vertisols, the presence of Acacia–Balanites woodlands, and mean maximum daily temperature (Thomson et al. 1999; Elnaiem et al. 1999; Elnaiem 2011). Research has also documented associations between environmental factors and VL that may contribute to outbreaks of the disease including relative humidity (Elnaiem et al. 1997; Salomon et al. 2012), precipitation (Hoogstraal and Heyneman 1969; Gebre-Michael et al. 2004), and normalized difference vegetation index (NDVI) (Hoogstraal and Heyneman 1969; Elnaiem et al. 1997, 2003;

FIGURE 11.1
Leishmania parasites enter the bodies of humans and other animals through bites from the sandfly (*Phlebotomus orientalis*).

Gebre-Michael et al. 2004; Rajesh and Sanjay 2013). Additionally, Ashford and Thomson (1991) suggested the possibility of a connection between a prolonged inundation (flooding) event in the 1960s and the corresponding 10-year drop in VL within the northern states of South Sudan. Although the importance of environmental variables in relation to the transmission dynamics of VL has been established, the lack of *in situ* data within the study region has led to inconclusive results regarding these relationships. We exploited the advantages of sustained and controlled Earth monitoring *via* Earth observations by the National Aeronautics and Space Administration (NASA). Using these data, we have shown the relationship between environmental factors and the spatiotemporal distribution of VL in the northern states (Sweeney et al. 2014), therefore demonstrating how Earth observations can be used for mapping risks of leishmaniasis.

The following sections present the methodology we have developed, which uses remote sensing to monitor climate variability, environmental conditions, and their impacts on the dynamics of infectious diseases. We also show how remotely sensed data can be accessed and evaluated and how they can be integrated into the decision process for mapping risks, creating EWS, and evaluating the impacts of control measures.

Climate and Environmental Factors: How Do They Help?

To date, much of the debate has centered on attribution of past changes in disease rates to climate change and the use of scenario-based models to project future changes in risk for specific diseases. Although these can give useful indications, the unavoidable uncertainty in such analyses, as well as contingency on other socioeconomic and public health determinants in the past or future, limit their utility as decision-support tools. For operational health agencies, the most pressing need is the strengthening of current disease control efforts to bring down current disease rates and manage short-term climate risks, which will, in turn, increase resilience to long-term climate change. The WHO and partner agencies are working through a range of programs to (1) ensure political support and financial investment in preventive and curative interventions to bring down current disease burdens; (2) promote a comprehensive approach to climate risk management; (3) support applied research, through definition of global and regional research agendas and targeted research initiatives on priority diseases and population groups (Campbell-Lendrum et al. 2015).

In this context, the International Research Institute for Climate and Society (IRI) develops research to first understand the mechanisms driving changes in transmission of diseases. We usually try to understand the relationship between diseases and the climate by creating spatial and temporal stratification of the diseases and population at risk (i.e., risk mapping). If a relationship exists between the diseases and the climate, we then estimate the seasonality of the disease and timing of intervention. We then develop frameworks for EWS to monitor in real time and forecast the risks of diseases transmission based on climate and environmental factors (i.e., creating EWS). Finally, once decision makers have put in place control measures to mitigate the problem, climate variability is considered to assess the efficacy of control measures (i.e., evaluation stage of mitigation measures).

Risk Maps

A risk map can aid in understanding the relationship between diseases and the climate. In the following example, we show how a risk map was produced to understand whether there is a relationship between climate and malaria in Eritrea (Figure 11.2). From this analysis, we can glean that malaria incidence exists across a rather sharp spatial gradient in Eritrea and that this variation is driven by climatic factors. As depicted by the areas in red, we can see that malaria incidence is high in the western part of Eritrea and peaks in September and October immediately after the rainy season, which

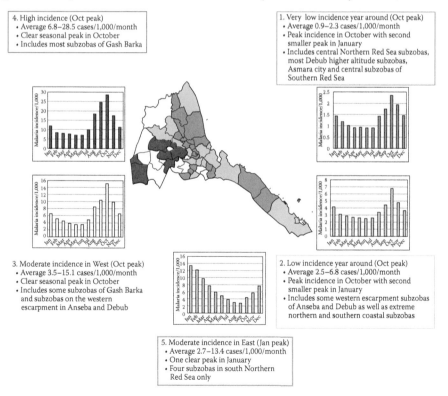

4. High incidence (Oct peak)
  • Average 6.8–28.5 cases/1,000/month
  • Clear seasonal peak in October
  • Includes most subzobas of Gash Barka

1. Very low incidence year around (Oct peak)
  • Average 0.9–2.3 cases/1,000/month
  • Peak incidence in October with second smaller peak in January
  • Includes central Northern Red Sea subzobas, most Debub higher altitude subzobas, Asmara city and central subzobas of Southern Red Sea

3. Moderate incidence in West (Oct peak)
  • Average 3.5–15.1 cases/1,000/month
  • Clear seasonal peak in October
  • Includes some subzobas of Gash Barka and subzobas on the western escarpment in Anseba and Debub

2. Low incidence year around (Oct peak)
  • Average 2.5–6.8 cases/1,000/month
  • Peak incidence in October with second smaller peak in January
  • Includes some western escarpment subzobas of Anseba and Debub as well as extreme northern and southern coastal subzobas

5. Moderate incidence in East (Jan peak)
  • Average 2.7–13.4 cases/1,000/month
  • One clear peak in January
  • Four subzobas in south Northern Red Sea only

FIGURE 11.2
A climate risk map showing the spatiotemporal stratification of malaria incidences in Eritrea at district level.

occurs in July and August. Conversely, in the highlands of Eritrea (central green part), malaria incidence is low because of the low temperatures at high altitudes. On the east coast, there are some areas where malaria peaks in January because of rainfall that occurs occasionally in December.

An analysis of this type showing how climate and environmental factors are related in time and space should typically be the first step in analyzing any type of climate-related issue. Once (if) a relationship has been demonstrated, it is then worthwhile trying to develop an EWS to monitor and forecast the conditions that may trigger a problem.

Early Warning System

The WHO has developed a framework for creating an EWS for malaria (Da Silva 2004). The framework is composed of four components (Figure 11.3):

1. Vulnerability assessment, including the assessment of current control measures, any problems related to resistance developed by the mosquitoes or the plasmodium, socioeconomic factors, such as migration of population, and so on.

2. Climate forecasting, allowing for forecasting, 3–6 months in advance, of the probability of an increase in precipitation or in temperature, weather conditions that may lead to an increase in risk for an outbreak of malaria.

3. Monitoring of climate and environmental factors, including monitoring of precipitation, temperature, and the presence of vegetation or water bodies that would influence the development of mosquitoes.

4. *Case surveillance:* Monitoring of malaria cases is performed at the hospital level and managed at the central level by the Ministry of Health.

This EWS can be also used as a framework for floods, natural disasters, droughts, and other events that have a relationship with climate and environmental factors.

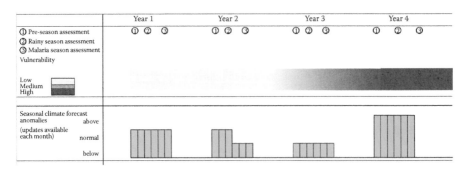

FIGURE 11.3
Early warning system developed for the World Health Organization.

TABLE 11.1

Possible Outcomes if Climate Information Is Not Incorporated into Malaria Impact Evaluations

	Changes in observed malaria following intervention (relative to baseline)	
Changes in climate suitability for malaria transmission following intervention (relative to baseline)	Decrease ⬇	No change or increase ⬆
Increase ⬆	May underestimate impact of intervention	May obscure impact of intervention
No change (average)	No effect	
Decrease ⬇	May overestimate impact of intervention	May obscure extent of failure of intervention

Source: From Dinku, T., Kanemba, A., Platzer B., and Thomson, M.C. (2014b). *Leveraging the Climate for Improved Malaria Control in Tanzania.* Earthzine.

Evaluation of Control Measures

The PMI uses the RBM partnership–approved methodology to evaluate whether the deployed interventions have had an impact on malaria morbidity and mortality. The methodology requires consideration of contextual (potentially confounding) factors that affect the epidemiology of malaria when using all-cause mortality as the measure of impact. These factors include increases in household income, better vaccination coverage, improvements in living conditions, and so on. Although the RBM methodology provides guidance on how to consider certain confounding factors when determining their potential impact on mortality, the effect of climate on malaria prevalence, and therefore mortality, is much less clear. Table 11.1 lists potential errors when not considering climate in malaria evaluations.

In order to conduct the analysis for the above three components, availability of decision-relevant climate and environmental information about the past, recent trends, current conditions, likely future trajectories, and associated impacts is a prerequisite for climate-informed decision-making.

Accessing Quality Data through Earth Observations

When working on vector-borne diseases, decision makers and researchers often face a lack of quality data required for optimal targeting of the intervention and surveillance. The results/decisions are critical as they impact on the lives of many people: "Bad data create bad policies."

Climate data and information—whether station- or satellite-generated—can increasingly be accessed freely online. Station data can typically be obtained from a country's national meteorological service (NMS). Depending on the quality control processes performed by the NMS, these data may be of high or low quality. However, station data are not always available especially in Africa. Some of the station data provided by the NMS are freely available through the Global Telecommunication System but often lack the spatial resolution needed.

Satellites provide raw data that are continuously archived and cover large areas of the globe. In order for decision makers to access, visualize, or manipulate these data, an interface is necessary. In many cases, the raw data may be free, but not all interfaces allow free access to their archived data. Sources for satellite-generated climate data are varied, and a selection is provided below, with further details—including Web addresses—provided in Table 11.2. The following are likely to be the most useful of the freely available satellite-based estimates.

Precipitation

- The *Global Precipitation Climatology Project* (GPCP) combines satellite and station data. The monthly data extend from 1979 onwards, while the daily product is from 1996 to present. This product has a low spatial resolution but is of interest when creating long time series to understand trends in past precipitation.[*]

- The *Climate Prediction Center (CPC) Merged Analysis of Precipitation* (CMAP) combines satellite and station data.[†] This product is very similar to the GPCP but has some differences due to different algorithms used to estimate precipitation.

- The *CPC MORPHing technique* (CMORPH) provides global precipitation estimates at very high spatial and temporal resolutions.[‡] This product is suitable for real-time monitoring of rainfall, provided a long history is not required, as data are only available from January 1998 on.

- The *Tropical Rainfall Measurement Mission* (TRMM) provides estimates of precipitation in the tropics. Monthly aggregates improve the quality of the data. They are available from January 1998 onward, but there is a delay of about a month or more in updating the data.

---

[*] GPCP provides global monthly 2.5° and daily 1° rainfall estimates.

[†] CMAP provides products at a spatial resolution of 2.5° with 5-day and monthly aggregations since 1979.

[‡] CMORPH provides precipitation analyses at 8-km spatial resolution and 30-minute temporal resolution.

TABLE 11.2

List of URLs and Use of Different Derived Earth Observation Products

			Precipitation Data
Satellite	Rainfall estimated from Famine Early Warning Systems Network through IRI Data Library	http://iridl.ldeo.columbia.edu/maproom/Health/Regional/Africa/Malaria/MEWS/index.html	Provides data as well as an analysis tool that allows its users to analyze time-series anomalies and download the results in one click (see Figure 11.3).
Satellite	GPCP from NASA	http://precip.gsfc.nasa.gov/gpcp_v2_comb.html http://iridl.ldeo.columbia.edu/SOURCES/.NASA/.GPCP/.V1DD/ http://iridl.ldeo.columbia.edu/SOURCES/.NASA/.GPCP/.V2/.satellite-gauge/	Global monthly 2.5° and daily 1° rainfall estimates. The monthly data extend from 1979 to present, while the daily product is from 1996 to present. Combines precipitation information available from different microwave sensors and special sensor microwave imager data from the Defense Meteorological Satellite Program (USA) and the infrared from geostationary satellites and secondarily from polar-orbiting satellites. Gauge data are assembled and analyzed by the Global Precipitation Climatology Centre. Daily data at 1° spatial resolution, which is around 100 km at the equator:
Satellite	CMAP from the NOAA-CPC	http://www.esrl.noaa.gov/psd/data/gridded/data.cmap.html http://iridl.ldeo.columbia.edu/SOURCES/.NOAA/.NCEP/.CPC/.Merged_Analysis/	Global precipitation at a spatial resolution of 2.5° with pentadal (5-day) and monthly aggregations since 1979. Useful for trend analysis in the past. Produces global precipitation in which observations from rain gauges are merged with precipitation estimates from several satellite-based algorithms and is very similar to that of GPCP.
Satellite	CMORPH of the NOAA-CPC	http://www.cpc.ncep.noaa.gov/products/janowiak/cmorph_description.html http://iridl.ldeo.columbia.edu/SOURCES/.NOAA/.NCEP/.CPC/.CMORPH/	Global precipitation analyses at very high spatial (8-km) and temporal (30-minute) resolutions, updated everyday in almost real time, making it a good candidate for real-time monitoring of rainfall. Data are available starting from December 2002. It uses precipitation estimates that have been derived exclusively from low-orbiter satellite microwave observations and whose features are transported *via* spatial propagation information that is obtained entirely from geostationary satellite IR data.

(Continued)

TABLE 11.2 (*Continued*)

List of URLs and Use of Different Derived Earth Observation Products

		Precipitation Data	
Satellite	TRMM from NASA and the National Aeronautics and Japan Aerospace Exploration Agency	http://precip.gsfc.nasa.gov/ trmm_comb.html http://iridl.ldeo.columbia.edu/ SOURCES/.NASA/.GES-DAAC/. TRMM_L3/.TRMM_3B42/.v7/. daily/.precipitation/	The data sets cover the period January 1998 to September 2014 (with about a month delay) at 0.25°. This product is of good quality and high temporal spatial resolution. It aims to improve the quantitative knowledge of the three-dimensional distribution of precipitation in the tropics. TRMM has a passive microwave radiometer, the first active space-borne precipitation radar, and a visible-infrared scanner, plus other instruments. It optimally merges microwave and IR rain estimates to produce 3-hourly precipitation fields at quarter-degree spatial resolution. Then, the 3-hourly products are aggregated to monthly and merged with gauge data over land to produce the best-estimated monthly precipitation field.
Satellite	RFE of the NOAA-CPC	http://www.cpc.ncep.noaa.gov/ products/fews/RFE2.0_desc.html http://iridl.ldeo.columbia.edu/ SOURCES/.NOAA/.NCEP/. CPC/.FEWS/	Specific to Africa, it provides accurate rainfall totals and is assumed to give the true rainfall near each station. The product is available at 11-km spatial resolution on a daily and 10-day basis and is operationally used for the Famine Early Warning System and Malaria Early Warning System. RFE2 uses microwave estimates in addition to continuing the use of cloud-top temperature and station rainfall data that forms the basis of RFE1. Meteosatgeo stationary satellite infrared data are acquired in 30-minute intervals, and areas depicting cloud-top temperatures of < 235 K are used to estimate convective rainfall.

(Continued)

TABLE 11.2 (Continued)

List of URLs and Use of Different Derived Earth Observation Products

			Precipitation Data
Temperature data			
Satellite	LST from MODIS (Aqua and Terra)	http://iridl.ldeo.columbia.edu/ SOURCES/.USGS/.LandDAAC/. MODIS/.1km/.8day/.version_005/ Africa: http://iridl.ldeo.columbia. edu/SOURCES/.USGS/. LandDAAC/.MODIS/.1km/.8day/. version005/.Aqua/ South America: http://iridl.ldeo. columbia.edu/SOURCES/.USGS/. LandDAAC/. MODIS/.1km/.8day/. version_005/.Terra/	For Africa and South America, the spatial resolution is 1 km and temporal resolution is 8 days. Maps are available for daytime and night-time images. Data are available from July 2002 to present for Africa and from March 2000 to present for South America. The data consist of 8-day composite land-surface temperature maps of continental Africa (derived from the MODIS sensor onboard the Aqua satellite) and map of South America (derived from the MODIS sensor onboard the Terra satellite).
	MODIS LST	Minimum air temperature: http:// iridl.ldeo.columbia.edu/maproom/ Health/Regional/Africa/Malaria/ LSTmin/index.html Maximum air temperature: http:// iridl.ldeo.columbia.edu/maproom/ Health/Regional/Africa/Malaria/ TMR/index.html	Provides a good estimation of the minimum and maximum air temperatures inferred from that minimum LST.
Vegetation cover			
Satellite	MODIS NDVI	http://iridl.ldeo.columbia.edu/ maproom/Health/Regional/ Africa/Malaria/NDVI/index. html?sregion=EAF	Provides a good estimation of vegetation abundance at a spatial resolution of 250 m.

(Continued)

TABLE 11.2 (*Continued*)

List of URLs and Use of Different Derived Earth Observation Products

			Precipitation Data
Satellite	MODIS EVI	http://iridl.ldeo.columbia.edu/maproom/Health/Regional/Africa/Malaria/NDVI/index.html?sregion=.EAF& var=EVI	Using a combination of middle-infrared, NIR, and red channels, it is possible to map water bodies and vegetation in sparse areas (Pekelet al. 2011) and to create new indices such as the GVMI (Ceccato et al. 2002) to map vegetation moisture content.
Satellite	MODIS reflectance for monitoring water bodies and bare soils	http://iridl.ldeo.columbia.edu/maproom/Health/Regional/Africa/Malaria/NDVI/index.html?sregion=.EAF& var=reflectance	
Satellite	Global NDVI of the NOAA-AVHRR	http://glcf.umd.edu/data/gimms/ http://iridl.ldeo.columbia.edu/SOURCES/.UMD/.GLCF/.GIMMS/.NDVIg/.global/.ndvi/	The GIMMS data set is an NDVI product available for a 25-year period spanning 1981–2006 at 8-km spatial resolution, every 15 days. The data set is derived from imagery obtained from the AVHRR instrument onboard the NOAA satellite series 7, 9, 11, 14, 16, and 17. This is an NDVI data set that has been corrected for calibration, view geometry, volcanic aerosols, and other effects not related to vegetation change.
Satellite	Vegetation indices from Terra MODIS and single channels (blue–red NIR–SWIR) of the NOAA-AVHRR	http://iridl.ldeo.columbia.edu/SOURCES/.USGS/.LandDAAC/.MODIS/ http://iridl.ldeo.columbia.edu/maproom/index.html	Provides consistent spatial and temporal comparisons of global vegetation conditions that can be used to monitor photosynthetic activity. The data are available at 250-m spatial resolution every 16 days from April 2000 to present. Two MODIS VIs, the NDVI and the EVI, are produced globally over land at 16-day compositing periods. Whereas the NDVI is chlorophyll-sensitive, the EVI is more responsive to canopy structural variations, including leaf area index, canopy type, plant physiognomy, and canopy architecture. The two vegetation indices complement each other in global vegetation studies and improve upon detection of vegetation changes and extraction of canopy biophysical parameters.

(Continued)

TABLE 11.2 (Continued)

List of URLs and Use of Different Derived Earth Observation Products

			Precipitation Data
Water bodies and inundation products			
Satellite	Landsat 5, 7, and 8 QuikSCAT, ERS scatterometer, SSM/I, and AMSR-E active/ passive microwave data sets	http://landsat.usgs.gov/ http://iridl.ldeo.columbia.edu/ SOURCES/.NASA/.JPL/. wetlands/	Landsat provides high spatial resolution (30 m) every 16 days. Those global products are available at 25-km spatial resolution on a daily basis.

Note: AVHRR, advanced very high resolution radiometer; CMAP, CPC Merged Analysis of Precipitation; CMORPH, CPC MORPHing technique; CPC, Climate Prediction Center; EVI, enhanced vegetation index; GIMMS, Global Inventory Modelling and Mapping Studies; GPCP, Global Precipitation Climatology Project; IR, infrared; IRI, International Research Institute for Climate and Society; LST, land-surface temperature; MODIS, moderate-resolution imaging spectroradiometer; NASA, National Aeronautics and Space Administration; NDVI, normalized difference vegetation index; NIR, near-infrared; NOAA, National Oceanic and Atmospheric Administration; RFE, African Rainfall Estimation; SWIR, shortwave infrared; TRMM, Tropical Rainfall Measurement Mission.

The product is of good quality if high spatial detail is required and real-time information is not critical.[*]

- The *Global Precipitation Measurement* (GPM) provides estimates of precipitation globally. They are available from March 2014 to present. The GPM is an extension of the TRMM rain-sensing package.[†]

- The *African Rainfall Estimate* (RFE) combines satellite and station data specifically for Africa. The data are available from 1995 and are useful for high spatial resolution but not for daily observations.[‡]

- The *Enhancing National Climate Services* (ENACTS) program combines all available rain gauge data from the national meteorology agencies of Ethiopia, Gambia, Ghana Madagascar, Mali, Rwanda, Tanzania, and Zambia, with satellite data for the last 30 years at high spatial resolution.[§] The ENACTS program generates the best data set at the national level and is expected to expand to other countries in Africa.

- *Climate Hazards Group Infrared Precipitation with Station* (CHIRPS) data are produced by the University of California, Santa Barbara, using a similar technique developed to create the ENACTS data but using fewer rain gauges.[¶]

Temperature

In Africa, the spatial distribution of weather stations is often limited and the dissemination of temperature data is variable, therefore limiting their use for real-time applications. Compensation for this paucity of information may be obtained by using satellite-based methods. The estimation of near-surface air temperature (Ta) is useful for a wide range of applications in health. It affects the transmission of malaria (Ceccato et al. 2012) in the highlands of East Africa. Air temperature is commonly obtained from synoptic measurements in weather stations. However, the derivation of Ta from the land-surface temperature (Ts) derived from satellite is far from straightforward. Studies have shown that it is possible to retrieve high-resolution Ta data from the moderate-resolution imaging spectroradiometer (MODIS) Ts products over different ecosystems in Africa (Ceccato et al. 2010; Vancutsem et al. 2010).

---

[*] TRMM provides precipitation data at 0.25°.
[†] GPM provides precipitation data at 0.1°.
[‡] RFE has an 11-km spatial resolution and a 10-day temporal resolution.
[§] ENACTS uses a spatial resolution of 10 km, and 10-day and monthly products are available for the last 30 years. The approach is now being considered for expansion into other countries in Kenya, Uganda, and West Africa.
[¶] The CHIRPS data cover the African continent at 5 km on a daily basis for the last 30 years and at 5 km every 10 days and on a monthly basis for the entire globe. The data are available *via* the IRI Data Library at: http://iridl.ldeo.columbia.edu/SOURCES/.UCSB/.CHIRPS/.v1p8/.

For temperature-based data, the following data sets are recommended:

- *Land-surface temperature* (LST) from MODIS provides land-surface temperature estimates. The data are available from July 2002 for Africa and from March 2000 for South America at a spatial resolution of 8 km. Separate estimates for daytime and nighttime temperatures are available. Maximum and minimum air temperature estimates can be derived from the land-surface temperatures.

- The ENACTS program has generated minimum and maximum temperatures for the eight countries listed above. This is accomplished by combining quality-controlled station data from all available stations with climate reanalysis products. The resulting data sets go back to 1961 at a spatial resolution of about 4 km.

Vegetation

Remote sensing can be used to distinguish vegetated areas from bare soils and other surface covers. Various vegetative properties can be gleaned from indices such as the NDVI, including but not limited to leaf area index, biomass, greenness, and chlorophyll. However, quantitative analyses are highly sensitive to the context of the study location, and relationships should be assessed prudently.

Practitioners can access data on vegetation cover through the following sources:

- *Global NDVI* is available from 1981 to 2006. The data set has been shown to be valid in representing vegetation patterns in certain regions (but not everywhere) and should be used with caution (Ceccato 2005b).

- *Terra MODIS NDVI* and enhanced vegetation index (EVI) are available for 16-day periods from April 2000 at 250-m resolution. The NDVI is an updated extension to the Global NDVI. The EVI is another index used to estimate vegetation that can complement the NDVI.

Water Bodies and Inundation Products

Using MODIS and Landsat reflectance channels in the middle infrared, near-infrared, and red ranges, it is possible to map small water bodies that are breeding sites for mosquitoes. Using a technique developed by Pekel et al. (2011), it is possible to automatically map the water bodies by transforming the red–green–blue color space (represented by the middle infrared, near-infrared and red channels) into a hue–saturation–value space that decouples chromaticity and luminance.

Global maps of inundated area fraction are derived at the 25-km scale from remote sensing observations from multiple satellite sources, focusing on data sets from active/passive microwave instruments (ERS scatterometer, QuikSCAT, SSM/I, and AMSR-E). Comparison and validation of global 25-km resolution data sets with fine-scale SAR-based regional wetland distributions is employed to ensure self-consistency. Those products are used to map flood events and their impacts on malaria and leishmaniasis in South Sudan (Sweeney et al. 2014).

Practitioners can access data on water bodies through the following sources:

- *Terra MODIS middle-infrared, near-infrared, and red reflectances* are available for 16-day periods from April 2000 onward at 250-m resolution.
- *Landsat middle-infrared, near-infrared, and red reflectances* are available every 16 days at 30-m spatial resolution.
- *Inundation fraction* products are available for daily, 6-day, and 10-day periods for the entire globe at 25-km spatial resolution (McDonald 2011).

Improving Data Quality and Accessibility

To address the spatial and temporal gaps in climate data as well as the lack of quality-controlled data, approaches are being developed based on the idea of "merging" station data with satellite and modelled data. Some of these methods are taking advantage of cooperation among climate scientists, meteorologists, and decision makers to develop a platform in which the now-quality-improved data can be accessed, manipulated, and integrated into the programs of national-level stakeholders and international partners, from inside the country and abroad.

ENACTS Approach

The IRI has been working with national meteorological agencies in Africa (Ethiopia, Gambia, Ghana, Madagascar, Mali, Rwanda, Tanzania, and Zambia) to improve the availability, access, and use of climate information by national decision makers and their international partners. The approach—ENACTS—is based on three pillars (Dinku et al. 2011, 2014a, 2014b).

Pillar 1: *Improving Data Availability.* Availability of climate data is improved by combining quality-controlled station data from the national observation network with satellite estimates for rainfall and elevation maps and reanalysis products for temperature. The final products are data sets with 30 or more years of rainfall and temperature for a 4-km grid across the country.

Pillar 2: *Enhancing Access to Climate Information.* Access to information products is enhanced by making information products available online. This is accomplished by customizing and installing the very powerful IRI Data Library and developing an online mapping service that provides user-friendly tools for the analysis, visualization, and download of climate information products.

Pillar 3: *Promoting the Use of Climate Information.* Use of climate information is promoted by engaging and collaborating directly with potential users.

By integrating ground-based observations with proxy satellite and modelled data, the ENACTS products and services overcome issues of data scarcity and poor quality, introducing quality-assessed and spatially complete data services into national meteorological agencies to serve stakeholder needs. One of the strengths of ENACTS is that it harnesses all local observational data, incorporating high definition information that globally produced or modelled products rarely access. The resulting spatially and temporally continuous data sets allow for the characterization of climate risks at a local scale. ENACTS enables analysis of climate data at multiple scales to enhance malaria control and elimination decisions. It uses detailed historical climate data to understand natural variability in temperature and rainfall over national, regional, and district scales.

CHIRPS Approach

A similar approach has been developed by CHIRPS. CHIRPS is a 30+ year quasi-global rainfall data set. Spanning 50°S–50°N (and all longitudes), starting in 1981 to near-present, CHIRPS incorporates 0.05° resolution satellite imagery with *in situ* station data to create gridded rainfall time series for trend analysis and seasonal drought monitoring. As of May 1, 2014, version 1.8 of CHIRPS is complete and available to the public. For detailed information on CHIRPS, refer to the paper on the US Geological Survey (USGS) website (http://pubs.usgs.gov/ds/832/pdf/ds832.pdf).

Since 1999, USGS and Climate Hazard Group scientists, supported by funding from the US Agency for International Development (USAID), NASA, and the National Oceanic and Atmospheric Administration (NOAA), have been developing techniques for producing rainfall maps, especially where surface data are sparse. CHIRPS was created in collaboration with scientists at the USGS Earth Resources Observation and Science Center in order to deliver reliable, up-to-date, and more complete data sets for a number of early warning objectives (such as trend analysis and seasonal drought monitoring).

Early research focused on combining models of terrain-induced precipitation enhancement with interpolated station data. More recently, new resources of satellite observations such as gridded satellite-based precipitation estimates from NASA and NOAA have been leveraged to build high-resolution (0.05°) gridded precipitation climatologies. When applied to satellite-based precipitation fields, these improved climatologies can remove systematic bias, a key technique in the production of the 1981 to near-present CHIRPS data set. The creation of CHIRPS has supported drought-monitoring efforts by the USAID Famine Early Warning Systems Network.

Two CHIRPS products are produced operationally: a rapid preliminary version and a later final version. The preliminary CHIRPS product is available, for the entire domain, 2 days after the end of a pentad (on the 2nd, 7th, 12th, 17th, 22nd, and 27th). The preliminary CHIRPS uses only a single station source, GTS. The final CHIRPS product takes advantage of several other station sources and is complete sometime after the 15th of the following month. Final monthly, dekad, pentad, and daily products are calculated at that time.

Data Accessibility

Over the past 30 years, the field of remote sensing has grown to include numerous national, intergovernmental, and private organizations that freely provide user-friendly high spatial and temporal resolution data sets. However, the ease of access should not be mistaken for ease of analysis as the data sets are still complex and require complex evaluation, especially when applied to decision-making.

The IRI has developed various tools to improve data accessibility and analysis for decision makers and interdisciplinary researchers alike. A Climate Data Library was built as an integrated knowledge system to support the use of climate and environmental information in climate-sensitive health decision-making. Initiated as an aid to climate scientists to do exploratory data analysis, it has expanded to provide a platform for interdisciplinary researchers focused on topics related to climate impacts on society.

IRI Data Library

As its name suggests, the IRI Climate Data Library is organized as a library: a collection of both locally held and remotely held data sets, designed to make the data more accessible for the library's users. Data sets in the library come from many different sources and many different "data cultures" in many different formats. By *data set*, we refer to a collection of data organized as multidimensional dependent variables, independent variables, and sub-data sets, along with the metadata (particularly metadata on purpose and use) that makes it possible to interpret the data in a meaningful manner.

The IRI Climate Data Library can be used *via* two distinct mechanisms that are designed to serve different communities. Expert Mode serves the needs of operational practitioners and researchers that have an in-depth knowledge of the functionality of the system and are able to customize it to their own specific needs. The Data Library programming language (Ingrid) can be used by advanced users to develop custom functions and perform tailored analyses. This functionality is widely used around the world by climate researchers, as Expert Mode allows users with programming skills a very extensive level of personalized functionality. Online tutorials, examples, and function definitions are part of the Data Library.

Map Rooms

In contrast to Expert Mode, the Map Rooms provide easy access to point-and-click map-based user interfaces that are built on Data Library infrastructure. The Map Rooms are the result of collaborative negotiations around information needs and make specific data and products for a region or time period available for a specific purpose to specific users and decision makers. The data and maps in these Map Rooms are available for quick and easy download to the user's desktop. For example, in 2006, in collaboration with the Desert Locust Information System of the Food and Agricultural Organization, the IRI developed an interface to estimate environmental conditions conducive to the development of swarms of desert locusts. The interface continues to be available as an online "clickable map," which allows desert locust officers (often working in remote areas) to visualize, analyze, and automatically download MODIS images and rainfall estimates derived from satellite measurements into their RAMSES-GIS software (Ceccato et al. 2007b). Problems with downloading very large images through the Internet, especially in areas in the world where the Internet is slow, such as northern Niger, are avoided by using the Data Library, which permits quickly zooming into the spatial and temporal area of interest and downloading only smaller region(s) of interest (Ceccato et al. 2006).

IRI Climate Data Library Archives and Near-Real-Time Updates

Global climate observations by ground stations, satellites, and modeled estimates of climatic conditions compose the vast majority of the Data Library's data archive. An extensive menu of maps and analysis used to monitor current global and regional climate, as well as historical data, are available from a wide range of sources including NASA, NOAA, CRU-UEA, WMO, ECMWF, GISS, and so on. From the Map Rooms, it is possible to readily access and download the publicly available data sets being viewed, including station, atmospheric, and oceanic observations and analyses, model-based analyses, and forecasts, as well as land-surface and vegetation information.

The near-real-time data sets are updated by automated software that retrieves the data as soon as it is available on the originating site. For instance, MODIS satellite data will be available in the IRI Climate Data Library within a day after processing is complete at the NASA data center. Examples of data and products that are of interest to the health community held in the data library are indicted in Table 11.1.

Downloading Data Library Data and Products

A Data Library user can download both images and data onto his or her desktop workstation. Data can be downloaded in standard ASCII and binary formats, Excel and R tabular formats, GIS formats, netCDF files, and directly to application software (such as GrADS and MATLAB®) that support the OPeNDAP data transfer protocol (Cornillon 2003). Over the last decade, OPeNDAP has emerged as a community standard for machine-to-machine data access and transfer and is widely used where data sharing is involved, for example, with the climate change scenarios developed as part of the Coupled Model Intercomparison Project for the Intergovernmental Panel on Climate Change (Meehl et al. 2007).

Images, including maps, produced in the Data Library can be delivered to the user's desktop in standard graphics formats like PostScript, JPEG, and PDF. The maps can also be made available in WMS, KML, and GIS formats that feed directly into applications such as Google Earth, Google Maps, or ArcGIS. Any analysis or data download done by the user is represented in a URL that can be saved to the user's desktop. This URL can be shared with collaborators to repeat the analysis. The URL can be incorporated into a script that is run periodically when either environmental or public health data sets are updated.

The IRI Data Library has enabled decision makers to have fast and easy access to the different Earth Observation products mentioned in the section "Improving Data Quality and Accessibility" and to analyze the data to understand the seasonality and trends of climate in relation to health.

Analysis

Often, climate information has to be analyzed and understood in terms of trends and anomalies to be integrated into decision processes for health.

Trends

Trends can be on a monthly basis, of interannual variability, decadal (10-year) variability, and long-term trends (10–100 years) as in the case of climate change. Figure 11.4 shows an example of time-series analysis for the last 100 years in the Sahel region of Africa. It is possible to see the inter-annual variability of rainfall (red line), the decadal variability (blue line), and the long-term trend (black line). The drought that the Sahel experienced in 1970–1980 is clearly visible, and a recovery of the rainfall is visible during the last 10 years. This particularity of the climate trends is of great importance when decision makers have to take measures for climate adaptation.

An example is shown in Figure 11.4, which indicates how rainfall in the Sahel region of Africa has changed over the last 100 years. From this figure, it is possible to gauge the following three components of climate variability:

1. *Interannual (red line)*: How climate can shift from year to year. Just as the weather today can differ considerably from that of yesterday, so the climate this year can be very different from last year. In fact, in nearly all places, the difference in rainfall from one year to the next is much larger than any changes that might be expected from climate change. The largest part of climate variability that will need to be managed by practitioners occurs at timescales of approximately 1–10 years—particularly when considering rainfall.

2. *Decadal (blue line)*: How climate can shift over periods of about 10–30 years. In some parts of the globe, there may be clusters of wet or dry years, possibly resulting in prolonged periods of drought or flooding. For example, the drought that the Sahel experienced during the 1970s and 1980s is clearly noted by the dip in the blue line,

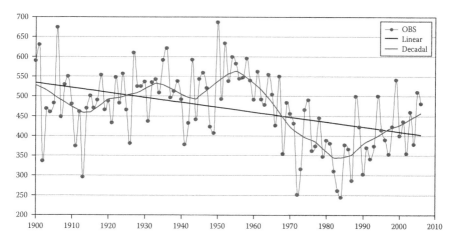

FIGURE 11.4
Observed annual rainfall (mm) in the Sahel, 1900–2006.

while a recovery to more normal levels of rainfall is visible from the mid-1980s to around 2010. This component of climate variability can be very important in adaptation planning: it is possible that short-term climate trends (perhaps over the coming 5–10 years) could be inconsistent with the long-term trend. For example, it is possible for an area to experience a period of a few years of steady or slowly cooling temperatures even if a strong warming trend is anticipated over the next century. Similarly, strong recent warming trends may be much faster than trends that can be expected for the coming decades.

3. *Long-term trend (black line)*: How climate can shift over the long term (beyond 30 years). The most important contributor at this timescale is the impact from climate change. Here, the long-term trend shows a decrease in rainfall over 100 years. However, long-term trends are much more evident in temperatures than in rainfall and are likely to be less relevant to those working in fragile contexts.

In East Africa, a similar trend has been observed where the last decade has experienced a deficit in rainfall. As a measure of climate adaptation, farmers in Ethiopia have started building dams and small reservoirs to collect and harvest water. This process has been very efficient in many regions but has also favored the development of breeding sites for mosquitoes with implications for increase in vector-borne disease transmission such as in East India (Baeza et al. 2013).

Anomalies

One of the most commonly used methods for assessing how the current climate deviates from "normal" is to create an anomaly by subtracting the historical average from the value in question. For example, the average rainfall for 21–31 January from 1980 to 2010 (say 135 mm) can be subtracted from the total for 21–31 January 2015 (say 200 mm), to create a positive "anomaly" of 65 mm (see Figure 11.5). It could then be determined that rainfall for 21–31 January 2015 was 65 mm above average. However climate anomalies are presented, monitoring is an important step in the identification of the early onset of severe conditions, especially of slow-onset hazards such as droughts. In combination with forecasting, monitoring can provide advanced warning of imminent hazardous conditions.

Figure 11.5 depicts rainfall anomalies over Africa. On the map, brown coloring indicates areas with below-average rainfall (negative anomalies) for the period 1–10 December 2014, and blue coloring indicates above-average rainfall (positive anomalies) for the same period. The darker the color, the greater the anomaly. For more information on climate anomalies, see Box 11.1.

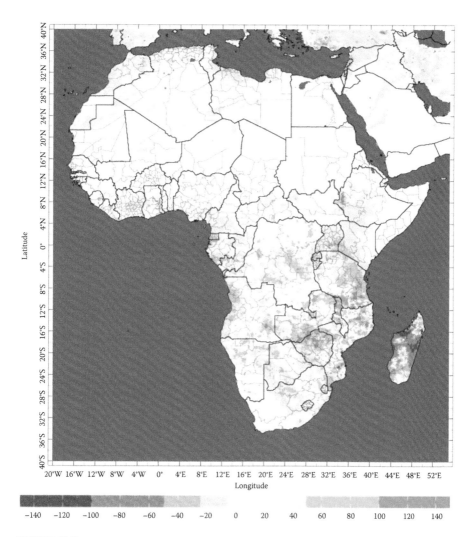

FIGURE 11.5
Rainfall anomalies (mm) over Africa.

BOX 11.1   CLIMATE ANOMALIES

Climate anomalies—the difference in climate between two time periods—can be expressed in different ways.

Percent of average—divide the observed rainfall by the average rainfall, and then express the result as a percentage. For example, if 200 mm of rainfall is received in a given period when average rainfall is 135 mm, the region received almost 50% more rain than typically falls during that time of year. When compared to anomalies, percentages may be easier to understand, but they can still be misleading. For example, in months or seasons that are typically dry, even a small amount of extra rainfall can translate into a large percentage change*: an area that receives 15 mm of rain, but typically receives only 5 mm of rain, would be receiving 300% of its average rainfall, but 15 mm is not much rain by almost any standard and is unlikely to have a major impact. In contrast, if an area that typically receives 500 mm instead receives 300% of its average rainfall, the results are likely to be devastating.

Categories—These are defined in terms of their historical frequency rather than in comparison to the average. The categories are defined by thresholds, which set upper and lower limits to the category. The thresholds may be defined in terms of the proportion of years in the historical record that had less rain or colder temperatures. For example, in seasonal forecasting three categories are commonly defined so that each category typically occurs once in every 3 years. In this case, *below normal* is defined so that historically one-third of years had less rainfall than the upper limit for this category.

---

* This percentage option is not applicable for temperature.

Inundation Products for Leishmaniasis

Surface inundation data sets developed at the City College of New York, as part of a NASA MEaSUREs (Making Earth System Data Records for Use in Research Environments) project supporting assembly of global scale Earth System Data Record (ESDR) and characterizing spatiotemporal attributes of inundated wetlands, developed new products to examine inundation dynamics (McDonald et al. 2011). We employed the Surface Water Microwave Product Series component of that ESDR derived from the QuikSCAT and AMSR-E active/passive microwave data sets to examine the period 2004–2009 for South Sudan (Schroeder et al. 2010). These three environmental variables (NDVI, precipitation, and inundation) aided in determining whether wet or dry years were more conducive to the transmission of leishmaniasis.

Inundation during April, May, June, and July also exhibited a strong inverse relationship with VL cases in SONDJ. This relationship was best explored when comparing the VL case data of a specific medical center to the inundation anomalies. Results are typified by the Lankien Medical Center analysis, where below-average inundation during April displays an inverse relationship with VL cases in the following SONDJ (Figure 11.6).

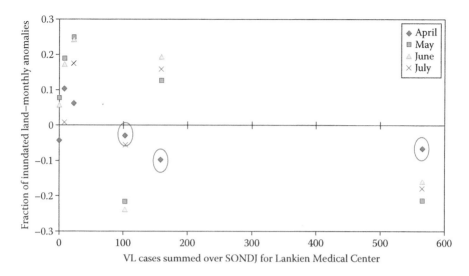

FIGURE 11.6
Monthly inundation anomalies (April–July) and VL cases summed over September through January for Lankien Medical Center in Jonglei State.

Drought may lead to below-average inundation, which could allow for soils to maintain their fissures, resulting in a sustained breeding season for sandflies (Quate 1964). Above-average precipitation and inundation might have the inverse effect, thus eliminating their breeding sites within the soil.

Water Bodies Products for Malaria

Using Landsat images at 30-m spatial resolution, it is possible to map small water bodies where mosquitoes will breed and transmit diseases such as malaria, dengue fever, chikungunya, and West Nile fever. By combining the middle-infrared channel (which is sensitive to water absorption), the near-infrared channel (which is sensitive to bare soil and vegetation canopy), and the red channel (which is sensitive to chlorophyll absorption), it is possible to map water bodies in blue, vegetation in green, and bare soils in brown (Pekel et al. 2011). In Figures 11.7 and 11.8, we overlay malaria data clustered by locations according to time series of malaria cases. Green dots represent villages with low malaria transmission, yellow and orange dots represent villages with medium to high malaria transmission, and red dots represent villages with very high malaria transmission. Villages with low malaria transmission are located either in the highland areas where temperature is the limiting factor or in the dry valley (Rift Valley). Villages with high malaria transmission are located in the dry valley where green vegetation and water bodies are present. The farmers have built dams, which are irrigation infrastructures creating good conditions for mosquitoes to breed (Figure 11.8).

FIGURE 11.7
Landsat image over the Rift Valley in Ethiopia.

FIGURE 11.8
Landsat image over the region of Metehara (8.82°N, 39.92°E) in Ethiopia.

Temperature for Malaria

Using Aqua MODIS land-surface temperature during the nighttime, we mapped the minimum air temperature in the highlands of Ethiopia every 8 days. The MODIS LST can be overlaid in ArcGIS with altitude lines showing that for the same altitude, the temperature can be either favorable for mosquitoes breeding (above 16°C) or unfavorable (below 16°C). Figure 11.9 shows the MODIS LST in the highland of Ethiopia close to the Rift Valley. This information is important for the Ministry of Health to target control measures to fight malaria.

Time series of MODIS LST can show how the temperature varies in the highlands and therefore impacts the transmission of malaria. Integrating precipitation and temperature into a vectorial capacity model (VCAP) allows Ministries of Health in Africa to assess the risk of malaria transmission in the epidemic zones of Africa (Ceccato et al. 2012).

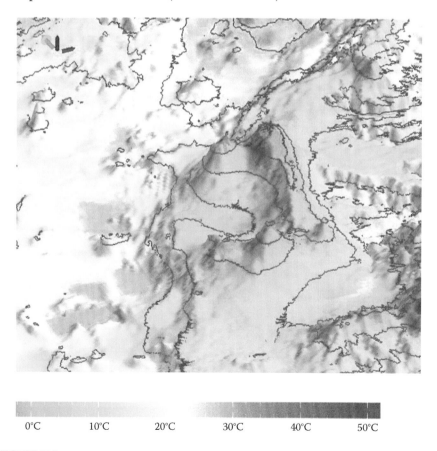

| 0°C | 10°C | 20°C | 30°C | 40°C | 50°C |

FIGURE 11.9
Moderate-resolution imaging spectroradiometer land-surface temperature during the night showing difference of temperatures according to the elevation in the highlands of Ethiopia.

The VCAP product (Figure 11.10) is made available on a regular basis for the period January 2004 to present on the IRI MapRoom website: http://iridl.ldeo.columbia.edu/maproom/Health/Regional/Africa/Malaria/VCAP/index.html

This online VCAP is used by control services and researchers. Validation of the VCAP product was performed using malaria data provided by (1) the National Malaria Control Program, Ministry of Health, Eritrea, and (2) the Service de Lutte contre le Paludisme, Antananarivo, Madagascar (Ceccato et al. 2012).

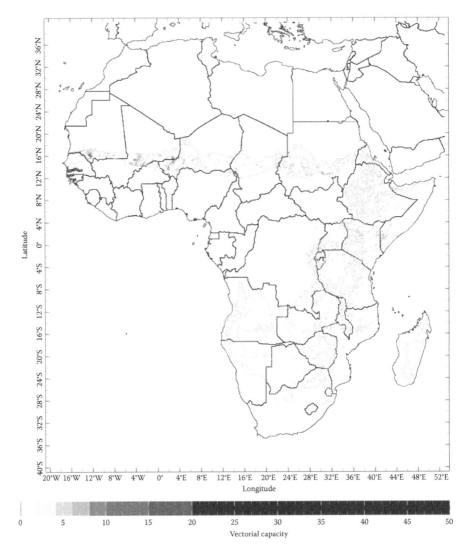

FIGURE 11.10
Vectorial capacity map produced by the US Geological Survey EROS Center.

Conclusion

During the last 30 years, much progress has been made in incorporating remote sensing and GIS into decision processes that can help Ministries of Health and researchers in fighting vector-borne diseases. The examples provided in this chapter show how climatic and environmental factors can be monitored using remote sensing and integrated into decision-making process for mapping risks, creating EWS, and evaluating the impacts of control measures. Until recently, image and processing costs prevented local decision makers from implementing remote sensing decision-support systems on a large scale. More recently, computer processing, data storage facilities, and easy access to remotely sensed products have become available at low cost, and high spatial resolution images have become accessible free of charge. Processing tools are also being made available to the user community at no cost (e.g., IRI Data Library). These developments have paved the way toward making countries more receptive to the implementation of remote sensing systems.

After 30 years of research and development to create the capability to control vector-borne diseases using remote sensing technologies, the pieces are finally falling into place to support global implementation of such technologies. Comprehensive and integrated EWS are available to minimize the impact of deadly diseases and the barriers to implementation, namely cost and data management capabilities, have disappeared. Data and good intentions alone, however, are not sufficient. Many countries also need assistance in the process of technology transfer and in structuring their national information systems and decision-making processes, if they are to derive full benefit from this exceedingly powerful technology.

References

Al-Hamdan, M.Z., Crosson, W.L., Economou, S.A., Estes, M.G., Estes, S.M., Hemmings, S.N., Kent, S.T., et al. (2014). Environmental public health applications using remotely sensed data. *Geocarto International*, 29(1): 85–98.

Anyamba, A., Chretien, J.P., Small, J., Tucker, C.J., Formenty, P.B., Richardson, J.H., Britch, S.C., et al. (2009). Prediction of a rift valley fever outbreak. *Proceedings of the National Academy of Sciences of the United States of America*, 106(3): 955–959.

Aregawi, M., Lynch, M., Bekele, W., Kebede, H., Jima, D., Taffese, H.S., Yenehun, M.A., et al. (2014). Measure of trends in malaria cases and deaths at hospitals, and the effect of antimalarial interventions, 2001–2011, Ethiopia. *PloS One*, 9(1): e106359.

Ashford, R.W., and Thomson, M. (1991). Visceral leishmaniasis in Sudan. A delayed development disaster. *Annals of Tropical Medicine and Parasitology*, 85: 571–572.

Baeza, A., Bouma, M.J., Dhiman, R., Baskerville, E., Ceccato, P., Yadav, R., and Pascual, M. (2013). Long-lasting transition towards sustainable elimination of desert malaria under irrigation development. *Proceedings of the National Academy of Sciences of the United States of America*, 110(37): 15157–15162, http://dx.doi. org/: 10.1073/pnas.1305728110.

Beck, L.R., Rodriguez, M.H., Dister, S.W., Rodriguez, A.D., Washino, R.K., Roberts, D.R. and Spanner M.A. (1997) Assessment of a remote sensing-based model for predicting malaria transmission risk in villages of Chiapas, Mexico. *American Journal of Tropical Medicine and Hygiene*, 56(1): 99–106.

Bhunia, G.S., Kesari, S., Chatterjee, N., Mandal, R., Kumar, V., and Das, P. (2012). Seasonal relationship between normalized difference vegetation index and abundance of the *Phlebotomus* kala-azar vector in an endemic focus in Bihar, India. *Geospatial Health*, 7: 51–62.

Bhunia, G.S., Kumar, V., Kumar, A.J., Das, P., and Kesari, S. (2010). The use of remote sensing in the identification of the eco-environmental factors associated with the risk of human visceral leishmaniasis (kala-azar) on the Gangetic plain, in north-eastern India. *Annals of Tropical Medicine and Parasitology*, 104: 35–53.

Buczak, A.L., Koshute, P.T., Babin, S.M., Feighner, B.H., and Lewis, S.H. (2012). A data-driven epidemiological prediction method for dengue outbreaks using local and remote sensing data. *BMC Medical Informatics and Decision Making*, 12(1): 124.

Campbell-Lendrum, D., Manga, L., Bagayoko, M., and Sommerfeld, J. (2015). Climate change and vector-borne diseases: What are the implications for public health research and policy? *Philosophical Transactions of the Royal Society B*, 370: 20130552. http://dx.doi.org/10.1098/rstb.2013.0552.

Ceccato, P., Gobron, N., Flasse, S., Pinty, B., and Tarantola, S. (2002). Designing a spectral index to estimate vegetation water content from remote sensing data: Part 1: Theoretical approach. *Remote Sensing of Environment*, 82(2–3): 188–197.

Ceccato, P., Connor, S.J., Jeanne, I., and Thomson, M.C. (2005a). Application of Geographical information system and remote sensing technologies for assessing and monitoring Malaria risk. *Parassitologia*, 47: 81–96.

Ceccato, P. (2005b). Operational Early Warning System Using SPOT-VGT and TERRA-MODIS to Predict Desert Locust Outbreaks. In *Proceedings of the 2nd VEGETATION International Users Conference*, 24–26 March 2004, Antwerpen. Editors: Veroustraete, F., Bartholome, E., Verstraeten, W.W. Luxembourg. Luxembourg: Office for Official Publication of the European Communities, ISBN 92-894-9004-7, EUR 21552 EN, 475 p.

Ceccato, P., Bell, M.A., Blumenthal, M.B., Connor, S.J., Dinku, T., Grover-Kopec, E.K., Ropelewski, C.F., and Thomson, M.C. (2006). Use of remote sensing for monitoring climate variability for integrated early warning systems: Applications for human diseases and desert Locust management. *International Geoscience and Remote Sensing Symposium (IGARSS)*, 270–274, Article number 4241221.

Ceccato, P., Ghebremeskel, T., Jaiteh, M., Graves, P.M., Levy, M., Ghebreselassie, S., Ogbamariam, A., et al. (2007a). Malaria stratification, climate and epidemic early warning in Eritrea. *American Journal of Tropical Medicine and Hygiene*, 77: 61–68.

Ceccato, P., Cressman, K., Giannini, A., and Trzaska, S. (2007b). The desert Locust Upsurge in West Africa (2003–2005): Information on the desert Locust early warning system, and the prospects for seasonal climate forecasting. *International Journal of Pest Management*, 53(1): 7–13.

Ceccato, P., Vancutsem, C., and Temimi, M. (2010). Monitoring air and land surface temperatures from remotely sensed data for climate-human health applications. *International Geoscience and Remote Sensing Symposium (IGARSS)*, 178–180, Article number 5649810.

Ceccato, P., Vancutsem, C., Klaver, R., Rowland, J., and Connor, S.J. (2012). A vectorial capacity product to monitor changing malaria transmission potential in epidemic regions of Africa. *Journal of Tropical Medicine*, 2012, Article ID 595948, 6 pp. http://dx.doi.org/10.1155/2012/595948.

Cornillon, P. (2003). OPeNDAP: Accessing data in a distributed, heterogeneous environment. *Data Science Journal*, 2: 164–174.

DaSilva J, Garanganga B, Teveredzi V, Marx SM, Mason SJ, and Connor SJ. (2004). Improving epidemic malaria planning, preparedness and response in Southern Africa. *Malaria Journal*, 3: 37.

Dinku, T., Ceccato, P., and Connor, S.J. (2011). Challenges to satellite rainfall estimation over mountainous and arid parts of East Africa. *International Journal of Remote Sensing*, 32(21): 5965–5979.

Dinku, T., Block, P., Sharoff, J., Hailemarian, K., Osgood, D., del Corral, J., Cousin, R. and Thomson, M. (2014a). Bridging critical gaps in climate services and applications in Africa. *Earth Perspectives*, 1(15): 1–13.

Dinku, T., Kanemba, A., Platzer B., and Thomson, M.C. (2014b). *Leveraging the Climate for Improved Malaria Control in Tanzania.* Earthzine. http://www.earthzine.org/2014/02/15/

Elnaiem, D.A., Hassan, H.K., and Ward, R.D. (1997). Phlebotomine sandflies in a focus of visceral leishmaniasis in a border area of eastern Sudan. *Annals of Tropical Medicine and Parasitology*, 91(3): 307–318.

Elnaiem, D.A., Hassan, H.K., and Ward, R.D. (1999). Associations of *Phlebotomus orientalis* and other sandflies with vegetation types in eastern Sudan focus of kala-azar. *Medical and Veterinary Entomology*, (13): 198–203.

Elnaiem, D.E.A. (2011). Ecology and control of the sand fly vectors of *Leishmania donovani* in East Africa, with special emphasis on *Phlebotomus orientalis*. *Journal of Vector Ecology*, 36(s1): S23–S31.

Elnaiem, D.E.A., Schorscher, J., Bendall, A., Obsomer, V., Osman, M.E., Mekkawi, A.M., and Thomson, M.C. (2003). Risk mapping of visceral leishmaniasis: The role of local variation in rainfall and altitude on the presence and incidence of kala-azar in eastern Sudan. *American Journal of Tropical Medicine and Hygiene*, 68(1): 10–17.

Fontaine, R.E., Najjar, A.E. and J.S. Prince (1961). The 1958 malaria epidemic in Ethiopia. *American Journal of Tropical Medicine and Hygiene*, 10: 795–803.

Gebre-Michael, T., Malone, J.B., Balkew, M., Ali, A., Berhe, N., Hailu, A., and Herzi, A.A. (2004). Mapping the potential distribution of *Phlebotomus martini* and *P. orientalis* (Diptera: Psychodidae), vectors of kala-azar in East Africa by use of geographic information systems. *Acta Tropica*, 90: 73–86.

Gerstl, S., Amsalu, R., and Ritmeijer, K. (2006). Accessibility of diagnostic and treatment centres for visceral leishmaniasis in Gedaref State, northern Sudan. *Tropical Medicine & International Health*, 11(2): 167–175.

Herbreteau, V., Demoraes, F., Khaungaew, W., and Souris, M. (2006). Use of geographic information system and remote sensing for assessing environment influence on Leptospirosis incidence, Phrae province Thailand. *International Journal of Geomatics*, 2(4): 43–50.

Hoogstraal, H., and Heyneman, D. (1969). Leishmaniasis in the Sudan Republic. *The American Journal of Tropical Medicine and Hygiene*, 18(6): 1091-1210.

International Soil Reference and Information Centre (ISRIC) – World Soil Information, 2013. Soil property maps of Africa at 1 km. Available for download at www. isric.org.

Kitron, U., Clennon, J.A., Cecere, M.C., Gürtler, R.E., King, C.H., and Vazquez-Prokopec, G. (2006). Upscale or downscale: Applications of fine scale remotely sensed data to Chagas disease in Argentina and schistosomiasis in Kenya. *Geospatial Health*, 1(1): 49–58.

Kolaczinski, J.H., Hope, A., Ruiz, J.A., Rumunu, J., Richer, M., and Seaman, J. (2008). Kala-azar epidemiology and control, southern Sudan. *Emerging infectious diseases*, 14(4): 664.

Linthicum, K.J., Anyamba, A., Tucker, C.J., Kelley, P.W., Myers, M.F., and Peters, C.J. (1999). Climate and satellite indicators to forecast Rift Valley fever epidemics in Kenya. *Science*, 285(5426): 397–400.

Machault, V., Yébakima, A., Etienne, M., Vignolles, C., Palany, Ph., Tourre, Y.M., Guérécheau, M., et al. (2014). Mapping entomological dengue risk levels in Martinique using high-resolution remote-sensing environmental data. *ISPRS International Journal of Geo-Information*, 3(4): 1352–1371, http://dx.doi. org/10.3390/ijgi3041352.

McDonald, K.C., Chapman, B., Podest, E., Schroeder, R., Flores, S., Willacy, K., Moghaddam, M., et al. (2011). Monitoring inundated wetlands ecosystems with satellite microwave remote sensing in support of earth system science research. *Conference Paper 34th International Symposium on Remote Sensing of Environment— The GEOSS Era: Towards Operational Environmental Monitoring*, 4 p.

Meehl, G.A., Covey, C., Delworth, T., Latif, M., McAvaney, B., Mitchell, J.F.B., Stouffer, R.J., et al. (2007). The WCRP CMIP3 multi-model dataset: A new era in climate change research. *Bulletin of the American Meteorological Society*, 88: 1383–1394.

Moreno-Madriñán, M.J., Crosson, W.L., Eisen, L., Estes, S.M., Estes Jr., M.G., Hayden, M., Hemmings, S.N., et al. (2014). Correlating remote sensing data with the abundance of Pupae of the Dengue Virus Mosquito Vector, *Aedes aegypti*, in Central Mexico. *ISPRS International Journal of Geo-Information*, 3(2): 732–749.

Omumbo, J., Ouma, J., Rapuoda, B., Craig, M.H., Le Sueur, D., and Snow, R.W. (1998). Mapping malaria transmission intensity using geographical information systems (GIS): An example from Kenya. *Annals of Tropical Medicine and Parasitology*, 1: 7–21.

Pekel, J.F., Ceccato, P., Vancutsem, C., Cressman, K., Vanbogaert, E., and Defourny, P. (2011). Development and application of multi-temporal colorimetric transformation to monitor vegetation in the Desert Locust habitat. *IEEE Journal of Selected Topics in Applied Earth Observations and Remote Sensing*, 4(2): 318–326.

Quate, L.W. (1964). Phlebotomus sandflies of the Paloich Area in the Sudan. *Journal of Medical Entomology*, 1(3): 213–268.

Rajesh, K., and Sanjay, K. (2013). Change in global climate and prevalence of Visceral Leishmaniasis. *International Journal of Scientific and Research Publications*, 3(1): 1–2.

Roux, E., Venâcio, A.F., Girres, J.-F., and Romaña, C.A. (2011). Spatial patterns and eco-epidemiological systems—part I: Multi-scale spatial modelling of the occurrence of Chagas disease. *Geospatial Health*, 6(1): 41–51.

Salomon, O.D., Quintana, M.G., Mastrangelo, A.V., and Fernandez, M.S. (2012). Leishmaniasis and climate change—Case study: Argentina. *Journal of Tropical Medicine*, 2012: 601242, http://dx.doi.org/10.1155/2012/601242.

Schorscher J.A, and Goris M. (1992). Incrimination of Phlebotomus (Larroussius) orientalis as a vector of visceral leishmaniasis in western Upper Nile Province, southern Sudan. *Transactions of the Royal Society of Tropical Medicine and Hygiene*, 86(6): 622–623.

Schroeder, R.M.A., Rawlins, K.C. McDonald, E., Podest, R., Zimmermann and Kueppers, M. (2010). Satellite microwave remote sensing of North Eurasian Inundation dynamics: Development of coarse-resolution products and comparison with high-resolution synthetic Aperture Radar data. *Environmental Research Letters*, special issue on Northern Hemisphere high latitude climate and environmental change, 5: 015003 (7 pp.). http://dx.doi.org/10.1088/1748-9326/5/1/015003.

Seaman, J., Mercer, A.J., and Sondorp, E. (1996). The epidemic of visceral leishmaniasis in western Upper Nile, southern Sudan: Course and impact from 1984 to 1994. *International Journal of Epidemiology*, 25(4): 862–871.

Simoonga, C., Utzinger, J., Brooker, S., Vounatsou, P., Appleton, C.C., Stensgaard, A.S., Olsen, A., et al. (2009). Remote sensing, geographical information system and spatial analysis for schistosomiasis epidemiology and ecology in Africa. *Parasitology*, 136(13): 1683–1693.

Skouloudis, A.N., and Rickerby, D.G. (2015). In-situ and remote sensing netwroks for environmental monitoring and global assessment of Leptospirosis outbreaks. *Procedia Engineering*, 107: 194–204.

Sweeney, A., Kruczkiewicz, A., Reid, C., Seaman, J., Abubakar, A., Ritmeijer, K., Doggale, C., et al. (2014). Utilizing NASA earth observations to explore the relationship between environmental factors and visceral leishmaniasis in the Northern States of the Republic of South Sudan. *Earthzine IEEE*, 16(11): 1–15.

Thomson, M.C., Elnaiem, D.A., Ashford, R.W., and Connor, S.J. (1999). Towards a kala-azar risk map for Sudan: Mapping the potential distribution of *Phlebotomus orientalis* using digital data of environmental variables. *Tropical Medicine and International Health*, 4(2): 105–113.

Vancutsem, C., Ceccato, P., Dinku, T., and Connor, S.J. (2010). Evaluation of MODIS Land surface temperature data to estimate air temperature in different ecosystems over Africa. *Remote Sensing of Environment*, 114(2): 449–465.

Walz, Y., Wegmann, M., Dech, S., Raso, G., and Utzinger, J. (2015). Risk profiling of schistosomiasis using remote sensing: Approaches, challenges and outlook. *Parasites & Vectors,* 8: 163, http://dx.doi.org/10.1186/s13071-015-0732-6.

Witt, C.J., Richards, A.L., Masuoka, P.M., Foley, D.H., Buczak, A.L., Musila, L.A., Richardson, J.H., et al. (2011). The AFHSC-Division of GEIS Operations Predictive Surveillance Program: A multidisciplinary approach for the early detection and response to disease outbreaks. *BMC Public Health*, 11(Suppl 2): S10.

World Health Organization (WHO). (2013). Neglected tropical diseases: The 17 neglected tropical diseases. Retrieved from http://www.who.int/neglected_diseases/diseases/en/

Yeshiwondim, A.K., Gopal, S., Hailemariam, A.T., Dengela D.O., and Patel, H.P. (2009). Spatial analysis of malaria incidence at the village level in areas with unstable transmission in Ethiopia. *International Journal of Health Geographics*, 8: 5, http://dx.doi.org/10.1186/1476-072X-8-5.

12

Scale in Disease Transmission, Surveillance, and Modeling

Guillaume Chabot-Couture

CONTENTS

Certain infectious diseases spread over continents, whereas others are found only in remote foci areas. For example, the influenza virus can infect people anywhere in the world; vector-borne diseases like malaria, dengue, and yellow fever can be carried from the tropics to temperate zones by infected travelers, who act as reservoirs. In contrast with these widespread diseases, poliomyelitis and Guinea worm are now found only in a few areas of Africa and Asia, because of long-running global campaigns to eradicate them.

As a starting point, we define the scale of an infectious disease as the geographic distance and the time period over which significant changes in prevalence and/or incidence take place [1]. Though infectious diseases can vary over widely different spans of time and space, scale can be discussed in terms of three main factors: the human mixing scale (a function of proximity and mobility), the scale of environmental determinants of transmission (e.g., airborne, waterborne, or vector-borne transmission), and the scale of disease control efforts (e.g., vaccination, treatment distribution, or disease prevention awareness).

Infectious disease control programs seek first to prevent and then reduce (if not completely eliminate) the burden of diseases. For such programs, the collection of accurate information on the geography and temporal features of disease incidence is crucial to understanding the determinants of transmission. Disease surveillance can be of varying intensities, from a collection of passively reported cases to active case searching by health workers. With regard to surveillance, the deployment and use of geographical information systems, Global Positioning System (GPS) technology, and remote sensing data can have an important beneficial impact: armed with a better understanding of disease transmission dynamics, control efforts can be better targeted towards populations most at risk, and outbreak response can happen more quickly.

In this context, the modeling and simulation of infectious disease transmission and incidence using computers is a rapidly developing tool that can support public health programs. These disease models can combine many different types of data, from clinical trials to disease incidence trends, into disease outbreak forecasts; they can also project the impact of new and untested interventions.

There are many relevant diseases that could be used to discuss the scale of disease transmission and the three factors that determine it. In this chapter, we discuss the human mixing scale, with examples from global influenza transmission; the scale of environmental factors of transmission, with examples from malaria and dracunculiasis (Guinea worm disease); and finally the scale of disease control efforts, using examples from global campaigns to eradicate Guinea worm and poliomyelitis.

Human Mixing Scale and Global Influenza Transmission

The scale of a global disease like influenza follows the scale of human interactions; it is a product of the scale of population density and population connectivity. The flu virus is an interesting and commonly experienced example because each year it travels the globe, spreading rapidly across workplaces, towns, and countries. Infections can happen throughout the year but are concentrated into one or more flu seasons. The duration of a flu season is typically 2–3 months, but the timing of flu epidemics can vary by country. For example, in the United States incidence typically peaks around the New Year, in Europe it peaks around the end of February, in South Africa it peaks in July, and in New Zealand it typically peaks in August [2]. Away from the tropics, flu seasonality is related to the increasing survival of flu virus particles in conditions of low humidity and low temperature. In the tropics, flu seasonality is related to periods of heavy rain [3,4]. In addition to climactic conditions, indoor crowding is often cited as a possible contributor to seasonal flu epidemics; holidays and the end of school terms often coincide with the beginning of epidemics. However, the local person-to-person scale of this second mechanism is very different from the region-wide spatial extends of climate zones. There is heterogeneity in transmission, and transmission can work at multiple timescales and spatial scales. Within a city or a workplace, people can become infected over multiple weeks as the disease propagates, even if the infectious period is on average 1 week [5]. In the end, not everyone catches the flu every year.

Our ability to measure (and study) the spread of the flu depends on the sensitivity of our surveillance efforts. Influenza surveillance typically involves a network of hospital or health centers that regularly send updates about the incidence of influenza-like illnesses and severe acute respiratory infections to a national laboratory. This national laboratory in turn shares epidemiological data and specimens with the World Health Organization (WHO) as part of the WHO Global Influenza Surveillance and Response System (GISRS). The GISRS (formerly GISN) was established in 1952 and includes 142 national laboratories in 112 WHO member states. The WHO Collaborating Centers, Essential Regulatory Laboratories, and H5 Reference Laboratories then perform advanced antigenic and genetic analyses on the samples the national laboratories share with them. These data are then fed into the WHO FluNet, a tool to share data and describe current global influenza trends [6]. Although the sensitivity of influenza surveillance can vary broadly between countries, depending on the strength of their health-care system, the data collected are nonetheless very useful. For example, the genetic data about the currently circulating strains of influenza are used to select which virus strains should compose the influenza vaccine, year after year. Because the strains circulating in the Southern Hemisphere are different from those circulating in the Northern Hemisphere, different vaccines are made to achieve better vaccine efficacy [7]. The flu surveillance network also monitors the emergence of new and more virulent virus strains, like the one that caused the H1N1 pandemic in 2009 [8].

Global epidemics, like that of H1N1 influenza or SARS, show how important connectivity is to disease dispersal. The further apart two cities are in terms of travel time, the more disconnected they become; rapid transit facilities travel and thus increases the flow of travelers. In this case, the geography of administrative boundaries and physical distances is important, but two cities can be well connected by air travel even when they are across the world from each other. Connectivity is also dependent on the number of travelers involved: major cities are better connected than rural cities. Connectivity is also asymmetrical: people from small cities tend to travel more often to larger cities than the reverse. Across a network of multiple cities, the connectivity of any two cities will be driven primarily by the path that has the shortest transit time.

Recent work analyzing the arrival time, in cities around the world, of the H1N1 flu pandemic in 2009 and of the SARS epidemic in 2003 has shown that a measure of effective distance (connectivity) between the origin of an epidemic and a given city, based on the principles outlined above, is linearly related to arrival time of the disease in that city (see Ref. [9] for the mathematical definition of *effective distance*). By contrast, it was shown that the relationship between geographical distance from the origin of the epidemic and the arrival time in a given city is much less clear [9]. Similarly, it has been shown that commuter flows within a country are the dominant source of local connectivity [10]. Overall, the more synchronous outbreaks are in different cities, the better connected these cities are [11].

The transmission patterns of a disease can also be studied using the viral genetic data of the infections. As the virus travels and infects people, it progressively mutates and these mutations can be analyzed to reconstruct the transmission pathways [12,13], sometimes even when only a fraction of cases are observed [14].

High-quality census data can help to map human populations with high resolution, but such data are often available only in developed countries; in less-developed countries where census data can be less reliable or outdated, there are important uncertainties in the mapping of population [15,16]. One approach to mapping population anonymously and cost-effectively has been to use cell phone data [17]. There is also ongoing work to use increasingly affordable and easy-to-use GPS receivers, geographic information systems (GIS), and remote sensing to map populations in the most affected areas [18–21].

In the developed world, population movement patterns have been studied actively [22–27], finding a long tail in the distribution of distances travelled as measured through the call records of cell phones, the movement of dollar bills, and survey data. By contrast, there is much less data on human mobility in the developing world and as a result it is less well understood [28–32]. A recent study of the population distribution in Africa found that the average travel time to the nearest city with a population 50,000 or more was 3.5 hours, which implies low geographic accessibility of rural areas to urban centers [33]. This lack of connectivity can limit disease transmission within a population but also provide reservoirs for the disease to persist for longer periods of time [34].

In order to limit the impact of the flu, a new flu vaccine is made available every year to combat the most recently circulating strains [35]. For certain high-risk populations such as pregnant women, the Strategic Advisory Group of Experts of the WHO also recommends a reduction in social contact and the use of antivirals. The influenza antivirals (oseltamivir and zanamivir) have been shown to reduce the symptoms and the duration of illness [36]. During the 2009 H1N1 pandemic, the WHO distributed 3 million doses of antivirals to 72 countries in an effort to reduce the burden of the disease [37].

In the fight against influenza, modeling can help, for example by studying the impact of prioritizing vaccine use, using antivirals, and/or deploying nonpharmaceutical interventions such as school closures in slowing down epidemics and reducing the burden of the disease [38]. Because the level of human connectivity around the world is large, it is a hard task to contain flu outbreaks. Modeling the network of cities and the flow of people between them can be used to predict the course of flu outbreaks with good agreement with past epidemics [39–41]. For example, modeling has shown that travel restrictions would need to be greater than 99% effective in order to delay outbreaks by more than 2–3 weeks [42]. However, the ability of models to make accurate predictions about upcoming epidemics depends on the data available to calibrate them. It is thus important to make available high resolution and high-quality disease surveillance data.

As the world continues to improve its surveillance of influenza, its mapping of population and their movements, and its deployment of influenza control measures, there may come a time when the human and economic cost of influenza is reduced significantly. In 1999, it was estimated that the flu cost between US $70 and $130 billion per year, just in the United States [43].

Environmental Suitability and Vector Diseases

In contrast to the flu, which can be transmitted everywhere humans come into contact, vector-borne diseases depend on their environment to transmit. This can put strong constraints on the scale of disease transmission: where the disease is found and what temporal patterns it follows [44]. In recent years, the geographical spread of vector diseases like chikungunya (transmitted by the *Aedes albopictus* and *Aedes aegypti* mosquitoes) and Zika (transmitted by *Aedes* genus mosquitoes) has increased and moved increasingly northward [45]. Understanding the environmental suitability of these vectors is an important part of predicting where the next disease outbreaks will occur and mitigate their impact.

For example, malaria is transmitted by *Anopheles* genus mosquitoes. The *Plasmodium* parasite that causes malaria lives part of its life cycle within the *Anopheles* mosquito and part of it within the human host. There are five types of malaria parasites: *Plasmodium falciparum, Plasmodium vivax, Plasmodium malariae, Plasmodium ovale, and Plasmodium knowlesi. Plasmodium falciparum* and *P. vivax* are the two most common malaria parasites, and *P. falciparum* is the most deadly. While in the mosquito, the rate at which the parasite develops and matures depends on air temperature in the mosquito habitat; if the temperature is too low, the maturation of the parasite cannot be completed during the lifetime of the mosquito, and malaria transmission is not possible [46–48]. Thus even within malaria-endemic countries, areas at higher altitude can be malaria-free or experience infrequent epidemics: average air temperature decreases with increasing altitude, as is the case in the highlands of Madagascar, Kenya, and Uganda [49,50]. With climate change, increasing temperature may lead to increasing malaria prevalence [51], although it is often difficult to separate the impact of climate change over the past decades from changes in malaria control [52] and increasing urbanization.

Similarly, the transmission of malaria is not possible in areas where the mosquito cannot live and reproduce; if the temperature is too high or rainfall is insufficient, malaria circulation cannot be sustained in the long term [44]. Thus, mapping the location of the malaria burden can help to uncover the determinants of transmission, identify hotspots of transmission (highlighting the heterogeneity of the risk), help to track the progress made by malaria control efforts, and further target these efforts. For example, geospatial modeling of the 2006 Zambia national malaria indicator survey showed that the variability in malaria prevalence depends on temperature, rainfall, distance to water bodies, land cover, altitude, rural/urban setting, and on interventions (spraying mosquito insecticides on indoor walls and distributing mosquito-killing bed nets) [53]. A similar effort in Kenya identified rural/urban location, temperature, precipitation, altitude, and distance to water bodies as strongly related to parasite prevalence. An analysis of these data showed a high level of local heterogeneity in the parasite prevalence [54]. Finally, in Ghana, an analysis of land cover data using remote sensing data showed that areas that were swamp-like, near villages, and within banana/plantain production were associated with a higher risk of malaria [55].

An extensive constellation of Earth-observing satellites is currently delivering information about rainfall, air temperature, humidity, vegetation cover, urban extent, and other environment-related data, for example, Landsat, GPM, Aqua, Terra, AURA, CloudSat, Tropical Rainfall Measuring Mission, NOAA-N, and GOES. Most importantly, the data collected by these instruments are available free of charge for academic use [56]. In many countries, this information, when combined with additional variables and interpreted within a computer model of malaria, can be used to understand the extent and intensity of malaria transmission [44,57,58]. Unfortunately, even with the impressive advances in the resolution of the available remote sensing data, there are important heterogeneities in the force of malaria transmission that will not be possible to measure from space. In order to gain better resolution, aerial imagery and ground measurements will be crucial in identifying features of transmission that would not have been apparent otherwise. For example, in Tanzania, an urban malaria control program identified mosquito breeding sites within Dar es Salaam using both stereoscopic 1-m aerial imagery and ground validation [59].

Outside of research projects and well-instrumented control programs, malaria surveillance most often relies on the reporting of suspected and confirmed cases by public and private health-care providers. Unfortunately, routine reporting of malaria cases can be uneven and incomplete. This is especially problematic in areas where malaria is endemic. For example, a recent study of malaria surveillance in Vietnam found that the public health information system had missed 50%–90% of cases found by passive case detection and 80%–95% of cases found by active case detection, predominantly due to people seeking care in the private sector [60]. Moreover, in 2014, the WHO estimated that only about one-tenth of the 198 million malaria cases that occurred worldwide were detected and reported through national malaria surveillance systems [61].

Often, data from independent cluster surveys, such as the Demographic and Health Survey and the Malaria Indicator Survey, are used to evaluate the performance of the health system in tracking, testing, and treating malaria patients [62] and help to compensate for data gaps, for example, where routine surveillance data are incomplete. One approach to overcoming the limitations of reported data has been to estimate malaria incidence based on data about death rates by age and a model for malaria-specific mortality [63]. Another approach has been to estimate incidence *a priori*, based on interpolating the results of active case searching surveys, geographic information on the boundary of malaria-endemic areas, and detailed maps of population [47]. Going forward, significant improvements in malaria surveillance will be necessary to better monitor the impact of control, elimination, and eventually eradication efforts [64]. As part of that expansion, active case detection will be useful for detecting the last chains of transmission in an area, especially when asymptomatic carriers play an important role in transmission [65].

In areas of high malaria prevalence, it may sometimes be sufficient for a person to be diagnosed on the basis of clinical symptoms (e.g., fever, sweats, chills, and headaches). In contrast, in low prevalence areas, rapid diagnostic tests (RDTs), microscopic observation of red blood cells, and polymerase chain reaction (PCR) are used to confirm parasitemia. The sensitivity and specificity of RDTs are typically excellent (around 90% and 95%, respectively) when the parasite count is above 100 per microliter of blood [66–68]. Moreover, many RDTs can differentiate *P. falciparum* from the other types of malaria, and the treatment can be modified as a result. The cost of RDTs is decreasing, and their usage is becoming more common. Furthermore, in the future it may become possible to perform routine in-the-field genetic analysis of a malaria infection, for example, using RT-PCR or DNA microarrays, in order to determine drug sensitivity.

The geographic scale of infections is different among the four different malaria parasites. The global scale of these parasites has been the subject of much study, most recently by the Malaria Atlas Project, which has been compiling data on parasite prevalence and clinical episodes, as well as incorporating climate suitability information and expert knowledge [46,47,69,70]. Globally, the most deadly parasite is *P. falciparum*, which can be found in Africa, Asia, and the Americas [47,48,71,72]. *Plasmodium vivax* is the second most deadly type, and it is found in Asia, Africa, and the Americas [48,73,74]. *Plasmodium ovale* is much less common, typically accounting for less than 5% of malaria infections; it is predominantly found in Africa and some parts of Asia [75,76]. *Plasmodium malariae* also receives limited attention because it is responsible for only a small fraction of malaria infections worldwide; it can be found in Africa, Asia, and parts of the Americas [77,78]. Finally, *P. knowlesi* is a primate malaria parasite that can also infect humans; it is found primarily in Southeast Asia [79,80]. Concurrently with parasite prevalence, the Malaria Atlas Project has also mapped the prevalence of different malaria vectors around the world [81]. For example, in Figure 12.1 [48,71,82], a map of estimated clinical burden of *P. falciparum* malaria is compared with a map of environmental suitability for the parasite. In Africa, high clinical burden

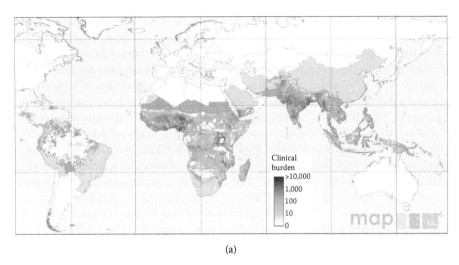

(a)

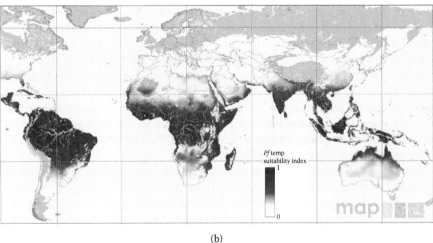

(b)

FIGURE 12.1

Comparing clinical burden of *Plasmodium falciparum* malaria in 2007 (number of clinical cases in people of all ages per year per 5×5 km) (a), with the environmental suitability of that parasite in 2010 (from unsuitable to fully suitable) (b), we see that transmission is less extensive than what could be sustained by the environment. In Africa, high clinical burden correlates well with a high suitability index. By comparison, in South and central America, in India, in Indonesia, and in the Northern coast of Australia, the clinical burden of P. *falciparum malaria* is less than what could be sustained. Maps are adapted from Home–Malaria Atlas Project. (n.d.). http://www.map.ox.ac.uk/, and detail about the methods can be found in Snow RW, et al. 2005. *Nature* 434: 214–217 and Hay SI, et al. 2010. *PLoS Med* 7: e1000290.

correlates well with a high suitability index. By comparison, in South and Central America, India, Indonesia, and the northern coast of Australia, the clinical burden of *P. falciparum* malaria is less than what could be sustained given the suitability index due to efficient control efforts. Overall, the collection of regular, geographically extensive data on parasite prevalence, in addition to disease surveillance efforts, is an important part of monitoring the impact of control efforts and changing epidemiological trends.

In Africa, the range of malaria transmission intensity is wide. For example, the entomological inoculation rate (EIR), the rate at which a human is bitten by an infected mosquito per year, varies from less than 1 to greater than 1,000 [83–85]. The force of transmission is also highly heterogeneous [86]; for example, the EIR can vary between <1 and 120 for villages within the same district [87], and mosquito density can vary appreciably even within a single village [88–90]. The result is that some children experience a few episodes of malaria per year, while other children within the same village experience many episodes. These variations in exposure are due in part to variations in distance to mosquito breeding sites, proximity to water bodies, household structural features, human behavior, and genetic factors [91].

In each country affected by malaria, the National Malaria Control Program (NMCP) develops a customized malaria control strategy. This strategy typically includes the distribution of long-lasting insecticide-treated nets, antimalarial drugs, diagnostic test materials, and (in specific settings) indoor residual spraying campaigns with insecticides. There is a large number of international organizations involved in the fight against malaria and working to help NMCPs achieve better outcomes. This help can come in the form of financial support, strategic support, research support, and high-level coordination across multiple efforts.

Over the past 15 years, there have been strong decreases in the incidence of malaria worldwide, which can be primarily attributed to effective malaria control [72]. From 2000 to 2015, the number of malaria cases worldwide is estimated to have decreased from 262 to 214 million, and the number of malaria deaths is estimated to have decreased from 839,000 to 438,000 [62]. In 2000, the fraction of people in sub-Saharan Africa sleeping under an insecticide-treated net was less than 2%, while in 2015 that fraction had increased to 55% [92]. This increase has been primarily driven by making these bed nets available at a subsidized price or free of charge, but more nets must be distributed in order to achieve universal coverage, at least for children under the age of 5 years. Based on household survey data from Africa, there has also been an increase in the fraction of malaria infections treated with artemisinin combination therapy (ACT), from less than 1% in 2005 to an estimated 16% in 2014 [62]. However, the majority of patients who seek care are treated with other drugs.

As efforts to control malaria continue to increase, the correct use of antimalarial drugs will be important to limit the appearance of drug-resistant parasites. Currently, as described in the 2015 World Malaria Report [62], drug-resistant parasites can be found in a number of areas around the world.

For example, in South America, West Africa, Southeast Asia, and Oceania, *P. falciparum* resistance has been reported to chloroquine, quinine, sulfadoxine-pyrimethamine, and mefloquine. Furthermore, resistance to ACT has been detected in the Greater Mekong subregion of Southeast Asia. In many areas along the Cambodia–Thailand border, *P. falciparum* has become resistant to most available antimalarial medicines. Therapeutic efficacy studies remain the gold standard for guiding drug policy, and they should be undertaken every 2 years. Studies of first- or second-line antimalarial treatments were completed in 66% of countries where *P. falciparum* efficacy studies were feasible. If the world were to lose artemisinin as a drug against malaria, as it did for chloroquine many years ago [93], more expensive drugs would be needed for treatment, and when those fail more people would die of malaria. Going forward, tracking and mapping the drug resistance of the different malaria parasites will help national control programs make decisions on what drugs their front-line health workers should be using to treat malaria patients.

Insecticide resistance in malaria vectors has been reported in 60 of 78 reporting countries around the world since 2010. Of these, 49 have reported resistance to two or more insecticide classes. The most commonly reported resistance is to pyrethroids, which is the insecticide most frequently used by malaria vector control efforts. In order to track the progress of this resistance globally, the WHO has established a system that gathers information from NMCPs, the African Network for Vector Resistance, the Malaria Atlas Project, the President's Malaria Initiative, and the published literature. In these studies, insecticide resistance was defined as less than 90% mosquito mortality on bioassays. In 2014, some 97 countries reported undertaking insecticide resistance monitoring. However, only 52 of these countries provided the WHO with resistance data, suggesting that many countries do not currently monitor insecticide resistance [62]. In order to fight this growing insecticide resistance, the WHO drafted guidelines for NMCPs that recommend the rotation of insecticides used, indoor residual spraying, and the introduction of long-lasting bed nets treated with insecticides other than pyrethroids, when they become available [94].

Malaria transmission dynamics can be complex. There are multiple types of malaria parasites, malaria mosquitoes, varying mosquito habitats and climate conditions, multiple types of vector and disease control efforts, and varying levels of drug resistance and insecticide resistance. Disease modeling software can be helpful in tackling this complexity, to help better understand the transmission dynamics and to study malaria control policies, ongoing or proposed.

Within malaria-infected regions, distributing antimalarial drugs to everyone can be an effective strategy when the prevalence of infections is very high, but in intermediate or low transmission settings it can lead to overtreatment and increase the risk of promoting drug resistance. In conditions of low to intermediate transmission, modeling has suggested that local malaria elimination could be achieved by repeated mass treatment, whereas in higher transmission areas it would need to be combined with other interventions to achieve long-term reduction in incidence [95]. Modeling can also help to estimate the impact of different test and treat strategies, such as case response to local drug administration to target hotspots of transmission and compare them with mass drug administration campaigns (the distribution of antimalarial drugs to all people in a given age range within a given area) [96]. An efficient strategy for malaria control would take advantage of the intrinsic heterogeneity of transmission.

Computer simulations of malaria models can be used to inform and compose an eradication strategy, incorporating the impact of novel tools, questions of technical feasibility, operational feasibility, and variability related to environmental conditions (different types of climates and different types of mosquitoes) [97]. For example, models can provide guidance as to what target performance a malaria vaccine should attain in order to be game-changing across different transmission settings [98]. Models can also help in studying the many spatial scales and heterogeneous landscapes important to the dynamics of transmission [99]. Malaria models can already be connected to weather forecasts and used to build early warning systems to predict whether the upcoming malaria season will be more or less intense than average [100,101].

Going forward, malaria elimination will need a strategy that incorporates more data about the dynamics of parasite populations, the movement of human populations, and higher resolution data about the environment. For example, because humans can spread malaria far further than mosquitoes can, understanding human movements will help to prevent or mitigate the reintroduction of the parasite from malaria-infected countries to malaria-free countries [31,102]. Recently, the mapping of human movements in Kenya, using mobile phone data, helped to identify sources and sinks of malaria transmission [103]. Remote sensing measurements of nighttime lights were also shown to be related to seasonal population migrations in Niger, which in turn explained the strong seasonal forcing of measles epidemics in the country [104].

Of all transmissible human diseases, the extent and scale of malaria transmission is among the best understood, but the extent of most infectious diseases is poorly known [16]. This is especially true for neglected tropical diseases, although that situation is improving [105–107]. Guinea worm is an exception to this rule. Its extent is well understood as a result of three decades of effort to eradicate it; cases remain only in a few dozen villages in Africa [108].

Guinea Worm Eradication

Since 1986, there has been a global campaign to eradicate Guinea worm disease (also referred to as *dracunculiasis,* or *dracontiasis*) [109–113]. Evidence suggests that this disease has existed for millennia. Before it became a target for eradication, it could be found from India to Egypt, and from Mali to Ethiopia [114].

Elements of the biology of this parasite enable transmission over short and long geographical scales, and the transmission cycle involves short and long times-cales. Recently, as the number of cases globally has been decreasing towards zero, the Carter Center and the programs to eradicate Guinea worm in Chad, South Sudan, Mali, and Ethiopia have increased their use of GIS and GPS to track down and effectively respond to the few remaining disease hot spots.

Dracunculiasis is the result of an infection by the *Dracunculus medinensis* worm. The lifecycle of this worm includes a long maturation (11–12 months) within humans, during which female worms are fertilized by male worms, grow to be around 70–120 cm in length, and then migrate towards the feet of the host. At the end of this incubation period, the female worm creates a lesion in the skin of its host and exits the lesion over a period of a few weeks, at a rate of a few centimeters per day. As the worm exits, the host feels a localized burning sensation and seeks to put the lesion in water. When the worm senses water, it ejects larvae in large quantities. Once in the water, the Guinea worm's first-stage larvae are eaten by water fleas (copepods) and go through their first matura-tion step, lasting 2–3 weeks. Following this maturation, a person can ingest the microscopic water fleas by drinking the contaminated water. The last stage of the maturation takes place within the human, completing the cycle [115].

In order for larvae maturation to take place inside the copepods, the water must be warm enough (20°C–30°C). The ponds and water bodies where the larvae are introduced must be stagnant and small enough for another human to have a chance to ingest the water; rivers or lakes are believed to not support transmission [116]. Thus, transmission tends to take place in either dry areas with a rainy season or rainy areas with a dry season, where small seasonal ponds exist for people to drink water from [116,117].

There is no vaccine, no cure, no treatment, and no acquired immunity to Guinea worm disease [118]. People young or old, male or female can be affected. The same person can be infected by multiple worms at the same time, and that person can be reinfected every year [119,120]. Transmission can be interrupted if safe water is provided. This can be done by providing cov-ered wells, teaching people to filter their water, keeping infected people out of the drinking water, or treating infected water with temephos (a chemical agent that kills the copepods) [121–124]. Containment of the exiting worms is an important way to interrupt transmission. Best practices involve identifying cases within 24 hours of emergence, bandaging the wound, and preventing the case from getting in contact with water (typically by isolating them) [125].

The number of infected areas in the world has been steadily decreasing since the beginning of the twentieth century with increasing access to safe drinking water. Before the disease was well understood and efforts were made to control it, Guinea worm had been reported in Senegal, Mauritania, Mali, Ivory Coast, Burkina Faso [126], Ghana [127–129], Benin, Togo, Niger, Nigeria [111,122], Cameroon, Chad [130], Central African Republic, South Sudan, Sudan, Ethiopia, Uganda, Kenya, Yemen, Iran, Uzbekistan, Kazakhstan, Pakistan [131], and India [108,132–134]. The southern republics of the USSR, Uzbekistan and

Kazakhstan eliminated the disease through focused efforts by 1930 [135–137]. Iran eliminated Guinea worm in the 1970s [132]. India began a Guinea worm control campaign in 1980, at the beginning of the decade of water safety, and then achieved countrywide elimination in 1996.

In 1986, it was estimated that there were 3.5 million cases of Guinea worm annually [108,138]. In 2015, only four countries reported the disease (South Sudan, Chad, Mali, and Ethiopia), and global incidence had reached an all-time low of 22 cases [108]. Figure 12.2 shows the date for the last recorded Guinea worm case, where data are available, for countries in Africa and Asia [132,135–137,139]. Angola and the Democratic Republic of the Congo have not been certified free of dracunculiasis, even though no cases have been reported from these countries for decades, because they have not demonstrated sufficient disease surveillance capacity to show that Guinea worm is not currently circulating.

Guinea worm is a very local infectious disease. Important risk factors are recent incidence of the disease nearby in the previous year and the presence of stagnant ponds used to source drinking water [116]. Nonetheless, the year-long incubation period allows for cases to be found very far from where the worm

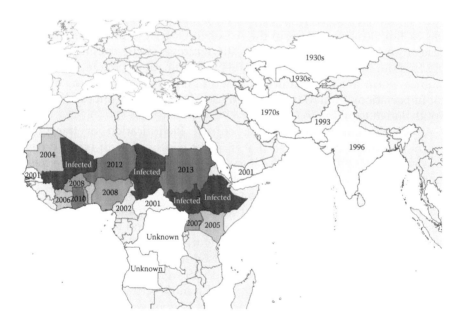

FIGURE 12.2
Map of last recorded Guinea worm case in Africa and Asia (Data from Ruiz-Tiben E. 2006. *Adv Parasitol* 61: 275–309; Fedchenko AP. 1971. *Am J Trop Med Hyg* 20: 511–523; Abdiev FT. 2001. *Med Parazitol (Mosk)*: 60–61; LMII of MP. 1999. Dracunculiasis eradication in Uzbekistan: country report/prepared by the L.M. Isaev Institute of Medical Parasitology; and WHO. (n.d.). Dracunculiasis certification status of countries—Data by country, http://apps.who.int/gho/data/node.main.A1633.) Of the four remaining infected countries, only small pockets of transmission remain. No record could be found of past cases for countries colored green, and they have been certified free of Guinea worm by the WHO.

was acquired. During the twentieth century, the industrialization of Nigeria and the introduction of train lines contributed to the dispersal of Guinea worm from rural areas to villages that had never been exposed [122,140]. Historically, cases have been found in the United States, in France, and in Great Britain that could be traced back to previous travel to Africa or other infected areas [141,142]. During the nineteenth century, Guinea worm was observed among slaves in Brazil, Cuba, Haiti, Grenada, and many other American countries, with their infections tracing back to Africa. Once the slave trade stopped, so did the circulation of Guinea worm in the Americas [142].

Because transmission includes both long- and short-range components, the surveillance for Guinea worm is a mix of active surveillance for villages near recently infected villages and passive surveillance for all other areas. In active surveillance villages, there is a village volunteer who watches for Guinea worm infections and helps to contain the disease if it is detected. Passive surveillance is based on raising awareness for the disease in the population and in distributing rewards for confirmed cases. The combined result of these surveillance efforts has been approximately 70% containment, even though some infected areas can be very remote (e.g., [143]). In certain situations, uncontained cases can lead to a rapid rise in incidence. For example, in Tanzikratene village in the Ansongo District of Mali, the number of cases rose from 3 in 2013 to 29 in 2014 within a single community drinking from an infected water source [144,145]. Sometimes, the introduction of Guinea worm in a community simply dies off. For example, this was seen for importations from Mali to Niger in 2012 [146], and for importations from South Sudan to Sudan in 2013 [147].

As eradication is approached, correct identification of the guinea worm becomes important so that it is not confused with other types of worms [148–151]. In 2014, laboratory confirmation of all worms extracted from patients became a requirement, and the majority of worm specimens were found not to be Guinea worm [152].

As the eradication campaign approaches the last chains of transmission, it becomes increasingly important to detect all cases and understand the last remaining patterns of transmission. New tools like GIS, mapping, and satellite imagery are increasingly being used for this purpose. For example, efforts to leverage satellite imagery and GIS were promising in helping to manage data and identify the location of remote settlements [153]. Landsat imagery helped to show that in areas where Guinea worm control had taken place agricultural productivity had increased, as measured by the normalized difference vegetation index [154]. More recently, GIS has been used to study the spatial distribution of a resurgence of cases in Niger [155]. The location of cases and villages under surveillance has been tracked year after year in South Sudan, showing narrowing geographical transmission [156]. In Chad, the peculiar epidemiology of the worm, involving both human and dogs, has been studied by comparing the location of the cases and their proximity to different types of water sources [130,157].

Historically and in most situations, the extent of Guinea worm infections is the extent of the cases of the disease; virtually all infections become detectable cases when the worm exits the host. However, the recent detection of Guinea worm in dogs, baboons, and other animals is a possible exception to this rule and it could further expand the extent of the disease. By contrast, assessing the scale of poliomyelitis transmission can be much more difficult, as cases represent less than 1% of all infections. As a result, the poliovirus can circulate unseen for years, especially in areas where surveillance sensitivity is poor.

Poliomyelitis Eradication

The story of poliomyelitis is one of changing scale. At the beginning of the twentieth century, when standards of hygiene improved in industrializing countries, poliomyelitis appeared as an epidemic disease with rising incidence, while in less developed countries incidence was lower and transmission was more stable [158,159]. A century later, polio is facing eradication and can be found only in a handful of countries. Today, the scale of poliomyelitis transmission reflects the few remaining areas without adequate vaccination coverage. As the efforts of the global polio eradication initiative (GPEI) continue and increasingly benefit from new technologies to track the disease and distribute the vaccines, we may witness its complete and final disappearance.

Poliomyelitis is a paralytic disease caused by poliovirus infection. This infectious disease is typically acquired in childhood, and about 99.5% of the time it leads to mild or flu-like symptoms. In rare cases (less than 0.5%), the poliovirus enters the nervous system, attacks motor neurons, and causes acute flaccid paralysis (AFP), for example, of one leg, one arm, or even worse of muscles we use for breathing. The poliovirus is excreted in feces and pharyngeal secretions, and transmitted person-to-person primarily by the hand-to-hand-to-mouth channel. Transmission is most intense within families [160,161], but polio can circulate across cities, countries, and between continents [162].

Outbreaks of infantile paralysis seem to have appeared in Scandinavian countries, the United States, and other developed countries in the 1880s [158]. Evidence suggests that before that time poliovirus infection took place while children were passively protected by their mother's antibodies. This passive protection typically disappears between 6 and 12 months of age. Within populations where the force of infection was high, the incidence of paralytic polio was low because children would be exposed and develop immunity while protected by their maternal antibodies. By contrast, within populations where the force of infection was low, children would be exposed later in life and be susceptible to paralysis, resulting in higher incidence of paralysis [163]. This would explain why, in the beginning of the twentieth century, poliomyelitis increased in incidence in the industrializing world, while the incidence appeared to be low in less-developed regions, like Cuba and Brazil [164].

Poliomyelitis incidence is seasonal. Before polio was eliminated from the United States, the peak of poliovirus transmission took place in the summer [165,166]. The seasonal variation in incidence was strong in New England, whereas in Hawaii it was much less pronounced, probably because of the milder climate. In South America, the areas that eliminated polio first were temperate and less humid, suggesting that the virus survives on surfaces for shorter periods in dry/temperate conditions [158].

When, in the 1950s, Salk and Sabin invented their vaccines against polio and these vaccines started being used by public health programs around the world, the incidence of poliomyelitis started to decline and its scale became increasingly tied to the quality of vaccination efforts. For example, in 1973, the United States achieved the elimination of endemic wild polioviruses. By using the Sabin vaccine in routine pediatric immunization and mass immunization drives, population immunity became sufficiently high so that virus circulation could not persist. Around the United States, herd immunity was achieved [167,168].

In 1980, following the successful eradication of smallpox [169], the WHO sought to identify other infectious diseases for eradication. At that time, all countries had introduced the Sabin and/or the Salk vaccine in their recommended pediatric immunization schedule. Worldwide, there were 112 countries reporting polio cases, and between 250,000 and 600,000 [158,170,171] polio cases, with an often-cited estimate of 350,000 incident cases in 1988 [172]. In the 1980s, the Pan American Health Organization made rapid progress in eliminating poliomyelitis from all their member states. In 1988, the World Health Assembly decided to follow this momentum and extend this elimination goal to all countries, calling for complete eradication of the poliovirus by the year 2000.

As shown in Figure 12.3 [172–174], the number of countries reporting poliomyelitis cases decreased steadily during the 1980s and this trend accelerated in the late 1990s as the deadline for eradication approached. In 2000, the number of polio-infected countries had decreased to 31, of which 20 had never interrupted poliovirus circulation (endemic), and there were 2,971 cases reported worldwide. In the 2000s, the number of endemic countries continued to decrease, but there were also frequent outbreaks in polio-free countries, which slowed down the global progress towards eradication. In 2015, Nigeria was removed from the list of polio-endemic countries, leaving Afghanistan and Pakistan as the only remaining countries. If wild poliovirus circulation is interrupted in those countries, then poliomyelitis will be on track to become the second human disease to ever be eradicated.

Like influenza, poliomyelitis was an infectious disease found globally. In the history of polio eradication, there are multiple examples of polio being transmitted by travelers across large distances. For example, in 1992, an outbreak of wild poliovirus type 3 took place in the Netherlands within religious

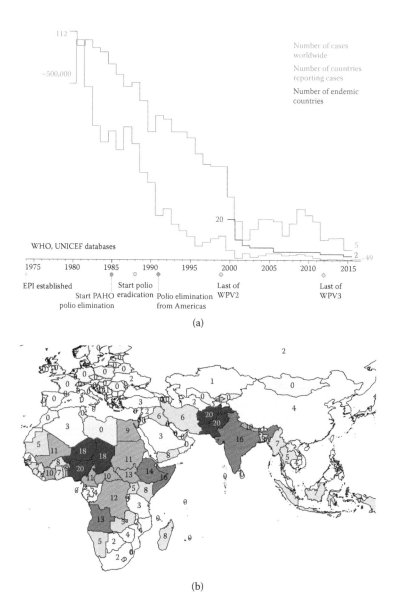

FIGURE 12.3

(a) Timeline of the global polio eradication, showing the number of poliomyelitis cases world-wide, the number of countries reporting cases, and differentiating this from the number of endemic countries. (b) Number of years reporting polio cases, wild or vaccine derived, between 1996 and 2015 (Data from Global Polio Eradication Initiative > Data and monitoring. www. polioeradication.org/Dataandmonitoring; WHO–AFP/polio data. (n.d.). https://CaseCount. aspx; WHO—Data, statistics and graphics by subject. http://apps.who.int/immunization_en/ index.html). Nigeria, Afghanistan, Pakistan, and India are preeminent on this map because they were endemic throughout the 2000s. Countries neighboring Nigeria also experienced frequent outbreaks and importations, especially Chad and Niger.

communities who refuse vaccination [175]. Within a few weeks, the virus had spread within this religious community throughout the country. From there, it travelled to a similar religious community in Canada [176]. Genetic sequencing of the outbreak strain showed that the virus had originated in a geographically remote region of India, although during that epidemic, the poliovirus did not propagate from Canada to the United States as it had in 1979 [177,178]. Between 1979 and 1995, the Netherlands detected numerous importations of wild poliovirus type 1 and type 3 from the Mediterranean, India, and Indonesia [179]. A review of the literature on polio outbreaks between 1996 and 2012 revealed 22 outbreaks affecting 39 countries, with five outbreaks involving multiple countries [180].

The detection and tracking of pandemics requires a sensitive and broad-reaching disease surveillance system. By that measure, the global poliomyelitis surveillance network is one of the best and most developed surveillance networks in the world. For example, in 2014, the GPEI investigated 103,974 cases of AFP from 161 countries, in order to find 415 cases of poliomyelitis, 359 of which were due to wild polioviruses and 56 that were due to circulating vaccine-derived polioviruses [172]. Each AFP case was investigated by a health-care provider; case investigations included the patient's vaccination history and the collection of two stool samples. These stool samples were analyzed by one of the laboratories in the Global Polio Laboratory Network (GPLN) in order to detect the presence of poliovirus, and (if found) to determine the serotype and genetic sequence of the virus. Since its beginning, the GPLN has evolved to detect polioviruses more rapidly, more cheaply, and to become more sensitive in detecting vaccine-derived polioviruses, even when they differ from the vaccine stock by only a few nucleotides [181,182]. The success of the GPLN has also inspired the development of a similar network of laboratories for measles and rubella surveillance [183].

Preventing global wide-reaching outbreaks can be done by maintaining high polio immunity in polio-free countries that are connected to infected countries. For that purpose, the global polio eradication program coordinates mass vaccination efforts across continents, often as regional synchronized SIAs, so that global funds can be directed to countries that are most at risk of polio outbreaks and help to contain the poliovirus to as few countries as possible. The number of vaccination campaigns conducted in these polio-free countries is thus proportional to the fraction of their population that is susceptible to polio and the probability that poliovirus might be imported from infected countries [184,185]. The scale of these efforts is enormous. Every year, more than a billion doses of oral polio vaccine are distributed by the GPEI around the world to undervaccinated populations in order to achieve polio eradication.

More practically, as a country approaches polio elimination, it is increasingly difficult to detect areas with gaps in vaccination coverage before they become infected with the virus. Because of the low case-to-infection ratio, most polio infections remain undetected and hundreds of infections must take place before one or more paralytic cases can be detected. Vaccination coverage is also difficult to measure and, in areas where it is chronically low, data quality also has a tendency to be poor [186]. By using a statistical modeling framework, it is possible to construct better and more robust risk predictions. This is done by combining the information from multiple data sets, using linear and nonlinear combinations, and calibrating the balance of the individual indicators using external data, such as past outbreaks or multi-indicator cluster surveys. For example, such a model helped the Nigerian polio eradication program focus their resources on the areas that were most at risk, before they became infected with polio [187].

The systematic genetic sequencing of polioviruses found in lab-confirmed poliomyelitis cases enables the monitoring of poliovirus circulation. Understanding the pathways taken by the virus as it circulates can help focus eradication efforts: sources and sinks in the transmission network can be identified, and areas where links in the transmission chains have been missed can be inferred. When in 2014 wild poliovirus of type 1 was found in the sewers of Rio de Janeiro [188], it was genetically linked to an ongoing outbreak in Equatorial Guinea, which had started in Cameroon in 2013 [189]. Tracing the genetic history of that virus, it was shown that it had originated in Chad 2 years before and had perhaps travelled through the Central African Republic in order to reach Equatorial Guinea. Similarly, when wild poliovirus was found in sewers of Cairo in December 2012, it was linked back to Pakistan, not Nigeria. This link was further confirmed when cases were later found in Syria and Iraq [190], and the virus was also found in the sewers of Israel [191]. However, it is not always possible to determine the origin of the detected virus with good precision because a large number of infections within the transmission chain were not observed, unlike what can be done with flu or Ebola, where the majority of infections within a transmission chain can be observed [14].

Within countries with endemic transmission such as Pakistan, the movements of the poliovirus are tracked in space and time by AFP surveillance and sometimes also with environmental surveillance. In addition, a phylogenetic analysis can be used to make sense of transmission patterns. In the capsid region of the poliovirus genome, the rate of base pair mutation is well known and can be used as a molecular clock [192]. Thus, the genetic sequence of any polio case provides a measure of its genetic distance to other polio cases; two cases detected near each other in time and space typically will have a similar genetic sequence, unless multiple different strains of poliovirus are co-circulating [193]. Conversely, even for polioviruses detected in different regions, their genetic sequences may indicate that they are closely related. Generally, it is possible to make some inferences about the transmission patterns of the poliovirus between provinces and sometimes between cities [14]. From this genetic linking, the scale of poliovirus transmission, its mixing in time and space within a polio-infected area, can be made apparent.

In Pakistan, regular poliovirus transmission is observed across the whole country, from Karachi to Peshawar. Polioviruses are also carried back and forth over the Afghanistan border, due to the strong economic and social links between the two countries. Phylogenetic analyses of the polioviruses found along the border show not only how frequently the viruses cross the border but also how this movement is predominantly directional, from Pakistan towards Afghanistan [193–196]. Within Pakistan, the virus is concentrated in a number of reservoirs: regions where poliovirus is regularly found (as determined from the detection of poliovirus in AFP cases or in environmental surveillance samples) and that export these viruses. The Pakistan reservoirs are centered on the cities of Peshawar, Karachi, and at times Quetta [194–196]. In the analysis of phylogenetic trees (dendrograms), it is possible to cluster sequences in groups based on genetic similarity. In tracking these clusters, a reduction in the number of circulation clusters indicates that transmission is narrowing, even though the number of cases may not have changed significantly [195,197]. The provenance of virus clusters can help confirm or refute causal links in the epidemiology of polio. In Pakistan, it recently showed how multiple clusters were co-circulating around the country, even though it appeared that most cases were the result of a large outbreak within the federally administered tribal areas [196].

Environmental surveillance enhances our understanding of the geographical extent of transmission because it is sensitive to poliovirus infections, not only paralytic cases. The transmission of the poliovirus without detection of poliomyelitis cases is referred to as *silent circulation*. This can take place when population immunity is high or when the population is predominantly IPV-immunized, or when there are gaps in AFP surveillance [191,196,198]. Another symptom of surveillance gaps is the detection of "orphan" polioviruses: viruses whose most closely related sequences in the VP1 protein area are more than 2% different. The detection of such viruses suggests that some links of the genetic tree have been missed [199].

In many polio-infected areas, GPS-enabled devices have been used to geolocate cases and conduct field surveys. Since 2010, the polio eradication program in Nigeria has been using GIS and GPS technology to reach more children during vaccination campaigns [200,201], and these efforts have contributed to a significant improvement in the reach and completeness of vaccination coverage. This eventually led to Nigeria eliminating poliovirus circulation in 2015.

The extensive size of house-to-house vaccination campaigns make their logistics inherently challenging. In order to reach millions of children in 3–6 days (on average), a detailed and accurate plan (a microplan) of villages and settlements is developed. Microplans include actual or estimated target populations for each geographic location, the logistical needs (cold chain equipment, number of vaccinators and supervisors, vehicles, consumables, etc.), and travel routes (and mode of transport) to reach every location. Because it is central to the vaccination efforts, villages not on this microplan are often repeatedly missed, thus resulting in geospatially and temporally concentrated pockets of unvaccinated children. These pockets thus enable the poliovirus to circulate.

To address this persistent challenge, the Nigerian program located and mapped approximately 100,000 settlements in 10 northern high-risk states using high-resolution imagery and field GPS data collection. These maps were then used to distribute all the vaccination work across teams of volunteers, in each ward of each local government area in all 10 northern high-risk states. The result of this disciplined logistic was that once-overlooked villages were now officially documented and included in vaccination activities [201]. In order to identify the remaining gaps in coverage and assure inclusion of the chronically missed settlements, the Nigerian program equipped over 12,000 of their vaccination teams with GPS-enabled smartphones that collected time-stamped positions as they visited remote villages or dense urban neighborhoods. At the end of each day, a measure of geographic coverage was used to direct vaccination work towards areas that had been missed during the previous days' effort. Today, this extensive vaccinator tracking system is operational throughout Nigeria and provides near-real-time feedback to stakeholders around the world [201].

The efforts of the GPEI to track and monitor the poliovirus are exceptional. The data thus collected offer a window into the different scales of poliovirus transmission, from the narrow circulation within the two remaining endemic countries, to the rare remaining international outbreaks they potentiate. Systematically collecting data of such quality for other vaccine-preventable diseases would enable improvements in their control efforts and further reduce the global burden of communicable diseases.

Conclusion

In this chapter, the scale of transmission, surveillance, and control efforts for four different infectious diseases was described and discussed. It was shown how the three main factors that must be considered when assessing scale are the human population distribution and its movement, the biology of the disease and its relationship to the environment, and the control efforts by global health programs. The flu, as it travels around the world, follows the distribution of people and travels the way people travel. By contrast, vector-borne diseases like malaria transmit in places where their vector can live and reproduce. Sometimes, when a vector-borne disease is reintroduced into a disease-free area where local mosquitoes could sustain circulation, it is due to the efforts of local health officials that disease outbreaks are prevented. When they take a global scope, disease control efforts can achieve eradication, as was the case for smallpox and rinderpest.

Currently, the extent of poliovirus and Guinea worm transmission around the world is so constricted by eradication efforts that disease incidence has decreased by a factor of 10,000 over the last 30 years.

The availability of data is crucial to understand the role that each scale factor plays in the overall scale of a disease. Without conducting more epidemiological studies and clinical trials and without establishing more robust and far-reaching disease surveillance systems, our understanding of the scale of disease transmission will continue to be incomplete. At the onset of the polio and Guinea worm eradication campaigns, broad surveying took place to estimate the global burden of each disease. This was part of evaluating the technical and operational feasibility of setting eradication goals.

Today, efforts to map diseases of global health importance are expanding. The tools of digital epidemiology (remote sensing, GIS, low-cost handheld GPS receivers, and computer disease models) are increasingly enabling data collection on an unprecedented scale. If these efforts can be used to increase both the quality and quantity of data available to fight diseases, we may be able to radically reduce the morbidity and mortality that communicable diseases create around the world. And if we are persistent, we may even permanently wipe out some of these diseases from the planet.

Acknowledgments

The author would like to thank Philip Eckhoff, Edward Wenger, and Victoria Gammino for their critical feedback on this chapter.

References

1. Quattrochi DA, Goodchild MF. (1997). *Scale in remote sensing and GIS*. CRC Press. 432 p. Available: https://books.google.com/books?hl=en&lr=&id=uyfbQN0s_tgC&pgis=1. Accessed 18 January 2016.
2. WHO. (n.d.). Influenza update. Available: http://www.who.int/influenza/surveillance_monitoring/updates/latest_update_GIP_surveillance/en/. Accessed 1 January 2015.
3. Tamerius JD, Shaman J, Alonso WJ, Alonso WJ, Bloom-Feshbach K, et al. (2013). Environmental predictors of seasonal influenza epidemics across temperate and tropical climates. *PLoS Pathog* 9: e1003194.
4. Lowen AC, Mubareka S, Steel J, Palese P. (2007). Influenza virus transmission is dependent on relative humidity and temperature. *PLoS Pathog* 3: 1470–1476.
5. Centers for Disease Control and Prevention. (n.d.). – Influenza (flu). Available: http://www.cdc.gov/flu/index.htm. Accessed 1 January 2015.

6. World Health Organization. (n.d.). *Global Influenza Surveillance and Response System (GISRS)*. Available: http://www.who.int/influenza/gisrs_laboratory/en/. Accessed 13 November 2015.

7. World Health Organization. (n.d.). *WHO recommendations on the composition of influenza virus vaccines*. Available: http://www.who.int/influenza/vaccines/virus/recommendations/en/. Accessed 13 November 2015.

8. World Health Organization. (n.d.). World now at the start of 2009 influenza pandemic. *Dis Outbreak News*. Available: http://www.who.int/mediacentre/news/statements/2009/h1n1_pandemic_phase6_20090611/en/. Accessed 13 November 2015.

9. Brockmann D, Helbing D. (2013). The hidden geometry of complex, network-driven contagion phenomena. *Science* 342: 1337–1342. Available: http://www.sciencemag.org/content/342/6164/1337.full.

10. Balcan D, Colizza V, Gonçalves B, Hu H, Ramasco JJ, et al. (2009). Multiscale mobility networks and the spatial spreading of infectious diseases. *Proc Natl Acad Sci USA* 106: 21484–21489.

11. Viboud C, Bjørnstad ON, Smith DL, Simonsen L, Miller MA, et al. (2006). Synchrony, waves, and spatial hierarchies in the spread of influenza. *Science* 312: 447–451.

12. Cheng X, Tan Y, He M, Lam TT-Y, Lu X, et al. (2013). Epidemiological dynamics and phylogeography of influenza virus in southern China. *J Infect Dis* 207: 106–114.

13. Wallace RG, Hodac H, Lathrop RH, Fitch WM. (2007). A statistical phylogeography of influenza A H5N1. *Proc Natl Acad Sci USA* 104: 4473–4478.

14. Famulare M, Hu H. (2015). Extracting transmission networks from phylogeographic data for epidemic and endemic diseases: Ebola virus in Sierra Leone, 2009 H1N1 pandemic influenza and polio in Nigeria. *Int Health* 7: 130–138.

15. Tatem AJ, Campiz N, Gething PW, Snow RW, Linard C. (2011). The effects of spatial population dataset choice on estimates of population at risk of disease. *Popul Health Metr* 9: 4.

16. Hay SI, Battle KE, Pigott DM, Smith DL, Moyes CL, et al. (2013). Global mapping of infectious disease. *Philos Trans R Soc Lond B Biol Sci* 368: 20120250.

17. Deville P, Linard C, Martin S, Gilbert M, Stevens FR, et al. (2014). Dynamic population mapping using mobile phone data. *Proc Natl Acad Sci USA* 111: 15888–15893.

18. Chabot-Couture G, Seaman VY, Wenger J, Moonen B, Magill A. (2015). Advancing digital methods in the fight against communicable diseases. *Int Health* 7: 79–81.

19. Gaughan AE, Stevens FR, Linard C, Jia P, Tatem AJ. (2013). High resolution population distribution maps for Southeast Asia in 2010 and 2015. *PLoS One* 8: e55882.

20. Tatem AJ, Goetz SJ, Hay SI. (2008). Fifty years of earth-observation satellites. *Am Sci* 96: 390.

21. Tatem AJ. (2014). Mapping the denominator: spatial demography in the measurement of progress. *Int Health* 6: 153–155.

22. Song C, Qu Z, Blumm N, Barabási A-L. (2010). Limits of predictability in human mobility. *Science* 327: 1018–1021. doi:10.1126/science.1177170.

23. Brockmann D, Hufnagel L, Geisel T. (2006). The scaling laws of human travel. *Nature* 439: 462–465. doi:10.1038/nature04292.

24. Bagrow JP, Lin Y-R. (2012). Mesoscopic structure and social aspects of human mobility. *PLoS One* 7: e37676. doi:10.1371/journal.pone.0037676.
25. Gonzalez MC, Hidalgo CA, Barabasi A-L. (2008). Understanding individual human mobility patterns. *Nature* 453: 779–782. doi:10.1038/nature06958.
26. Simini F, González MC, Maritan A, Barabási A-L. (2012). A universal model for mobility and migration patterns. *Nature* 484: 96–100. doi:10.1038/nature10856.
27. Noulas A, Scellato S, Lambiotte R, Pontil M, Mascolo C. (2012). A tale of many cities: Universal patterns in human urban mobility. *PLoS One* 7. doi:10.1371/journal.pone.0037027.
28. Wesolowski A, Buckee CO, Pindolia DK, Eagle N, Smith DL, et al. (2013). The use of census migration sata to approximate human movement patterns across temporal scales. *PLoS One* 8. doi:10.1371/journal.pone.0052971.
29. Wesolowski A, Stresman G, Eagle N, Stevenson J, Owaga C, et al. (2014). Quantifying travel behavior for infectious disease research: a comparison of data from surveys and mobile phones. *Sci Rep* 4: 5678.
30. Tatem AJ. (2014). Mapping population and pathogen movements. *Int Health* 6(1): 5–11.
31. Tatem AJ, Smith DL. (2010). International population movements and regional *Plasmodium falciparum* malaria elimination strategies. *Proc Natl Acad Sci USA* 107: 12222–12227. doi:10.1073/pnas.1002971107.
32. Vazquez-Prokopec GM, Bisanzio D, Stoddard ST, Paz-Soldan V, Morrison AC, et al. (2013). Using GPS technology to quantify human mobility, dynamic contacts and infectious disease dynamics in a resource-poor urban environment. *PLoS One* 8: e58802.
33. Linard C, Gilbert M, Snow RW, Noor AM, Tatem AJ. (2012). Population distribution, settlement patterns and accessibility across Africa in 2010. *PLoS One* 7: e31743.
34. Anderson RM, May RM. (1992). *Infectious diseases of humans: Dynamics and control.* OUP, Oxford. 757 p.
35. Russell CA, Jones TC, Barr IG, Cox NJ, Garten RJ, et al. (2008). Influenza vaccine strain selection and recent studies on the global migration of seasonal influenza viruses. *Vaccine* 26 Suppl 4: D31–D34.
36. Jefferson T, Jones MA, Doshi P, Del Mar CB, Hama R, et al. (2014). Neuraminidase inhibitors for preventing and treating influenza in healthy adults and children. *Cochrane Database Syst Rev* 4: CD008965.
37. Fineberg HV. (2014). Pandemic preparedness and response—lessons from the H1N1 influenza of 2009. *N Engl J Med* 370: 1335–1342.
38. Halloran ME, Ferguson NM, Eubank S, Longini IM, Cummings DAT, et al. (2008). Modeling targeted layered containment of an influenza pandemic in the United States. *Proc Natl Acad Sci USA* 105: 4639–4644.
39. Merler S, Ajelli M. (2010). The role of population heterogeneity and human mobility in the spread of pandemic influenza. *Proc Biol Sci* 277: 557–565.
40. Colizza V, Barrat A, Barthélemy M, Vespignani A. (2007). Predictability and epidemic pathways in global outbreaks of infectious diseases: the SARS case study. *BMC Med* 5: 34.
41. Balcan D, Gonçalves B, Hu H, Ramasco JJ, Colizza V, et al. (2010). Modeling the spatial spread of infectious diseases: the global epidemic and mobility computational model. *J Comput Sci* 1: 132–145.

42. Ferguson NM, Cummings DAT, Fraser C, Cajka JC, Cooley PC, et al. (2006). Strategies for mitigating an influenza pandemic. *Nature* 442: 448–452. doi:10.1038/nature04795.

43. Meltzer M, Cox N, Fukuda K. (1999). The economic impact of pandemic influenza in the United States: priorities for intervention. *Emerg Infect Dis*. Available: http://wwwlive.who.int/entity/influenza_vaccines_plan/resources/ARTICLE_Economic_Impact_of_Pandemic_Influenza_in_the_US.pdf. Accessed 12 January 2016.

44. Chabot-Couture G, Nigmatulina K, Eckhoff P. (2014). An environmental data set for vector-borne disease modeling and epidemiology. *PLoS One* 9: e94741.

45. Bogoch II, Brady OJ, Kraemer MUG, German M, Creatore MI, et al. (2016). Anticipating the international spread of Zika virus from Brazil. *Lancet* 387: 335–336.

46. Hay SI, Guerra CA, Tatem AJ, Noor AM, Snow RW. (2004). The global distribution and population at risk of malaria: past, present, and future. *Lancet Infect Dis* 4: 327–336.

47. Snow RW, Guerra CA, Noor AM, Myint HY, Hay SI. (2005). The global distribution of clinical episodes of *Plasmodium falciparum* malaria. *Nature* 434: 214–217. doi:10.1038/nature03342.

48. Gething PW, Van Boeckel TP, Smith DL, Guerra CA, Patil AP, et al. (2011). Modelling the global constraints of temperature on transmission of *Plasmodium falciparum* and *P. vivax*. *Parasit Vectors* 4: 92. doi:10.1186/1756-3305-4-92.

49. Kristan M, Abeku TA, Beard J, Okia M, Rapuoda B, et al. (2008). Variations in entomological indices in relation to weather patterns and malaria incidence in East African highlands: implications for epidemic prevention and control. *Malar J* 7: 231.

50. Lindsay SW, Martens WJ. (1998). Malaria in the African highlands: past, present and future. *Bull World Health Organ* 76: 33–45.

51. Ruiz D, Brun C, Connor SJ, Omumbo JA, Lyon B, et al. (2014). Testing a multi-malaria-model ensemble against 30 years of data in the Kenyan highlands. *Malar J* 13: 206.

52. Gething PW, Smith DL, Patil AP, Tatem AJ, Snow RW, et al. (2010). Climate change and the global malaria recession. *Nature* 465: 342–345.

53. Riedel N, Vounatsou P, Miller JM, Gosoniu L, Chizema-Kawesha E, et al. (2010). Geographical patterns and predictors of malaria risk in Zambia: Bayesian geostatistical modelling of the 2006 Zambia national malaria indicator survey (ZMIS). *Malar J* 9: 37.

54. Noor AM, Gething PW, Alegana VA, Patil AP, Hay SI, et al. (2009). The risks of malaria infection in Kenya in 2009. *BMC Infect Dis* 9: 180.

55. Krefis AC, Schwarz NG, Nkrumah B, Acquah S, Loag W, et al. (2011). Spatial analysis of land cover determinants of malaria incidence in the Ashanti Region, Ghana. *PLoS One* 6: e17905.

56. NASA. (n.d.). *Earth data*. Available: https://earthdata.nasa.gov/. Accessed 1 January 2016.

57. Ceccato P, Connor SJ, Jeanne I, Thomson MC. (2005). Application of geographical information systems and remote sensing technologies for assessing and monitoring malaria risk. *Parassitologia* 47: 81–96.

58. Weiss DJ, Mappin B, Dalrymple U, Bhatt S, Cameron E, et al. (2015). Re-examining environmental correlates of *Plasmodium falciparum* malaria endemicity: a data-intensive variable selection approach. *Malar J* 14: 68.

59. De Castro MC, Yamagata Y, Mtasiwa D, Tanner M, Utzinger J, et al. (2004). Integrated urban malaria control: a case study in Dar es Salaam, Tanzania. *Am J Trop Med Hyg* 71: 103–117.

60. Erhart A, Thang ND, Xa NX, Thieu NQ, Hung LX, et al. (2007). Accuracy of the health information system on malaria surveillance in Vietnam. *Trans R Soc Trop Med Hyg* 101: 216–225.

61. World Health Organization. (n.d.) *Malaria Surveillance.* Available: http://www.who.int/malaria/areas/surveillance/en/. Accessed 20 November 2015.

62. World Health Organization. (2015). *World Malaria Report 2015.* Geneva: World Health Organization.

63. Snow R, Craig M, Deichmann U, Marsh K. (1999). Estimating mortality, morbidity and disability due to malaria among Africa's non-pregnant population. *Bull World Health Organ* 77: 624.

64. World Health Organization. (2012). Disease surveillance for malaria elimination: operational manual. *World Health Organization.* Available: http://www.who.int/malaria/publications/atoz/9789241503334/en/. Accessed 26 November 2015.

65. Sturrock HJW, Hsiang MS, Cohen JM, Smith DL, Greenhouse B, et al. (2013). Targeting asymptomatic malaria infections: active surveillance in control and elimination. *PLoS Med* 10: e1001467.

66. Wongsrichanalai C, Barcus MJ, Muth S, Sutamihardja A, Wernsdorfer WH. (2007). A review of malaria diagnostic tools: microscopy and rapid diagnostic test (RDT). *Am J Trop Med Hyg* 77: 119–127.

67. Stauffer WM, Cartwright CP, Olson DA, Juni BA, Taylor CM, et al. (2009). Diagnostic performance of rapid diagnostic tests versus blood smears for malaria in US clinical practice. *Clin Infect Dis* 49: 908–913.

68. Moody A. (2002). Rapid diagnostic tests for malaria parasites. *Clin Microbiol Rev* 15: 66–78.

69. Dalrymple U, Mappin B, Gething PW. (2015). Malaria mapping: understanding the global endemicity of falciparum and vivax malaria. *BMC Med* 13: 140.

70. Smith DL, Dushoff J, Snow RW, Hay SI. (2005). The entomological inoculation rate and *Plasmodium falciparum* infection in African children. *Nature* 438: 492–495. doi:10.1038/nature04024.

71. Hay SI, Okiro EA, Gething PW, Patil AP, Tatem AJ, et al. (2010). Estimating the global clinical burden of *Plasmodium falciparum* malaria in 2007. *PLoS Med* 7: e1000290.

72. Bhatt S, Weiss DJ, Cameron E, Bisanzio D, Mappin B, et al. (2015). The effect of malaria control on *Plasmodium falciparum* in Africa between 2000 and 2015. *Nature* 526: 207–211. doi:10.1038/nature15535.

73. Gething PW, Elyazar IRF, Moyes CL, Smith DL, Battle KE, et al. (2012). A long neglected world malaria map: *Plasmodium vivax* endemicity in 2010. *PLoS Negl Trop Dis* 6: e1814.

74. Howes RE, Reiner RC, Battle KE, Longbottom J, Mappin B, et al. (2015). *Plasmodium vivax* transmission in Africa. *PLoS Negl Trop Dis* 9: e0004222.

75. Collins WE, Jeffery GM. (1999). A retrospective examination of sporozoite- and trophozoite-induced infections with *Plasmodium falciparum*: development of parasitologic and clinical immunity during primary infection. *Am J Trop Med Hyg* 61: 4–19.

76. Sutherland CJ, Tanomsing N, Nolder D, Oguike M, Jennison C, et al. (2010). Two nonrecombining sympatric forms of the human malaria parasite *Plasmodium ovale* occur globally. *J Infect Dis* 201: 1544–1550.

77. Mueller I, Zimmerman PA, Reeder JC. (2007). *Plasmodium malariae* and *Plasmodium ovale*—the "bashful" malaria parasites. *Trends Parasitol* 23: 278–283.

78. Collins WE, Jeffery GM. (2007). *Plasmodium malariae*: parasite and disease. *Clin Microbiol Rev* 20: 579–592.

79. Cox-Singh J, Davis TME, Lee K-S, Shamsul SSG, Matusop A, et al. (2008). *Plasmodium knowlesi* malaria in humans is widely distributed and potentially life threatening. *Clin Infect Dis* 46: 165–171.

80. Chin W, Contacos PG, Coatney GR, Kimball HR. (1965). A naturally acquired quotidian-type malaria in man transferable to monkeys. *Science* 149: 865.

81. Sinka ME, Bangs MJ, Manguin S, Rubio-Palis Y, Chareonviriyaphap T, et al. (2012). A global map of dominant malaria vectors. *Parasit Vectors* 5: 69.

82. Malaria Atlas Project. Available: http://www.map.ox.ac.uk/. Accessed 11 September 2015.

83. Beier JC, Killeen GF, Githure JI. (1999). Short report: entomologic inoculation rates and *Plasmodium falciparum* malaria prevalence in Africa. *Am J Trop Med Hyg* 61: 109–113.

84. Kelly-Hope LA, McKenzie FE. (2009). The multiplicity of malaria transmission: a review of entomological inoculation rate measurements and methods across sub-Saharan Africa. *Malar J* 8: 19.

85. Hay SI, Guerra CA, Tatem AJ, Atkinson PM, Snow RW. (2005). Urbanization, malaria transmission and disease burden in Africa. *Nat Rev Microbiol* 3: 81–90.

86. Smith DL, Perkins TA, Reiner RC, Barker CM, Niu T, et al. (2014) Recasting the theory of mosquito-borne pathogen transmission dynamics and control. *Trans R Soc Trop Med Hyg* 108: 185–197.

87. Mbogo CM, Mwangangi JM, Nzovu J, Gu W, Yan G, et al. (2003). Spatial and temporal heterogeneity of Anopheles mosquitoes and *Plasmodium falciparum* transmission along the Kenyan coast. *Am J Trop Med Hyg* 68: 734–742.

88. Gaudart J, Poudiougou B, Dicko A, Ranque S, Toure O, et al. (2006). Space-time clustering of childhood malaria at the household level: a dynamic cohort in a Mali village. *BMC Public Health* 6: 286.

89. Smith T, Charlwood JD, Takken W, Tanner M, Spiegelhalter DJ. (1995). Mapping the densities of malaria vectors within a single village. *Acta Trop* 59: 1–18.

90. Ribeiro JM, Seulu F, Abose T, Kidane G, Teklehaimanot A. (1996). Temporal and spatial distribution of anopheline mosquitos in an Ethiopian village: implications for malaria control strategies. *Bull World Health Organ* 74: 299–305.

91. Bousema T, Griffin JT, Sauerwein RW, Smith DL, Churcher TS, et al. (2012). Hitting hotspots: spatial targeting of malaria for control and elimination. *PLoS Med* 9: e1001165.

92. Bhatt S, Weiss DJ, Mappin B, Dalrymple U, Cameron E, et al. (2015). Coverage and system efficiencies of insecticide-treated nets in Africa from 2000 to 2017. *Elife* 4: e09672.

93. D'Alessandro U, Buttiëns H. (2001). History and importance of antimalarial drug resistance. *Trop Med Int Health* 6: 845–848.

94. World Health Organization. (2012). Global plan for insecticide resistance management in malaria vectors. *World Health Organization*. Available: http://www.who.int/malaria/publications/atoz/gpirm/en/. Accessed 18 January 2016.

95. Okell LC, Griffin JT, Kleinschmidt I, Hollingsworth TD, Churcher TS, et al. (2011). The potential contribution of mass treatment to the control of *Plasmodium falciparum* malaria. *PLoS One* 6: e20179.

96. Gerardin J, Bever CA, Hamainza B, Miller JM, Eckhoff PA, et al. (2016). Optimal population-level infection detection strategies for malaria control and elimination in a spatial model of malaria transmission. *PLoS Comput Biol* 12: e1004707.

97. The MalEra Consultative Group on Modeling. (2011). A research agenda for malaria eradication: modeling. *PLoS Med* 8: e1000403.

98. Wenger EA, Eckhoff PA. (2013). A mathematical model of the impact of present and future malaria vaccines. *Malar J* 12: 126.

99. Eckhoff PA, Bever CA, Gerardin J, Wenger EA, Smith DL. (2015). From puddles to planet: modeling approaches to vector-borne diseases at varying resolution and scale. *Curr Opin Insect Sci* 10: 118–123.

100. Thomson MC, Doblas-Reyes FJ, Mason SJ, Hagedorn R, Connor SJ, et al. (2006). Malaria early warnings based on seasonal climate forecasts from multi-model ensembles. *Nature* 439: 576–579. doi:10.1038/nature04503.

101. Grover-Kopec E, Kawano M, Klaver R, Blumenthal B, Ceccato P, et al. (2005). An online operational rainfall-monitoring resource for epidemic malaria early warning systems in Africa. *Malar J* 4: 6.

102. Pindolia DK, Garcia AJ, Wesolowski A, Smith DL, Buckee CO, et al. (2012). Human movement data for malaria control and elimination strategic planning. *Malar J* 11: 205. doi:10.1186/1475-2875-11-205.

103. Wesolowski A, Eagle N, Tatem AJ, Smith DL, Noor AM, et al. (2012). Quantifying the impact of human mobility on malaria. *Science* 338: 267–270.

104. Bharti N, Tatem AJ, Ferrari MJ, Grais RF, Djibo A, et al. (2011). Explaining seasonal fluctuations of measles in Niger using nighttime lights imagery. *Science* 334: 1424–1427.

105. Pullan RL, Brooker SJ. (2012). The global limits and population at risk of soil-transmitted helminth infections in 2010. *Parasit Vectors* 5: 81.

106. Brooker S, Hotez PJ, Bundy DAP. (2010). The global atlas of helminth infection: mapping the way forward in neglected tropical disease control. *PLoS Negl Trop Dis* 4: e779.

107. Polack S, Brooker S, Kuper H, Mariotti S, Mabey D, et al. (2005). Mapping the global distribution of trachoma. *Bull World Health Organ* 83: 913–919.

108. Carter Center. (n.d.). Guinea worm. Available: http://www.cartercenter.org/health/guinea_worm/index.html. Accessed 30 August 2015.

109. Cairncross S, Tayeh A, Korkor AS. (2012). Why is dracunculiasis eradication taking so long? *Trends Parasitol* 28: 225–230.

110. Hopkins DR, Ruiz-Tiben E. (1990). Dracunculiasis eradication: target 1995. *Am J Trop Med Hyg* 43: 296–300.

111. Miri ES, Hopkins DR, Ruiz-Tiben E, Keana AS, Withers PC, et al. (2010). Nigeria's triumph: dracunculiasis eradicated. *Am J Trop Med Hyg* 83: 215–225.

112. Hopkins DR, Ruiz-Tiben E, Kaiser RL, Agle AN, Withers PC. (1993). Dracunculiasis eradication: beginning of the end. *Am J Trop Med Hyg* 49: 281–289.

113. Hopkins DR, Ruiz-Tiben E, Ruebush TK. (1997). Dracunculiasis eradication: almost a reality. *Am J Trop Med Hyg* 57: 252–259.

114. Adamson PB. (1988). Dracontiasis in antiquity. *Med Hist* 32: 204–209.

115. Muller R. (1971). Dracunculus and dracunculiasis. *Adv Parasitol* 9: 73–151.

116. Cairncross S, Muller R, Zagaria N. (2002). Dracunculiasis (Guinea worm disease) and the eradication initiative. *Clin Microbiol Rev* 15: 223–246.

117. Guiguemde TR. (1986). Climatic characteristics of endemic zones and epidemiologic modalities of dracunculosis in Africa. *Bull Soc Pathol Exot Filiales* 79: 89–95.

118. Centers for Disease Control and Prevention. (n.d.). Guinea worm. Available: http://www.cdc.gov/parasites/guineaworm/. Accessed 12 November 2015.

119. Biswas G, Sankara DP, Agua-Agum J, Maiga A. (2013). Dracunculiasis (guinea worm disease): eradication without a drug or a vaccine. *Philos Trans R Soc Lond B Biol Sci* 368: 20120146.

120. Reddy CR, Narasaiah IL, Parvathi G. (1969). Epidemiological studies on guineaworm infection. *Bull World Health Organ* 40: 521–529.

121. Watts SJ. (1986). The comparative study of patterns of guinea worm prevalence as a guide to control strategies. *Soc Sci Med* 23: 975–982.

122. Edungbola LD, Watts S. (1984). An outbreak of dracunculiasis in a peri-urban community of Ilorin, Kwara State, Nigeria. *Acta Trop* 41: 155–163.

123. Esrey SA, Potash JB, Roberts L, Shiff C. (1991). Effects of improved water supply and sanitation on ascariasis, diarrhoea, dracunculiasis, hookworm infection, schistosomiasis, and trachoma. *Bull World Health Organ* 69: 609–621.

124. Odoom JK, Ntim NAA, Sarkodie B, Addo J, Minta-Asare K, et al. (2014). Evaluation of AFP surveillance indicators in polio-free Ghana, 2009–2013. *BMC Public Health* 14: 687.

125. Hochberg N, Ruiz-Tiben E, Downs P, Fagan J, Maguire JH. (2008). The role of case containment centers in the eradication of dracunculiasis in Togo and Ghana. *Am J Trop Med Hyg* 79: 722–728.

126. Kambire SR, Kangoye LT, Hien R, Yameogo G, Hutin Y, et al. (1993). Dracunculiasis in Burkina Faso: results of a national survey. *J Trop Med Hyg* 96: 357–362.

127. Belcher DW, Wurapa FK, Ward WB, Lourie IM. (1975). Guinea worm in southern Ghana: its epidemiology and impact on agricultural productivity. *Am J Trop Med Hyg* 24: 243–249.

128. Tayeh A, Cairncross S, Maude GH. (1993). Water sources and other determinants of dracunculiasis in the northern region of Ghana. *J Helminthol* 67: 213–225.

129. Diamenu SK, Nyaku AA. (1998). Guinea worm disease—a chance for successful eradication in the Volta region, Ghana. *Soc Sci Med* 47: 405–410.

130. Eberhard ML, Ruiz-Tiben E, Hopkins DR, Farrell C, Toe F, et al. (2014). The peculiar epidemiology of dracunculiasis in Chad. *Am J Trop Med Hyg* 90: 61–70.

131. Hopkins DR, Azam M, Ruiz-Tiben E, Kappus KD. (1995). Eradication of dracunculiasis from Pakistan. *Lancet (London, England)* 346: 621–624.

132. Sahba GH, Arfaa F, Fardin A, Ardalan A. (1973). Studies on dracontiasis in Iran. *Am J Trop Med Hyg* 22: 343–347.

133. Ruiz-Tiben E, Hopkins DR. (2006). Dracunculiasis (guinea worm disease) eradication. *Adv Parasitol* 61: 275–309.

134. Johnson S, Joshi V. (1982). Dracontiasis in Rajasthan. VI. Epidemiology of dracontiasis in Barmer district, Western Rajasthan, India. *Int J Epidemiol* 11: 26–30.

135. Fedchenko AP. (1971). Concerning the structure and reproduction of the guinea worm (*Filaria medinensis* L.). *Am J Trop Med Hyg* 20: 511–523.

136. Abdiev FT, Shamgunova GS. (2001). Control of parasitic diseases in Uzbekistan. *Med Parazitol (Mosk)*: 60–61. Available: http://europepmc.org/abstract/med/11680381. Accessed 23 August 2015.

137. (Uzbekistan) LMII of MP, Elimination WHOCDE and (1999). Dracunculiasis eradication in Uzbekistan: country report/prepared by the L.M. Isaev Institute of Medical Parasitology, Ministry of Health, Republic of Uzbekistan, September 1998. Available: http://www.who.int/iris/handle/10665/66035. Accessed 30 August 2015.

138. Watts SJ. (1987). Dracunculiasis in Africa in 1986: its geographic extent, incidence, and at-risk population. *Am J Trop Med Hyg* 37: 119–125.

139. WHO. (n.d.). Dracunculiasis certification status of countries—Data by country. Available: http://apps.who.int/gho/data/node.main.A1633. Accessed 11 September 2015.

140. Watts SJ. (1984). Population mobility, urban development and dracunculiasis in Kwara State, Nigeria. *Soc Sci Med* 19: 471–473.

141. Carme B, Duda M, Datry A, Gentilini M. (1981). *Dracunculus medinensis* infestation following holidays in West Africa. Epidemiological implications (author's transl)]. *Nouv Presse Med* 10: 2711–2713.

142. Watts S. (2000). Dracunculiasis in the Caribbean and South America: a contribution to the history of dracunculiasis eradication. *Med Hist* 44: 227–250.

143. Centers for Disease Control and Prevention. (2016). Provisional count for 2015: 22 cases of Guinea worm disease worldwide. *Guinea Worm Wrap-up*. Available: http://www.cartercenter.org/resources/pdfs/news/health_publications/guinea_worm/wrap-up/238.pdf.

144. Centers for Disease Control and Prevention. (2013). Mali: not all missed opportunities were due to insecurity. *Guinea Worm Wrap-up*. Available: http://www.cartercenter.org/resources/pdfs/news/health_publications/guinea_worm/wrap-up/222.pdf.

145. Centers for Disease Control and Prevention. (2014). Mali: increase in cases and in case containment. *Guinea Worm Wrap-up*. Available: http://www.cartercenter.org/resources/pdfs/news/health_publications/guinea_worm/wrap-up/229.pdf.

146. Centers for Disease Control and Prevention. (2013). Mali's GWEP struggles during 2012. *Guinea Worm Wrap-up*. Available: http://www.cartercenter.org/resources/pdfs/news/health_publications/guinea_worm/wrap-up/216.pdf.

147. Centers for Disease Control and Prevention. (2014). Progress toward global eradication of dracunculiasis—January 2013–June 2014. *Morb Mortal Wkly Rep* 63: 1050.

148. Mbong EN, Sume GE, Danbe F, Kum WK, Mbi VO, et al. (2015). Not every worm wrapped around a stick is a guinea worm: a case of *Onchocerca volvulus* mimicking *Dracunculus medinensis*. *Parasit Vectors* 8: 374.

149. Eberhard ML, Melemoko G, Zee AK, Weisskopf MG, Ruiz-Tiben E. (2001). Misidentification of *Onchocerca volvulus* as guinea worm. *Ann Trop Med Parasitol* 95: 821–826.

150. Eberhard ML, Ruiz-Tiben E. (2014). Cutaneous emergence of Eustrongylides in two persons from South Sudan. *Am J Trop Med Hyg* 90: 315–317.

151. Watts S. (1998). An ancient scourge: The end of dracunculiasis in EGYPT. *Soc Sci Med* 46: 811–819.

152. Centers for Disease Control and Prevention. (2015). *Guinea Worm Wrap-up* Figure 9. Atlanta, GA: Centers for Disease Control and Prevention.

153. Clarke KC, Osleeb JP, Sherry JM, Meert JP, Larsson RW. (1991). The use of remote sensing and geographic information systems in UNICEF's dracunculiasis (Guinea worm) eradication effort. *Prev Vet Med* 11: 229–235.

154. Ahearn SC, De RC. (2007). Monitoring the effects of Dracunculiasis remediation on agricultural productivity using satellite data. *Int J Remote Sens* 17: 917–929.

155. Royal N. (2014). Of water and worms: Guinea worm re-emergence in Niger. *J Water Health* 12: 184–195.

156. Centers for Disease Control and Prevention. (2015). South Sudan closing in on the worm. *Guinea Worm Wrap-up*. Atlanta, GA: Centers for Disease Control and Prevention.

157. Centers for Disease Control and Prevention. (2015). Guinea worm race: 2015 will Chad be the last endemic country? *Guinea Worm Wrap-up*. Atlanta, GA: Centers for Disease Control and Prevention.

158. Nathanson N, Kew OM. (2010). From emergence to eradication: the epidemiology of poliomyelitis deconstructed. *Am J Epidemiol* 172: 1213–1229. doi:10.1093/aje/kwq320.

159. Nathanson N, Martin JR. (1979). The epidemiology of poliomyelitis: enigmas surrounding its appearance, epidemicity, and disappearance. *Am J Epidemiol* 110: 672–692.

160. Benyesh-Melnick M, Melnick JL, Rawls WE, Wimberly I, Oro JB, et al. (1967). Studies of the immunogenicity, communicability and genetic stability of oral poliovaccine administered during the winter. *Am J Epidemiol* 86: 112–136.

161. Kroon FP, Weiland HT, van Loon AM, van FR. (1995). Abortive and subclinical poliomyelitis in a family during the 1992 epidemic in The Netherlands. *Clin Infect Dis* 20: 454–456.

162. Wilder-Smith A, Leong W-Y, Lopez LF, Amaku M, Quam M, et al. (2015). Potential for international spread of wild poliovirus via travelers. *BMC Med* 13:133.

163. Paffenbarger RS, Wilson VO, Bodian D, Watt J. (1954). Spread of poliomyelitis an analysis of contact during epidemic periods. *Am J Hyg* 60: 63–82.

164. Lavinder CH, Freeman AW, Frost WH. (1918). *Epidemiologic studies of poliomyelitis in New York City and the northeastern United States during the year 1916*. US Government Printing Office.

165. Serfling RE, Sherman IL. (1953). Poliomyelitis distribution in the United States. *Public Heal Rep* 68: 453.

166. Martinez-Bakker M, King AA, Rohani P. (2015). Unraveling the transmission ecology of polio. *PLoS Biol* 13: e1002172.

167. Fine PEM. (1993). Herd immunity: history, theory, practice. *Epidemiol Rev* 15: 265–302.

168. Trevelyan B, Smallman-Raynor M, Cliff AD. (2005). The spatial dynamics of poliomyelitis in the United States: From epidemic emergence to vaccine-induced retreat, 1910–1971. *Ann Assoc Am Geogr* 95: 269–293.

169. Fenner F, Henderson DA, Arita I, Jezek Z, Ladnyi ID, et al. (1988). Smallpox and its eradication. Available: http://www.who.int/iris/handle/10665/39485. Accessed 11 September 2015.

170. Hinman AR, Foege WH, de Quadros CA, Patriarca PA, Orenstein WA, et al. (1987). The case for global eradication of poliomyelitis. *Bull World Health Organ* 65: 835–840.

171. Hull HF, Ward NA, Hull BP, Milstien JB, de Quadros C. (1994). Paralytic poliomyelitis: seasoned strategies, disappearing disease. *Lancet (London, England)* 343: 1331–1337.

172. Global Polio Eradication Initiative. (n.d.). Data and monitoring. Available: http://www.polioeradication.org/Dataandmonitoring.aspx. Accessed 11 September 2015.

173. WHO. (n.d.). AFP/polio data. Available: https://extranet.who.int/polis/public/CaseCount.aspx. Accessed 11 September 2015.
174. WHO | Data, statistics and graphics by subject—Polio case count (n.d.). Available: http://apps.who.int/immunization_monitoring/data/data_subject/en/index.html. Accessed 11 September 2015.
175. Oostvogel PM, van Wijngaarden JK, van der Avoort HG, Mulders MN, Conyn-van Spaendonck MA, et al. (1994). Poliomyelitis outbreak in an unvaccinated community in The Netherlands, 1992-93. *Lancet* 344: 665–670.
176. Drebot M, Mulders M, Campbell J, Kew O, Fonseca K, et al. (1997). Molecular detection of an importation of type 3 wild poliovirus into Canada from The Netherlands in 1993. *Appl Envir Microbiol* 63: 519–523.
177. Centers for Disease Control and Prevention. (1979). Follow-up on poliomyelitis—United States, Canada, Netherlands. *Morb Mortal Wkly Rep* 28: 345.
178. Furesz J, Armstrong RE, Contreras G. (1978). Viral and epidemiological links between poliomyelitis outbreaks in unprotected communities in Canada and the Netherlands. *Lancet (London, England)* 2: 1248.
179. Mulders MN, Reimerink JH, Koopmans MP, van Loon AM, van der Avoort HG. (1997). Genetic analysis of wild-type poliovirus importation into The Netherlands (1979–1995). *J Infect Dis* 176: 617–624.
180. Mach O, Tangermann RH, Wassilak SG, Singh S, Sutter RW. (2014). Outbreaks of paralytic poliomyelitis during 1996–2012: the changing epidemiology of a disease in the final stages of eradication. *J Infect Dis* 210 Suppl: S275–S282.
181. Kilpatrick DR, Yang C-F, Ching K, Vincent A, Iber J, et al. (2009). Rapid group-, serotype-, and vaccine strain-specific identification of poliovirus isolates by real-time reverse transcription-PCR using degenerate primers and probes containing deoxyinosine residues. *J Clin Microbiol* 47: 1939–1941.
182. CDC. (2009). Laboratory surveillance for wild and vaccine-derived polioviruses – worldwide, January 2008-June 2009. (2009). *Morb Mortal Wkly Rep* 58: 950–954.
183. World Health Organization. (n.d.). *Measles and Rubella laboratory network.* Available: http://www.who.int/immunization/monitoring_surveillance/burden/laboratory/measles/en/. Accessed 13 November 2015.
184. O'Reilly KM, Chauvin C, Aylward RB, Maher C, Okiror S, et al. (2011). A statistical model of the international spread of wild poliovirus in Africa used to predict and prevent outbreaks. *PLoS Med* 8: e1001109.
185. Lowther SA, Roesel S, O'Connor P, Landaverde M, Oblapenko G, et al. (2013). World Health Organization regional assessments of the risks of poliovirus outbreaks. *Risk Anal* 33: 664–679.
186. Brown AE, Okayasu H, Nzioki MM, Wadood MZ, Chabot-Couture G, et al. (2014). Lot quality assurance sampling to monitor supplemental immunization activity quality: an essential tool for improving performance in polio endemic countries. *J Infect Dis* 210 Suppl : S333–S340.
187. Upfill-Brown AM, Lyons HM, Pate MA, Shuaib F, Baig S, et al. (2014). Predictive spatial risk model of poliovirus to aid prioritization and hasten eradication in Nigeria. *BMC Med* 12: 92.
188. World Health Organization. (2014). Detection of poliovirus in sewage, Brazil. *Dis Outbreak News.* Available: http://www.who.int/csr/don/2014_6_23polio/en/. Accessed 14 September 2015.

189. World Health Organization. (2013). Wild poliovirus in Cameroon. *Dis Outbreak News*. Available: http://www.who.int/csr/don/2013_11_21/en/. Accessed 14 September 2015.

190. World Health Organization. (2014). Polio outbreak in the Middle East—update. *Dis Outbreak News*. Available: http://www.who.int/csr/don/2014_3_21polio/en/. Accessed 14 September 2015.

191. Anis E, Kopel E, Singer SR, Kaliner E, Moerman L, et al. (2013). Insidious reintroduction of wild poliovirus into israel, 2013. *Eurosurveillance* 18. Available: http://www.eurosurveillance.org/ViewArticle.aspx?ArticleId=20586. Accessed 23 January 2015.

192. Jorba J, Campagnoli R, De L, Kew O. (2008). Calibration of multiple poliovirus molecular clocks covering an extended evolutionary range. *J Virol* 82: 4429–4440.

193. Alam MM, Shaukat S, Sharif S, Angez M, Khurshid A, et al. (2014). Detection of multiple cocirculating wild poliovirus type 1 lineages through environmental surveillance: impact and progress during 2011–2013 in Pakistan. *J Infect Dis* 210 Suppl: S324–S332.

194. Shaukat S, Angez M, Alam MM, Sharif S, Khurshid A, et al. (2014). Molecular characterization and phylogenetic relationship of wild type 1 poliovirus strains circulating across Pakistan and Afghanistan bordering areas during 2010–2012. *PLoS One* 9: e107697.

195. Angez M, Shaukat S, Alam MM, Sharif S, Khurshid A, et al. (2012). Genetic relationships and epidemiological links between wild type 1 poliovirus isolates in Pakistan and Afghanistan. *Virol J* 9: 51.

196. Alam MM, Sharif S, Shaukat S, Angez M, Khurshid A, et al. (2015). Genomic surveillance elucidates persistent wild poliovirus transmission during 2013–2015 in major eservoir areas of Pakistan. *Clin Infect Dis*. Available: http://cid.oxfordjournals.org/content/early/2015/09/27/cid.civ831.abstract. Accessed 13 October 2015.

197. Burns CC, Shaw J, Jorba J, Bukbuk D, Adu F, et al. (2013). Multiple independent emergences of type 2 vaccine-derived polioviruses during a large outbreak in northern Nigeria. *J Virol* 87: 4907–4922.

198. Mangal TD, Aylward RB, Grassly NC. (2013). The potential impact of outine immunization with inactivated poliovirus vaccine on wild-type or vaccine-derived poliovirus outbreaks in a posteradication setting. *Am J Epidemiol* 178: 1579–1587.

199. Gumede N, Jorba J, Deshpande J, Pallansch M, Yogolelo R, et al. (2014). Phylogeny of imported and reestablished wild polioviruses in the Democratic Republic of the Congo from 2006 to 2011. *J Infect Dis* 210 Suppl: S361–S367.

200. Gammino VM, Nuhu A, Chenoweth P, Manneh F, Young RR, et al. (2014). Using geographic information systems to track polio vaccination team performance: pilot project report. *J Infect Dis* 210 Suppl: S98–S101.

201. Barau I, Zubairu M, Mwanza MN, Seaman VY. (2014). Improving polio vaccination coverage in Nigeria through the use of geographic information system technology. *J Infect Dis* 210 Suppl: S102–S110.

13

Remote Sensing and Socioeconomic Data Integration: Lessons from the NASA Socioeconomic Data and Applications Center

Alex de Sherbinin

CONTENTS

Introduction

The combination of remote sensing and socioeconomic data layers in geographic information system (GIS) packages and models for human-environment research has become commonplace in the past two decades [1, 2]. The possibilities of integrating spatial data only arose as technology made it possible to process large data files in a common spatial framework. In the predigital era, similar work was performed through acetate overlays of multiple map layers. Early GIS packages such as IDRISI and ArcINFO opened up new areas for research and applied uses of data from multiple sources.

A second thing that facilitated integration was the conversion of data among different formats, specifically, vector data (points, lines, and polygons) to raster formats, and raster data to point and polygon formats. While again the software made possible the conversion in both directions, there arose recognition that to properly convert vector data to raster formats required special algorithms [3] and that pregridding data products to facilitate the integration could be useful. This led to the development of the first globally gridded population product, Gridded Population of the World [3], which is now in its fourth incarnation. At the same time, there arose recognition that scale issues are important with the conversion, resampling, aggregation, and integration of spatial data sets from different knowledge domains [4], and their effects need to be understood.

Early work to rasterize vector data was sponsored by the National Aeronautics and Space Administration (NASA) Socioeconomic Data and Applications Center (SEDAC) in collaboration with Dr. Waldo Tobler and colleagues at the University of California, Santa Barbara. SEDAC was established in 1993 with the mission to develop data and applications that support the integration of socioeconomic data with the suite of NASA satellite Earth science data. SEDAC is one of 12 Distributed Active Archive Centers (DAACs) under the Earth Observing System Data and Information System, and it is unique in that it is the only NASA data center that provides spatial socioeconomic data complementary to the remote sensing data assets distributed by the other DAACs.

In an effort to track the impact of its work and the value of scientific data more generally, SEDAC tracks citations of its data in the scholarly literature. Although SEDAC data are used on their own and with a variety of non–remote sensing data sources, roughly one-quarter of citations refer to both SEDAC and remote sensing data, and a large number of those refer to actual cases of data integration—here defined as the combination of socioeconomic and remote sensing data in a spatial framework for analytical purposes. These citations occur in natural and social sciences, health, engineering, and multidisciplinary journals [5]. Within the universe of SEDAC data, SEDAC's gridded population data products, the Gridded Population of the World, version 3 (GPWv3) [6] and Global Rural-Urban Mapping Project, version 1 (GRUMP) [7] are most commonly cited, making up about 70% of all data set citations. GPWv3 is a population grid based solely on census inputs with a grid resolution of 2.5 arc-minutes,* and GRUMP is a collection of data products, including an urban extents mask, a lightly modeled population grid at 30 arc-seconds resolution in which population is reallocated from larger administrative units to urban areas, and a settlement points data set. Because of the preponderance of citations that demonstrate the integration of SEDAC's population data with remote sensing products, I focus primarily on this category of citations before turning to examples of the integration of other SEDAC data products such as poverty grids and infrastructure data. While this review is limited to SEDAC data products, the approaches

---

* GPWv4, released in 2015, has a resolution of 30 arc-seconds and a greater number of census input units (more than 12 million) than the prior versions.

to integration reviewed here are consistent with literature based on other gridded population data products such as LandScan and WorldPop, and the issues addressed in the "Discussion and Conclusions" section relating to data integration are relevant to those products as well.

The purpose of this chapter is to provide a timely review of data integration approaches in major application areas, including public health, natural hazards, poverty research, climate change vulnerability, biodiversity conservation, land use and land cover change, and urbanization and to discuss scale issues in data integration. Rather than grouping the papers by application area, I categorize the use of population data with remote sensing into five approaches and then provide examples of studies that fit into each category. I briefly summarize results in order to give a flavor for the kinds of findings that are made possible through data integration. I then turn to a brief summary of integration of remote sensing with other types of socioeconomic data ("Remote Sensing Integration with Other Kinds of Socioeconomic Data"), before discussing some methodological challenges that remain.

Data and Methods

This chapter draws on a universe of 381 peer-reviewed articles, reports, and conference papers published from 1994 to July 2015, all of which cited SEDAC socioeconomic data together with remote sensing data. The collection is more heavily weighted towards the latter part of this period; 324 out of 381 citations (85%) are from the year 2010 onwards. I selected articles in which there was a substantive integration of SEDAC data with remote sensing data (and not just an incidental reference to either) and that best represent the power of spatial data integration. This is not a systematic meta-analysis but rather a first-cut synthesis of common approaches to data integration.

Results

Based on an examination of the types of data integration found in the literature, I created a typology of remote sensing and population data integration based on the purpose for which the population data are used:

1. To *identify drivers* of deforestation/biodiversity loss/land cover change/fire activity
2. To *control* for the influence of population density in a regression model

3. To *assess exposure* of populations to hazards/infectious disease/pollutants

4. To *weight indicators* and indices for health/hazards research by the population that is exposed

5. To *mask areas* based on population density/human influence thresholds

The following subsections provide examples of each type of application. I then provide some examples of the integration of other types of SEDAC data.

Identifying Drivers

Population size, density, and change have often been cited as drivers of environmental problems [8]. In the classic DPSIR framework [9], driving forces (D) such as population change increase demand for goods, which contribute to pressures (P) such as pollutant emissions, a change in the state (S) of the environment, impacts (I) on human health or ecosystems, and ultimately policy responses (R). Growing populations have been held responsible for a host of ills, from deforestation to depletion of ocean resources [8]. There have been a range of studies using gridded population data and remote sensing to understand these drivers; here I focus on a few examples.

Using methods derived from GPW to develop a consistent time series population grid, Aide et al. [10] examined the population and other factors, such as soya production for export, that are driving processes of deforestation and reforestation in Latin America. They developed annual maps of change in woody vegetation and other land cover classes between 2001 and 2010 for each of 16,050 municipalities in Latin America and the Caribbean. In addition, they developed consistent time series population grids (based on a consistent set of census boundaries) and then used nonparametric regression analyses to determine which environmental or population variables best explained the variation in woody vegetation change. They found that deforestation in moist forests tended to occur in lowland areas with low population density, but woody cover change was *not* related to municipality-scale population change. The results emphasize the importance of quantifying deforestation and reforestation at multiple spatial scales and linking these changes with global drivers such as the global demand for food.

Van Asselen et al. [11] used the GPWv3 population density grid to identify drivers of wetland loss globally, finding that agricultural development is the main proximate cause of wetland conversion, whereas economic growth and population density are the most frequently identified underlying forces. Drivers of fire risk have also been an area of data integration. Hantson et al. [12] used satellite-derived moderate-resolution imaging spectroradiometer (MODIS) burned-area data (MCD45) to obtain global individual fire size data for 2002–2010, grouped together for each 2° grid. A global map of fire size distribution was produced by plotting the exponent of the power law. The drivers of the spatial trends in fire size distribution, including vegetation productivity, precipitation, population density, and net income, were analyzed using a generalized additive model. A global map of the fire size distribution, as approached by the power law, shows strong spatial patterns. These are associated both with climatic variables (precipitation and evapotranspiration) and with anthropogenic variables (cropland cover and population density). Archibald et al. [13] investigated the factors controlling the extent of fire in southern Africa using MODIS burned area maps and spatial data on the environmental factors thought to affect burnt area, including biophysical factors on the one hand and human factors on the other. A random forest regression tree model explained 68% of the variance in burnt area, and tree cover, rainfall in the previous 2 years, and rainfall seasonality were the most important predictors. However, human factors such as grazing, roads per unit area, population density, and cultivation fraction also played a role depending on climatic conditions.

Controlling for Population Factors in a Regression Model

In econometric models, it is often necessary to control for the influence of variables other than the ones of interest in order to ensure that the effect one is measuring is that of the independent variable of interest. For example, one may want to control for the influence of population density in a study of the impact of a certain conservation intervention on the number or change in population of some target species so as to isolate the impact of the intervention from known factors that influence population size. Nelson and Chomitz [14] used GRUMPv1 population settlement points [15] in combination with MODIS forest cover and active fire data to assess effectiveness of protected areas in the tropics. The settlement points along with distance to roads were used to identify drivers of forest fires. They found that tropical forests in protected areas have fewer fires than nonprotected forests when controlling for slope, rainfall, road and settlement proximity, and other factors affecting both deforestation and protected area placement.

De Sherbinin [16] used population density as a variable defining degree of urbanization in a model that sought to identify the geographic and biophysical correlates of child malnutrition in Africa. Elevation derived from NASA Shuttle Radar Topography Mission (SRTM) was also used in the model. The only biophysical variable that was found to be correlated with malnutrition, when controlling for population density and other metrics of urbanization, was drought frequency.

Assessing Exposure

Exposure mapping represents one of the biggest applications of population grids and is particularly found in public health and natural hazards literature. While many factors may exacerbate risk (age, health condition, etc.), a first cut is to understand simply who is exposed, since it is not possible to be at risk in the absence of exposure [17].

As an example of assessment of exposure to health risks, SEDAC's GPWv3 data were used in conjunction with aerosol optical depth (AOD) data from MISR and MODIS and the NASA GISS E2-PUCCINI general circulation model in a paper by Marlier et al. [18] on health risks from landscape fire emissions in Southeast Asia (Figure 13.1). They combined satellite-derived fire estimates and atmospheric modelling to quantify health effects from fire emissions in Southeast Asia from 1997 to 2006 and found that fire emissions contribute to

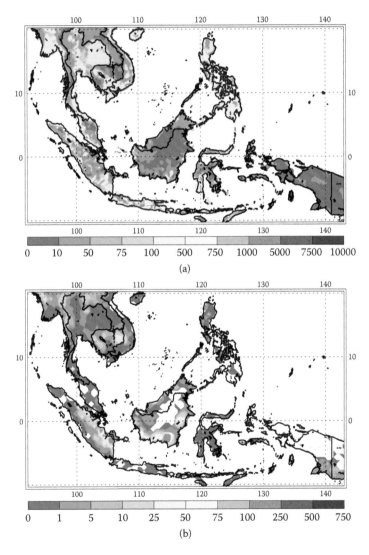

FIGURE 13.1
(a) 2005 population density, in persons per km², from Gridded Population of the World, version 3, and (b) 1997–2006 mean fire emissions, in g C/m² per month, from the Global Fire Emissions Database, version 3.

200 days per year of fine particulate matter $PM_{2.5}$ concentrations exceeding WHO interim health targets (50 μm^2) and an estimated 10,800-person annual increase in adult mortality during ENSO years with higher fire activity.

Chuvieco et al. [19] sought to determine the exposure of housing, among other things, to forest fires using GPWv3 and the MODIS thermal anomalies product. The value of houses was estimated according to the market prices of real estate and land, the level of economic development, and population density. Housing values are highest in the United States, all of Europe, eastern China, and southeastern Australia. The most fire-affected regions are found in the western United States and southeastern Australia.

Several studies have examined population exposure in the coastal zone using combinations of SRTM or ASTER Global Digital Elevation Model with SEDAC's population data sets. Studies have been conducted of population exposure to sea level rise [20], storm surge [21–23], and tsunamis [24]. De Sherbinin et al. [25] used ACE2 data, which are altimetry-adjusted SRTM data particularly suitable in densely forested areas, together with population density and change data from GPWv4 and projections to 2050 developed by colleagues at Baruch College, to assess the vulnerability to coastal storm surge along the West African coast. The results suggest that the region's population is highly exposed to coastal stressors and that Nigeria and Benin (and particularly Lagos and Cotonou) are particularly at risk, owing to a combination of very low-lying topography and rapid population growth due to coastal urbanization.

McGranahan et al. [20] used GRUMPv1 population grids to assess the population in the low-elevation coastal zone (LECZ), defined as the land between 0 and 10 m above mean sea level. The results suggest that roughly 10% of the world's population in the year 2000 and 13% of the urban population are exposed to sea level rise (SLR). A separate mapping of vulnerable populations in relation to the SRTM-derived 1- and 2-m LECZ was conducted for several small island states and major river deltas by Warner et al. [26] in an effort to identify areas that may experience future out-migration owing to climate stressors (Figure 13.2).

Lopez-Carr et al. [27] used GRUMPv1 time series population density to measure changes in population in areas of Africa that are particularly exposed to drought. The authors used the Climate Hazards Group Infrared Precipitation with Stations data, which blends infrared geostationary satellite observations with *in situ* station observations to produce monthly grids of precipitation. The results highlighted the Lake Victoria region and parts of Ethiopia, where high population growth and precipitation decline coincide.

Population-Weighting Indicators

In the development of indicators of exposure, it is important to weight the hazard by the population exposed in order to compare exposure across administrative units. De Sherbinin et al. [28] used MODIS and MISR AOD data to

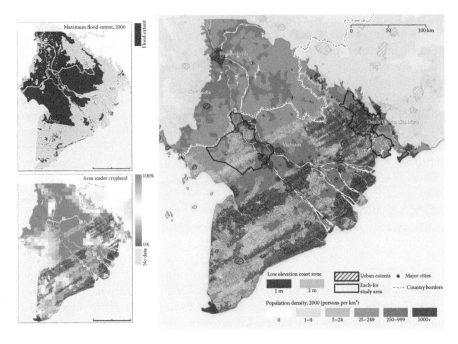

FIGURE 13.2
Population density and urban extents from Global Rural-Urban Mapping Project, version 1, in combination with Shuttle Radar Topography Mission–defined sea level rise in the Mekong Delta.

develop ground-level estimates of $PM_{2.5}$ and population-weighted the results to create indicators of exposure to $PM_{2.5}$. The purpose of the population-weighting was to ensure that high pollution levels in urban areas where larger populations are exposed to often-higher concentrations of air pollutants are represented more in the average $PM_{2.5}$ concentration per administrative unit. They first multiplied the $PM_{2.5}$ concentrations by population in that grid cell and then divided by the total population of the administrative region. The results, published in the 2014 Environmental Performance Index [29], suggested that China and India had the highest population-weighted $PM_{2.5}$ exposures of 48 and 32 micrograms per cubic meter, respectively. Nearly 70% of China's population and 40% of India's population reside in areas with annual $PM_{2.5}$ concentrations above the WHO threshold of 35 micrograms per cubic meter for unhealthy concentrations for sensitive groups.

Dilley et al. [30] estimated risk levels by combining hazard exposure with historical vulnerability for two indicators of elements at risk—population as measured by GPWv3 and gross domestic product (GDP) per unit area—for six major natural hazards: earthquakes, volcanoes, landslides, floods, drought, and cyclones. The landslide, drought, and cyclone data all included remote sensing inputs. By calculating the relative risks for each grid cell based on the population and GDP exposed, they were able to estimate risk levels at subnational scales.

Masking Areas Based on Population Density

Masking is necessary in geospatial analyses in order to remove from consideration areas that meet certain criteria. For example, researchers might mask out nonforested areas in a study of deforestation or mask out urban areas in a study of land cover change in rural areas. SEDAC population and urban area masks are commonly used for masking.

In a study of the link between deforestation and malaria risk by Guerra et al. [31], SEDAC's GRUMPv1 population grid was used in conjunction with satellite-derived estimates of tree cover from the Food and Agriculture Organization (FAO) Forest Resources Assessment. GRUMPv1 was used as a mask to both remove unpopulated areas and identify urban areas. The purpose was to exclude areas with human population densities that were considered too low or too high for malaria transmission. Areas with less than one person/km were deemed free of malaria risk, because human–vector contact in such areas would be sufficiently low to interrupt transmission. Population density thresholds were then defined, by region, as a proxy of urban agglomerations, to allow for the effect of urbanization on malaria transmission. Approximately 136 million people were found to be at risk of malaria within all tropical forests, and by coupling these numbers with the country specific rates of deforestation, it was possible to rank malaria endemic countries according to their potential for change in the population at risk of malaria as the result of deforestation.

De Sherbinin [16] used GRUMPv1 to mask out areas below two persons per square km population density in the aforementioned study assessing the geographical and biophysical correlates of child malnutrition in Africa. The rationale was to remove from consideration those areas that are thinly settled so that zonal statistics compiled for each of the correlates would more likely reflect conditions in areas where surveys of child malnutrition were conducted.

Remote Sensing Integration with Other Kinds of Socioeconomic Data

The integration of remote sensing with other (i.e., nonpopulation) types of socioeconomic data cannot be so easily categorized. Hence, I provide a review of some of the main lines of research.

The section "Assessing Exposure" provides a number of examples of data integration for exposure mapping. However, there is a recognition that exposure alone is insufficient to predict impacts, since populations are heterogeneous and differentially vulnerable to the same impact [32]. What is needed is an understanding of the *vulnerability* of the population to a given hazard.

Here, a number of studies have sought to assess vulnerability through the use of SEDAC spatial poverty data sets such as the Global Subnational Infant Mortality Rates (IMRs), v1, and Global Subnational Prevalence of Child Malnutrition, v1 [33, 34]. For example, de Sherbinin et al. (2015) [35] gave examples of the integration of multiple types of remote sensing and socioeconomic data sets for vulnerability mapping in West Africa, and Petrie et al. [36] used the child malnutrition grid with Defense Meteorological Satellite Program-Operational Linescan System (DMSP-OLS) nighttime lights data to map vulnerability in the Limpopo Basin. Kok et al. [37] used satellite-derived indicators such as grassland productivity, infrastructure density, and soil erosion, together with the Global IMRs data set, to identify patterns of vulnerability of smallholder farmers in drylands globally. They found clusters of severe vulnerability in the Sahel, the Horn of Africa, and Central Asia.

In other work, Bai et al. [38] compared poverty metrics from the aforementioned Center for International Earth Science Information Network (CIESIN) poverty data sets with indices of land degradation based on the Global Inventory Modeling and Mapping Studies (GIMMS) long-term greenness data from advanced very high resolution radiometer and MODIS [39] (Figure 13.3). They found that the correlation coefficients were 0.20 for both infant mortality and for underweight children and land degradation indices but concluded that a much more rigorous analysis is needed to tease out the underlying biophysical and social and economic drivers. Similarly, Gerber et al. [40] used the MODIS normalized difference vegetation index with the poverty mapping data to understand how land degradation is influenced by natural and anthropogenic factors, including socioeconomic conditions.

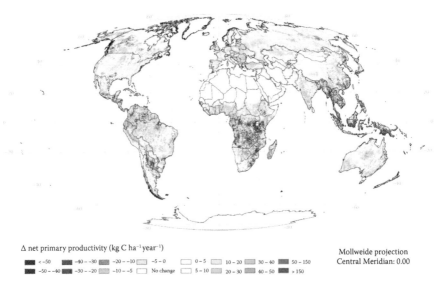

Δ net primary productivity (kg C ha^{-1}year^{-1})

| ■ < −50 | ■ −40 − −30 | ▨ −20 − −10 | ▧ −5 − 0 | □ 0 − 5 | ▨ 10 − 20 | ▨ 30 − 40 | ■ 50 − 150 |
| ■ −50 − −40 | ■ −30 − −20 | ▧ −10 − −5 | □ No change | ▧ 5 − 10 | ▨ 20 − 30 | ■ 40 − 50 | ■ > 150 |

Mollweide projection
Central Meridian: 0.00

FIGURE 13.3
Global change in net primary productivity, 1981–2006 (extreme deserts designated at no change).

Common Issues with Data Integration

In this section, I address some common issues encountered when integrating socioeconomic and remote sensing data with particular reference to scale issues. At a minimum, these issues need to be acknowledged in research findings; some need to be avoided while others can be addressed by taking corrective actions. In all cases, they will affect the validity of the research findings. This section does not address the many issues of linking people to pixels particular to local studies of land use and cover change; readers wishing a comprehensive exploration of these issues are advised to read Rindfuss et al. [41]. There is further discussion of many of these issues elsewhere in this volume, as well as in Goodchild and Quattrochi [42], Cao and Lam [4], and Atkinson and Tate [43].

Probably the biggest issue for data integration is the choice of scale, which ideally should be related to the scale of action—that is, the scales at which variation in spatially varying phenomena is best observed [4]. All too often, the choice of common scale for data integration is dictated by the measurement scale of available data—which could be the coarsest or finest resolution data set—rather than the operational scale. While coarser resolution (small scale) data sets can be resampled, that does not change the underlying or nominal scale. For example, climate projections may have grid cell sizes of 0.5–1 degree and may be resampled at higher resolution to integrate with higher resolution data, but the result is blocks of rasters with the same values. The same issue arises when converting small (map) scale vector data to raster formats, as is done for GPW. This is generally done using a proportional allocation algorithm to distribute polygon values to grid cells [44], yet the likelihood that grid cell values represent on-the-ground values declines the smaller the grid cell and the larger the administrative area (admin unit). Another way of saying this is that the larger the input units, generally, the higher the uncertainty. An important piece of information in uncertainty assessment is the average size of input units relative to the target grid cell size.*

Similarly, when spatially kriging point data in order to create a continuous surface, uncertainty in the resulting grid will depend on the density of the point measurements and the choice of grid cell size [43]. This is one reason that the Demographic and Health Survey (DHS) recommends using Bayesian kriging, since this method generates a surface of the spatial errors in the results: spatial errors are highest where measurements are most sparse. De Sherbinin et al. (2015) [35] did this for a Mali vulnerability map and were

---

* In GPWv4, this average input unit varies depending on the country. The average input unit resolution for very high development regions is 944 sq. km, whereas the low and medium human development countries have an average input resolution of 3,518 and 4,700 sq. km, respectively. Input unit size varies within countries, with unit sizes generally being smallest in densely settled urban areas.

able to incorporate the spatial errors in the DHS and climate data in the final map output, to give users a greater sense of the confidence that one could have in the results.

For the first two types of research identified in this review (examining drivers and spatial regression models), researchers are faced with the choice of whether to conduct statistical analyses on a pixel-by-pixel basis or aggregate to admin or other ecologically defined units. One artifact of conducting analyses on the grid cell basis is that it tends to artificially inflate the number of observations, N, which all things being equal will tend to increase the likelihood that correlations are statistically significant. Additional problems can arise when combining socioeconomic data that don't vary over large areas (e.g., population density grids based on admin1 or admin2 inputs) with continuously varying remote sensing data. In essence, these two data sets do not co-vary as one would expect in the real world. Population grids derived from dasymetric mapping using remote sensing or other ancillary data (see below) can be used, but then one must be careful that the remote sensing or other ancillary data do not create problems of endogeneity, where the same factors used to model the population grid also co-vary over space with the independent variable.

In contrast, researchers converting all grids through zonal statistics to admin or ecological units can run into the modifiable areal unit problem (MAUP) [4]. MAUP refers to the fact that the results of a statistical analysis can be substantially altered by the choice of areal units that are chosen as the unit of analysis—for example, enumeration areas or post codes or higher levels of aggregation such as counties or states/provinces. Values for almost all parameters—such as population count, density, or characteristics—will depend in part on the choice of unit, with larger units tending to have higher standard deviations while averaging out extremes in the data.

Whether running regressions on grid cells, administrative units, or ecologically defined units, researchers must address the issue of spatial autocorrelation. Spatial autocorrelation relates to Tobler's first law of geography, which simply holds that "everything is related to everything else, but near things are more related than distant things" [45]. In an ideal world, researchers would apply regression models that take into account spatial autocorrelation (i.e., spatial error models or geographically weighted regression), but as various authors have pointed out [16, 46], researchers who are not well versed in working with geographic data may apply traditional OLS regression techniques without regard for the effects of spatial covariation on model results (i.e., the fact that estimated coefficients are biased and that the standard errors are artificially deflated).

Researchers are often interested in using the continuous observations from remote sensing to adjust or create socioeconomic data. For example, population grids such as LandScan and WorldPop use dasymetric mapping to reallocate populations on the map based on relationships between physical parameters (as measured by radiances across multiple bands at sensor) and the census population results. However, these relationships vary by region, so one must be careful not to use, for example, the observed relationship between nighttime light radiance levels and population density in the United States or Europe to allocate populations to grid cells based on luminosity in Africa. The same issue can arise when remote sensing–derived parameters (e.g., lacunarity and irregular street patterns) are used to infer housing or population characteristics for urban slum and poverty mapping [47]. Patterns that are predictive in one country may not be predictive in another country.

Spatial mismatches and inconsistencies are also a perennial problem and again have to do with the scale of the different data sets being integrated. To generate an accurate estimate of the number of people in the LECZ [48], CIESIN had to first adjust the boundaries of census units to the remote sensing–derived coastline. At the small cartographic scale of many census geographies, census-provided boundaries would have shown populations already residing below sea level. There are also temporal inconsistencies, such as those that occur between regularly updated or near-real-time remote sensing observations and census data that could be more than a decade old in some regions. Finally, there are inconsistent definitions/measurements in socioeconomic data across countries. For example, in the production of the West Africa coastal vulnerability assessment, de Sherbinin et al. [35] developed a spatially disaggregated household wealth indicator based on DHS data that showed clear anomalies across borders owing to differences in the basket of goods used to measure wealth across countries. While SEDAC intends to develop grids of population by age and sex as part of GPWv4, development of a wider suite of census variable grids globally has been hindered by the lack of commonalities across countries.

Data integration implies some degree of interdisciplinarity. A problem inherent in interdisciplinary research is that data users may not be familiar with the science behind the data sets they use or may lack understanding on how they were constructed. For example, Williams [49] studied population growth in biodiversity hotpots used population grid time series to calculate change in population in hot spots, but the population projections were performed using national growth rates allocated equally over the entire country. Thus, the supposed population growth rates near hot spots were no different than the national growth rates. In other cases, population change across time periods in natural units such as river basins or ecosystems may be an artifact of changes in underlying census boundaries and the size of the admin input units rather than actual on-the-ground growth or decline in population size. Aide et al. [10] are among the few to have dealt with this issue in a systematic way by constructing a time series population database for Latin America with consistent boundaries over time.

A final issue for data integration and interdisciplinary studies more generally are the ways in which uncertainty is measured and addressed. Remote sensing scientists have highly refined methods for measuring and assessing the impact of errors in their measurements, for example, through classification accuracies and standard errors. Most spatial socioeconomic data do not come with corresponding error bars for the estimates contained in them.

Thus, characterizing the validity and accuracy of derived products can be challenging. De Sherbinin et al. [35] characterized the *spatial* errors in major input layers for their Mali vulnerability map, but they were unable to characterize the *measurement* errors for any of their data layers. The end result is that products derived from data integration contain high levels of uncertainty.

Discussion and Conclusions

This chapter provided examples of five types of population and remote sensing data integration and also provided examples of the integration of gridded poverty data layers with remote sensing data. The categorization emerged from the literature, creating a typology of data integration in which scale issues play out differently. In general, for masking and population exposure estimates, the issues of scale of the underlying population data will only affect the area to be masked or the population size estimate. For analysis of drivers and spatial regression, scale issues may have greater impacts on results. As a general rule, it can be said that the coarser the resolution of the underlying data, the greater the uncertainty in the results of any analysis. Yet there are limits, and the introduction of very high resolution remote sensing data and population mapping at 100-m resolution does not necessarily solve problems of data integration, since fundamentally the question of appropriate scale has to do with the scale of operation of the underlying phenomena and the geographic scale (or bounding box) of the study in question.

It is important for users to understand the fitness for use of the rapidly growing number of gridded population data sets for different kinds of applications. These include SEDAC's GPW and GRUMP, LandScan, WorldPop, and a commercial product developed by Esri. The modeled products typically seek to identify the daytime ambient population, as opposed to where people reside (and presumably spend their nights). Each data product has its strengths and weaknesses, and users need to be aware of the methods and underlying assumptions in order to select the appropriate product.

Gridded population and other socioeconomic data sets have become a critical ingredient for human dimensions research in the Anthropocene. The sample of applications here represents a small but representative fraction of the total research that has been made possible by the gridding of population and socioeconomic data. SEDAC has been a pioneer in these efforts, and it is gratifying to see a proliferation of efforts—including the History Database of the Global Environment, LandScan, European Forum for Geostatistics, WorldPop at the University of Southampton, Esri, and Internet.org—that are gridding population and socioeconomic products for a range of application areas. This will mean more fit-for-purpose data products that will facilitate integration with remote sensing and other data streams.

Acknowledgments

Support for the development of this chapter was provided under NASA contract NNG13HQ04C for the continued operation of SEDAC.

References

1. Council, N.N.R., *People and Pixels: Linking Remote Sensing and Social Science*. 1998, Washington, DC: National Academy Press.
2. Fox, J., et al., *People and the Environment: Approaches for Linking Household and Community Surveys to Remtoe Sensing and GIS*. 2003, Boston, MA: Kluwer Academic Publishers.
3. Tobler, W., et al., World population in a grid of spherical quadrilaterals. *International Journal of Population Geography*, 1997. **3**: 203–225.
4. Cao, C. and N. Lam, Understanding the Scale and Resolution Effects in Remote Sensing and GIS, in *Scale in Remote Sensing and GIS*, Editor. Dale A. Quattrochi, Michael F. Goodchild, Boca Raton, FL: CRC Press. pp. 57–72.
5. Chen, R.S., R. Downs, and J. Schumacher, Assessing the Interdisciplinary Use of Socioeconomic and Remote Sensing Data in the Earth Sciences, in *2013 Annual Meeting of the American Geophysical Union*, 2013, San Francisco, CA.
6. Balestri, S. and M.A. Maggioni, Blood diamonds, dirty gold and spatial spillovers measuring conflict dynamics in West Africa. *Peace Economics, Peace Science and Public Policy*, 2014. **20**(4): 551–564.
7. Giannetti, B.F., et al., A review of limitations of GDP and alternative indices to monitor human wellbeing and to manage eco-system functionality. *Journal of Cleaner Production*, 2015. **87**: 11–25.
8. de Sherbinin, A., et al., Population and environment. *Annual Reviews of Environmnent and Resources*, 2007. **32**: 345–373.
9. Agu, G, *The DPSIR Framework Used by the EEA*, 2007, Available from: http://ia2dec.pbe.eea.europa.eu/knowledge_base/Frameworks/doc101182.
10. Aide, T.M., et al., Deforestation and reforestation of Latin America and the Caribbean (2001–2010). *Biotropica*, 2013. **45**(2): 262–271.
11. van Asselen, S., et al., Drivers of wetland conversion: a global meta-analysis. *PLoS One*, 2013. **8**(11): e81292.
12. Hantson, S., S. Pueyo. and E. Chuvieco, Global fire size distribution is driven by human impact and climate. *Global Ecology and Biogeography*, 2015. **24**(1): 77–86.
13. Archibald, S., et al., What limits fire? An examination of drivers of burnt area in Southern Africa. *Global Change Biology*, 2009. **15**(3): 613–630.
14. Nelson, A. and K.M. Chomitz, Effectiveness of strict vs. multiple use protected areas in reducing tropical forest fires: a global analysis using matching methods. *PLoS One*, 2011. **6**(8): e22722.
15. van Donkelaar, A., et al., Use of satellite observations for long-term exposure assessment of global concentrations of fine particulate matter. *Environmental Health Perspectives*, 2015. **123**: 135–143.

16. de Sherbinin, A. The biophysical and geographical correlates of child malnutrition in Africa. *Population, Space and Place*, 2011. **17**(1): 27–46.

17. IPCC. *Special Report on Managing the Risks of Extreme Events and Disasters to Advance Climate Change Adaptation (SREX)*, University of Cambridge: Cambridge, UK.

18. Marlier, M.E., et al., El Niño and health risks from landscape fire emissions in southeast Asia. *Nature Climate Change*, 2013. **3**(5): 131–136.

19. Chuvieco, E., et al., Integration of ecological and socio-economic factors to assess global vulnerability to wildfire. *Global Ecology and Biogeography*, 2014. **23**(2): 245–258.

20. McGranahan, G., D. Balk. and B. Anderson. The rising tide: assessing the risks of climate change and human settlements in low elevation coastal zones. *Environment and Urbanization*, 2007. **19**(1): 17–37.

21. Brecht, H., et al., Sea-level rise and storm surges. *The Journal of Environment and Development*, 2012. **21**(1): 120–138.

22. Dasgupta, S., et al., Sea-Level Rise And Storm Surges: A Comparative Analysis of Impacts in Developgin Countries, Volume 1, in *World Bank Policy Research Working Paper*. 2009, The World Bank: Washington, DC. 43.

23. Ozcelik, C., Y. Gorokhovich. and S. Doocy. Storm surge modelling with geographic information systems: estimating areas and population affected by cyclone Nargis. *International Journal of Climatology*, 2012. **32**(1): 95–107.

24. Løvholt, F., et al., Tsunami hazard and exposure on the global scale. *Earth-Science Reviews*, 2012. **110**(1–4): 58–73.

25. Chucholl, C., Predicting the risk of introduction and establishment of an exotic aquarium animal in Europe: insights from one decade of Marmorkrebs (Crustacea, Astacida, Cambaridae) releases. *Management of Biological Invasions*, 2014. **5**(4): 309–318.

26. Warner, K., et al., *In Search of Shelter: Mapping the Effects of Climate Change on Human Migration and Develoment*. 2009, Nairobi, Kenya.

27. López-Carr, D., et al., A spatial analysis of population dynamics and climate change in Africa: potential vulnerability hot spots emerge where precipitation declines and demographic pressures coincide. *Population and Environment*, 2014. **35**(3): 323–339.

28. Galassi, G. and G. Spada. Sea–level rise in the Mediterranean Sea to 2050: roles of terrestrial ice melt, steric effects and glacial isostatic adjustment. *Global and Planetary Change*, 2014. **123**(Part A): 55–66.

29. Nelson, E., et al., Identifying the Opportunity Cost of Critical Habitat Designation under the U.S. Endangered Species Act, in *Economics Department Working Paper Series*. 2014, Bowdoin College. 27.

30. Dilley, M., et al., *Natural Disaster Hotspots: A Global Risk Analysis*. 2005, World Bank: Washington, DC. 145.

31. Guerra, C.A., R.W. Snow. and S.I. Hay. A global assessment of closed forests, deforestation and malaria risk. *Annals of Tropical Medicine & Parasitology*, 2006. **100**(3): 189–204.

32. Soares, M.B., A.S. Gagnon, and R.M. Doherty. Conceptual elements of climate change vulnerability assessments. *International Journal of Climate Change Strategies and Management*, 2012. 4(1): 6–35.

33. Center for International Earth Science Information Network—CIESIN—Columbia University. *Poverty Mapping Project: Global Subnational Prevalence of Child Malnutrition*. 2005, NASA Socioeconomic Data and Applications Center (SEDAC): Palisades, NY.

34. Center for International Earth Science Information Network—CIESIN—Columbia University. *Poverty Mapping Project: Global Subnational Infant Mortality Rates.* 2005, Palisades, NY: NASA Socioeconomic Data and Applications Center (SEDAC): Palisades, NY.

35. de Sherbinin, A., et al., Data Integration for Climate Vulnerability Mapping In West Africa. *ISPRS International Journal of Geo-Information*, 2015.

36. Petrie, B., et al., *Risk, Vulnerability and Resilience in the Limpopo River Basin.* 2014, RESILIM Program, USAID: Cape Town.

37. Kok, M.T.J., et al., *Quantitative Analysis of Patterns of Vulnerability to Global Environmental Change.* 2010, PBL Netherlands Environmental Assessment Agency the Hague, the Netherlands. 90.

38. Bai, Z., et al., Land Degradation and Ecosystem Services, in *Ecosystem Services and Carbon Sequestration in the Biosphere*, R. Lal et al., Editors. 2013, Springer: Netherlands. pp. 357–381.

39. Tucker, C.J., J.E. Pinzon, and M.E. Brown. *Global Inventory Modeling and Mapping Studies.* 2004, Global Land Cover Facility, University of Maryland: College Park, MD.

40. Gerber, N., E. Nkonya, and J. Braun. Land Degradation, Poverty and Marginality, in *Marginality*, J. von Braun and F.W. Gatzweiler, Editors. 2014, Springer: Netherlands. pp. 181–202.

41. Rindfuss, R., et al., Linking Pixels and People, in *Land Change Science*, G. Gutman et al., Editors. 2004, Kluwer Academic Publishers: Boston, MA. pp. 379–394.

42. Goodchild, M.F. and D.A. Quattrochi. Introduction: Scale, Multiscaling, Remote Sensing, and GIS, in *Scale in Remote Sensing and GIS*, D.A. Quattrochi and M.F. Goodchild, Editors. 1997, CRC Press: Boca Raton, FL. pp. 1–11.

43. Atkinson, P.A. and N.J. Tate. scale problems and geostatistical solutions: A review. *The Professional Geographer*, 2000. **52**(4): 607–623.

44. Balk, D., G. Yetman. and A. De Sherbinin. Construction of Gridded Population and Poverty Data Sets from Different Data Sources. in *European Forum for Geostatistics Conference*. 2010, EFGS: Tallinn, Estonia.

45. Tobler, W., A computer movie simulating urban growth in the Detroit region. *Economic Geography*, 1970. **46**(2): 234–240.

46. PR, V., et al., County child poverty rates in the US: A spatial regression approach. *Population Research and Policy Review*, 2006. **25**(4): 369–391.

47. Sliuzas, R., G. Mboup, and A. de Sherbinin, *Report on the Expert Group Meeting on Slum Identification Using Geo-Information Technology, 2008*, Enschede, The Netherlands, 21–23 May 2008.

48. Center for International Earth Science Information Network—CIESIN—Columbia University, *Low Elevation Coastal Zone (LECZ) Urban-Rural Population and Land Area Estimates, Version 2*. 2013, NASA Socioeconomic Data and Applications Center (SEDAC): Palisades, NY.

49. Williams, J. Humans and biodiversity: population and demographic trends in the hotspots. *Population & Environment*, 2013. **34**(4): 1–14.

Index

Printed and bound by CPI Group (UK) Ltd, Croydon, CR0 4YY

01/11/2024

01782617-0013